Lecture Notes in Mathematics

Edited by A. Dold and B. Eckı

701

Functional Analysis Methods in Numerical Analysis

Special Session, American Mathematical Society, St. Louis, Missouri 1977

Edited by M. Zuhair Nashed

Springer-Verlag
Berlin Heidelberg New York 1979

Editor
M. Zuhair Nashed
Department of Mathematical
Sciences
University of Delaware
501 Kirkbridge Office Building
Newark, DE 19711/USA

AMS Subject Classifications (1970): 65J05, 65F20, 65K05, 65L05, 65L10, 65L15, 41A15, 41A65, 45A05, 46N05, 47A50, 47A55, 47A99, 47H99, 49A25, 49A40, 49G20, 73E99

ISBN 3-540-09110-6 Springer-Verlag Berlin Heidelberg New York
ISBN 0-387-09110-6 Springer-Verlag New York Heidelberg Berlin

Printed in Germany

Printing and binding: Beltz Offsetdruck, Hemsbach/Bergstr.
2141/3140-543210

PREFACE

This volume essentially represents the Proceedings of a Special Session on the topic *Functional Analysis Methods in Numerical Analysis* which was held at the 1977 Annual Meeting of the *American Mathematical Society* in St. Louis, Missouri. The purpose of this session was to bring together some of the leading contributors to this field (from the United States) and to stimulate greater interaction between functional analysis and numerical analysis, as well as to provide a critical assessment of this interaction.

This is the first time that a Special Session on this topic is held at an annual meeting of the AMS. The organization of this session was received with flattering support by the invited speakers, and with great enthusiasm by the participants. In addition to invited talks, there were several contributed papers and an informal discussion of open problems and desirable directions of research in numerical functional analysis.

I should like to thank the Program Committee of the AMS for making this interesting event possible, and to express my appreciation to Professor A. Dold and Springer-Verlag for encouraging the publication of these proceedings which made it possible to highlight in a single volume some of the recent advances in this field and to capture the spirit of the Session. As its title indicates, the present volume provides some perspectives into applications of *methods of functional analysis* to *numerical analysis and approximation theory and methods*. It complements an earlier volume in the *Lecture Notes in Mathematics* entitled *Applications of Methods of Functional Analysis to Problems in Mechanics* [1] and several other volumes in the LNM series on approximation theory and numerical analysis which were proceedings of conferences held in Germany and elsewhere.

There was little interest in numerical functional analysis during its early development. The field was viewed with a mixture of suspicion and indifference. Functional analysts dismissed it as "too soft"; numerical analysts claimed it was "not useful". This situation has changed drastically over the past fifteen years. The reader interested in a retrospective assessment of the development and contributions of this field is invited to read the reviews [2] and [3], which also contain a bibliography of selected references. Today most of the advanced numerical analysis is best approached from a functional analysis point of view. A numerical functional analyst (or any of his relatives) no longer has to be on the defensive.

We conclude by imagining what would happen to numerical analysis if methods of functional analysis were banned. The error estimates industry would virtually collapse ... and last (and in this case least) this volume would become a collector's item.

M. Z. Nashed

[1] P. Germain and B. Naylorles, editors, Applications of Methods of Functional Analysis to Problems in Mechanics, Lecture Notes in Mathematics, Vol. 503, Springer-Verlag, Berlin-Heidelberg-New York.

[2] M. Z. Nashed, Bull. Amer. Math. Soc., 82 (1976), pp. 825-834.

[3] M. Z. Nashed, SIAM Review, 19 (1977), pp. 341-358.

Annual Meeting of the AMS in St. Louis, Missouri (27-31 January, 1977)
Special Session on Functional Analysis Methods in Numerical Analysis

Organizer: M. Zuhair Mashed

List of Invited Speakers and Contributors to the Session

Mieczyslaw Altman, Louisiana State University

Philip M. Anselone, Oregon State University

David L. Barrow, Texas A & M University

Edward K. Blum, University of Southern California

James H. Bramble, Cornell University

E. W. Cheney, University of Texas at Austin

John dePillis, University of California at Riverside and Santa Cruz

James V. Herod, Georgia Institute of Technology

Herbert B. Keller, California Institute of Technology

Peter D. Lax, Courant Institute - New York University

W. Robert Mann, University of North Carolina, Chapel Hill

M. Zuhair Nashed, University of Delaware

John W. Neuberger, North Texas State University

John E. Osborn, University of Maryland

W. V. Petryshyn, Rutgers University

Patricia M. Prenter, Colorado State University

Luois B. Rall, University of Wisconsin - Madison

G. W. Reddien, Vanderbilt University

Ridgway Scott, Brookhaven National Laboratory

W. Gilbert Strang, Massachusetts Institute of Technology

Richard A. Tapia, Rice University

Richard S. Varga, Kent State University

CONTENTS

A STRATEGY THEORY OF SOLVING EQUATIONS

by

Mieczyslaw Altman
Department of Mathematics
Louisiana State University
Baton Rouge, Louisiana 70803

Introduction. The strategy theory to be discussed is actually a part of the general theory of contractor directions developed by the author in a series of investigations (see [1 - 4]). We have shown there the structure of the general theory which is such that it takes all the benefits of the abstract differential calculus whereas no differentiation of any kind is involved in the general definition of contractor directions. Moreover, there are cases of operators which are not differentiable and which nevertheless can be handled by the method of contractor directions. This is, for instance, the case of contraction mappings. There is also another advantage of the general theory, namely, if the range of the operator involved in the equation is closed, then the method of contractor directions provides solvability conditions which are both necessary and sufficient. At the same time, a number of recent results obtained for nonlinear operators with closed ranges appear to be special cases of the general theory of contractor directions (see [1]). In general case, the theory mentioned above is also applicable to nonlinear operators without closed ranges. However, in this case a further development of the concept of contractor directions is necessary, that is, certain additional conditions are imposed on the sets of contractor directions. These requirements are not artificial restrictions, they are rather natural and essential hypotheses which can be verified in the case of the classical Newton-Kantorovich method. Although the method of contractor directions yields general existence theorems for nonlinear equations in Banach spaces under rather weak hypotheses, the main disadvantage is that the method does not provide a constructive tool for finding a solution to the equation. However, under an additional hypothesis, one can obtain an iterative procedure which converges to a solution, and yields an error estimate. The constructive aspect of the general method of contractor directions is the main purpose of this paper. A general iterative method is presented which contains the well-known Newton-Kantorovich method as a special case. In addition, new results are obtained for the Newton-Kantorovich method under Hölder continuity of the Fréchet derivative of the operator in question. Moreover, a general method is proposed which under appropriate hypotheses has such an advantage that while starting with an x_0 which is not good enough for the application of Kantorovich's theorem, we may arrive at an $x_n = \bar{x}_0$ which does satisfy the Kantorovich hypothesis and then the process turns automatically into the Newton-Kantorovich method yielding quadratic convergence.

As to the computational arithmetic, in addition to the standard ones, three different techniques are used which are based upon Lemma 4.1, Lemma 5.2 and Proposition A, Section 6, respectively.

1. Let $P\colon\ D(P) \subset X \to Y$ be a nonlinear mapping, where $D(P)$ is a vector space and X, Y are real or complex Banach spaces. Denote by $\mathbb{B}$ the class of increasing continuous functions B such that

(i) $B(0) = 0,\quad B(s) > 0$ for $s > 0$;

(ii) $\int_0^a s^{-1} B(s)ds < \infty$ for some positive a.

Let $g\colon\ X \to (0,\infty)$ be an arbitrary functional which is bounded on closed bounded sets of X.

Definition 1.1 [3]. Given $P: D(P) \subset X \to Y$, then $\Gamma_x(P) = \Gamma_x(P,q)$ is a set of contractor directions at $x \in D(P)$, which has the (B,q)-property, if for arbitrary $y \in \Gamma_x(P)$, there exist a positive number $\varepsilon = \varepsilon(x,y) \leq 1$ and an element $h \in X$ such that

(1.1) $$||P(x+\varepsilon h)-Px-\varepsilon y|| \leq q\ \varepsilon\ ||y||$$

(1.2) $$||h|| \leq B(g(x)\cdot||y||),$$

where $x + h \in D(P)$ and $q = q(P) < 1$ is some positive number which is independent of $x \in D(P)$.

Definition 1.2 [3]. A mapping $P: D(P) \subset X \to Y$ is said to have the property (α) if for each $x \in D(P)$, a set $\Gamma_x(P)$ of contractor directions with the (B,g)-property exists, which is dense in some ball with center 0 in Y.

Definition 1.3 [3]. The nonlinear mapping $P: D(P) \subset X \to Y$ is (B,g)-differentiable at $x \in D(P)$ if there is a dense subset $V \subset Y$ such that for arbitrary $y \in V$, there exists an element $h \in X$ which satisfies condition (1.2) and

(1.3) $$||P(x+\varepsilon h)-Px-\varepsilon y||/\varepsilon \to 0 \quad \text{as } \varepsilon \to 0+.$$

The (B,g)-differentiability obviously implies property (α).

Let $S = S(x_0,r)$ be an open ball with center $x_0 \in D(P)$ and radius r, and put $U = D(P) \cap \bar{S}$, where $\bar{S}$ is the closure of S.

Theorem 1.1 [3]. Suppose that the following hypotheses are satisfied:

a) $P: U \to Y$ is closed on U;

b) for each $x \in U_0 = D(P) \cap S$, we have that $-Px \in \Gamma_x(P)$ with the (B,g)-property, where $g(x) \equiv C$ (some positive constant);

c) $r \geq (1-q)^{-1} \int_{0}^{a} s^{-1} B(s)ds, \quad Ce^{1-q} ||Px_0|| \leq a.$

Then equation $Px = 0$ has a solution x in U.

2. STRATEGIC DIRECTIONS [3]

Definition 2.1. If $-Px \in \Gamma_x(P)$, i.e., $y = -Px$ satisfies condition (1.1), then the corresponding h is said to be a strategic direction for P at $x \in D(P)$. If, in addition, h satisfies condition (1.2) with $y = -Px$, then h is said to be a (B,g)-strategic direction for P at $x \in D(P)$. If for every $x \in X$, there exists a strategic direction $h = \delta(x)$, then δ: $D(P) \to X$ is called a strategic mapping for P or briefly a strategy for P. If, in addition $h = \delta(x)$ satisfies condition (1.2) with $y = -Px$ for every $x \in D(P)$, then δ is called a (B,g)-strategy for P.

Suppose that $h \in X$ is such that

$$(2.1) \qquad ||P(x+\varepsilon h)-(1-\varepsilon)Px||/\varepsilon \to 0 \quad \text{as } \varepsilon \to 0+.$$

Then $-Px \in \Gamma_x(P,q)$ for arbitrary positive $q < 1$, i.e., $-Px$ is a contractor direction for P at x and h is a strategic direction for P at x.

If for every $x \in D(P)$ there exists an element $h = \delta(x)$ satisfying condition (2.1), then δ: $D(P) \to X$ is a strategy for P. If, in addition, $h = \delta(x)$ satisfies condition (1.2) with $y = -Px$, then δ: $D(P) \to X$ is obviously a (B,g)-strategy for P.

By using the method of contract or directions, it is shown in [1] that if P: $D(P) \subset X \to Y$ is a nonlinear closed mapping with closed range and has a strategy, then P is a mapping onto Y.

A number of authors have obtained similar results by using different methods which are applicable only in the case of an operator with closed range (for references see [1]). At this point we should emphasize that the introduction of the concept of (B,g)-strategic direction is mainly responsible for the fact that the method of contractor directions is also applicable in the general case of a nonlinear operator without closed range. Moreover, this concept is also essential for the constructive version of the method of contractor directions.

3. THE ITERATIVE METHOD OF CONTRACTOR DIRECTIONS

Under an additional hypothesis, the local existence Theorem 1.1 which is based on the method of contractor directions, also yields an iterative method to be discussed below. Suppose that the hypotheses of Theorem 1.1 are fulfilled, i.e., in particular, the nonlinear operator P has a (B,g)-strategy δ: $U_o \to X$, where $g(x) \equiv C$ (some positive constant). Let us consider the following iterative procedure. Given x_o, put

$$x_{n+1} = x_n + \varepsilon_n h_n, \quad n = 0,1,\dots, \tag{3.1}$$

where $0 < \varepsilon_n \leq 1$ and $h_n = \delta(x_n)$ are chosen so as to satisfy conditions (1.1) and (1.2) with $x = x_n$ and $y = -Px_n$. The following theorem is stated in [3] without proof.

<u>Theorem 3.1</u>. Under the hypotheses of Theorem 1.1, suppose in addition, that $\{\varepsilon_n\}$ can be chosen so that

$$\sum_{n=1}^{\infty} \varepsilon_n = \infty. \tag{3.2}$$

Then the sequence $\{x_n\}$ determined by the iterative process (3.1) converges to a solution $x^* \in U$ of equation $Px = 0$, and the error estimate is

$$||x^*-x_n|| \leq (1-q)^{-1} \int_o^{a_n} s^{-1} B(s)ds, \tag{3.3}$$

where $a_n = e^{1-q} C ||Px_0|| e^{-(1-q)t_n}$, $t_n = \sum_{i=0}^{n-1} \varepsilon_i$.

Proof. We prove by induction that

$$(3.4_n) \qquad ||Px_n|| \le e^{-(1-q)t_n} ||Px_0||$$

$$(3.5_n) \qquad ||x_{n+1}-x_n|| \le (t_{n+1}-t_n)B(C||Px_0||e^{-(1-q)t_n})$$

where $t_{n+1} - t_n = \varepsilon_n$, $t_o = 0$. It follows from (3.4_n) and (1.1) with $x = x_n$, $y = Px_n$ and $\varepsilon = \varepsilon_n$ that, by virtue of (3.1),

$$\begin{aligned} ||Px_{n+1}|| &\le (1-\varepsilon_n) ||Px_n|| + q\varepsilon_n ||Px_n|| = \\ &= (1-(1-q)\varepsilon_n) ||Px_n|| < e^{-(1-q)\varepsilon_n} ||Px_n|| \le \\ &\le e^{-(1-q)t_{n+1}} ||Px_o||, \end{aligned}$$

that is, condition (3.4_{n+1}) is satisfied. Condition (3.5_n) results from (3.1), (3.4_n) and (1.2) with $h = h_n$, $g(x) \equiv C$ and $y = -Px_n$. As a consequence of the above, we obtain the following estimates.

$$\begin{aligned} ||x_m-x_n|| &\le \sum_{i=n}^{m-1} ||x_{i+1}-x_i|| \le \\ &\le \sum_{i=n}^{m-1} (t_{i+1}-t_i)B(C||Px_0||e^{-(1-q)t_i}) = \\ &= \sum_{i=n}^{m-1} (t_{i+1}-t_i)B(C\, e^{(1-q)(t_{i+1}-t_i)} ||Px_0||e^{-(1-q)t_{i+1}}) < \\ &< \sum_{i=n}^{m-1} (t_{i+1}-t_i)B(C\, e^{1-q} ||Px_0||e^{-(1-q)t_{i+1}}) \le \\ &\le \sum_{i=n}^{m-1} \int_{t_i}^{t_{i+1}} B(C\, e^{1-q} ||Px_0||e^{-(1-q)t})dt = \\ &= \int_{t_n}^{t_m} B(C\, e^{1-q} ||Px_0||e^{-(1-q)t})dt. \end{aligned}$$

Hence, we obtain the following estimate

$$(3.6) \qquad ||x_m-x_n|| \le \int_{t_n}^{t_m} B(C\, e^{1-q} ||Px_0||e^{-(1-q)t})dt.$$

By a simple computation we obtain from (3.6) that

$$(3.7) \qquad ||x_m - x_n|| \le (1+q)^{-1} \int_{a_m}^{a_n} s^{-1} B(s)ds,$$

where $a_i = C\, e^{1-q}\, ||Px_0||\, e^{-(1-q)t_i}$, $i = 0,1,\ldots$. By virtue of (ii), inequality (3.7) shows that $\{x_n\}$ is a Cauchy sequence. Hence, we get from (3.7) the estimate (3.3), by letting m go to infinity where x^* is the limit of the sequence $\{x_n\}$. As a consequence of (3.4), we obtain that

$$||Px_n|| \to 0 \text{ as } n \to \infty,$$

since $t_n \to \infty$ as $n \to \infty$, by virtue of (3.2). Since P is a closed operator, we infer that $Px^* = 0$. It remains to verify that

$$(3.8) \qquad ||x_n - x_0|| < r = (1-q)^{-1} \int_0^a s^{-1} B(s)ds, \quad a = C\, e^{1-q}\, ||Px_0||.$$

But this statement follows from (3.7), by putting $n = 0$ and replacing m by n, since $t_o = 0$. Thus, inequality (3.8) is satisfied for all $n = 0,1,\ldots$. Thus, the proof of Theorem 3.1 is completed.

Remark 3.1. The error estimate (3.3) shows that the rate of convergence of the sequence $\{x_n\}$ depends on the choice of the step length ε_n in (3.1), and the larger the ε_n the faster the convergence. The optimum rate of convergence is attained for $\varepsilon_n = 1$, $n = 0,1,\ldots$, if such a choice is feasible. In such an optimum case the error estimate (3.3) yields

$$(3.9) \qquad ||x^* - x_n|| \le (1-q)^{-1} \int_0^{a_n} s^{-1} B(s)ds,$$

where $a_n = e^{1-q}\, C\, ||Px_0||\, e^{-(1-q)n}$, $n = 0,1,\ldots$.

Evidently, the same formula (3.3) for the error estimate also shows that the smaller the $0 < q < 1$ the faster the convergence of the sequence $\{x_n\}$.

Suppose that the nonlinear operator $P\colon\ D(P) \subset X \to Y$ is closed and has a Gâteaux derivative $P'(x)$ which is additive and homogeneous, being defined in some open ball $S(x_o,r)$ with radius r and center x_o. We also assume that $P'(x)$ has a right inverse $P'(x)^{-1}$ for all $x \in S(x_o,r)$. The following theorem is a special case of Theorem 1.1.

Theorem 3.2. Suppose that the following hypotheses are satisfied:

a) $\qquad P\colon\ \bar{S}(x_o,r) \to Y$ is closed on $\bar{S}(x_o,r)$;

b) $\qquad ||P'(x)^{-1}Px|| \leq B(||Px||)$ for all $x \in S(x_o,r)$,

where B is some function from the class $\mathbb{B}$;

c) $\qquad r \geq (1-q)^{-1} \int_o^a s^{-1}\, B(s)ds,\quad a = e^{1-q}\,||Px_o||,$

where $q < 1$ is some positive number. Then the equation $Px = 0$ has a solution $x^* \in \bar{S}(x_o,r)$.

Theorem 3.3. Under the hypotheses of Theorem 3.2, consider the sequence (3.1) with $h_n = -P'(x_n)Px_n$. If $\{\varepsilon_n\}$ can be chosen so as to satisfy condition (3.2), then the sequence $\{x_n\}$ determined by the formula (3.1), i.e.,

$$x_{n+1} = x_n - \varepsilon_n\, P'(x_n)^{-1}\, Px_n,\quad n = 0,1,\ldots \tag{3.10}$$

converges to a solution x^* of equation $Px = 0$, and we have $x_n \in \bar{S}(x_o,r)$ for $n = 0,1,\ldots$, and $x^* \in \bar{S}(x_o,r)$, and the error estimate (3.3) holds.

Proof. The proof follows immediately from Theorem 3.1.

Evidently if the operator P has a nonsingular Fréchet derivative P' which satisfies condition b) of Theorem 3.2, then P has a (B,g)-strategy with $g(x) \equiv 1$. Such a strategy is called a Newton strategy, and the iterative method (3.10) is called the Newton-Kantorovich strategy with small steps. If $\varepsilon_n = \varepsilon \leq 1$ for $n = 0,1,\ldots$, then the method (3.10) is called the Newton-Kantorovich strategy with fixed steps. The classical Newton-Kantorovich method is such a method with $\varepsilon_n = 1$ for $n = 0,1,\ldots$. The Newton strategy satisfying condition b) of Theorem 3.2, where P' is the Gâteaux derivative (additive and homogeneous), is called the weak Newton strategy for P. It is clear that if P has a Gâteaux derivative $P'(x)$ which has a right inverse $P'(x)^{-1}$ for all $x \varepsilon S(x_0,r)$, then we have for every $x \varepsilon S(x_0,r)$

(3.11) $$||P(x-\varepsilon P'(x)^{-1}Px)-(1-\varepsilon)Px||/\varepsilon||Px|| \to 0 \text{ as } \varepsilon \to 0+.$$

Suppose that the convergence in (3.11) is uniform in $x \varepsilon S(x_0,r)$. Then for an arbitrary positive number $q < 1$, there exists a positive number $\varepsilon \leq 1$ such that the inequality

(3.12) $$||P(x-\varepsilon P'(x)^{-1}Px)-(1-\varepsilon)Px|| < q\,\varepsilon\,||Px||$$

is satisfied for all $x \varepsilon S(x_0,r)$. In this way we can obtain a Newton-Kantorovich method with a fixed step. Thus, we have the following:

<u>Theorem 3.4</u>. Suppose that the operator P has a weak Newton strategy satisfying condition b) of Theorem 3.2. Suppose, in addition, that the convergence in (3.11) is uniform. The iterative procedure

(3.13) $$x_{n+1} = x_n - \varepsilon P'(x_n)^{-1} Px_n , \quad n = 0,1,\ldots ,$$

where ε is determined by relation (3.12), converges to a solution $x^* \varepsilon \bar{S}(x_0,r)$, where r is defined by condition c) of Theorem 3.2. The sequence $\{x_n\}$ remains in $S(x_0,r)$ and the following error estimate holds:

(3.14) $$||x^*-x_n|| \leq (1-q)^{-1} \int_0^{a_n} s^{-1} B(s)ds,$$

where $a_n = e^{1-q} ||Px_0|| e^{-(1-q)n\varepsilon}$, $n = 0,1,\dots$.

Proof. The proof follows immediately from Theorem 3.3, where $\varepsilon_n = \varepsilon$, $n = 0,1,\dots$, ε being determined by relation (3.12).

Let us notice that Theorem 3.3 does not require the continuity of the derivative $P'(x)$.

Corollary 3.1. Suppose that the following conditions are satisfied: The operator P has a Fréchet derivative P' which is nonsingular and satisfies conditions b) and c) of Theorem 3.2. There exist positive constants C and $\alpha \leq 1$ such that

$$B(s) \leq C\, s^{1/(1+\alpha)} \text{ for } 0 \leq s. \tag{3.15}$$

The derivative P' is α-Hölder continuous with constant K, i.e.,

$$||P'(x)-P'(\bar{x})|| \leq K\, ||x-\bar{x}||^{\alpha} \text{ for all } x,\bar{x} \in S(x_0,r). \tag{3.16}$$

Then all hypotheses of Theorem 3.4 are fulfilled.

Proof. It suffices to prove that the convergence in (3.11) is uniform or relation (3.12) holds for some $0 < \varepsilon \leq 1$. In fact, we have, by virtue of (3.16) and (3.15), that

$$||P(x-\varepsilon\, P'(x)^{-1}Px)-(1-\varepsilon)Px|| \leq (1+\alpha)^{-1}\, K\, \varepsilon^{1+\alpha}\, [B||Px||)]^{1+\alpha} \leq$$

$$\leq (1+\alpha)^{-1}\, K\, \varepsilon^{1+\alpha}\, C^{1/(1+\alpha)}\, ||Px|| \leq q\, \varepsilon\, ||Px||$$

if $\varepsilon \leq 1$ is chosen so as to satisfy the inequality

$$(1+\alpha)^{-1}\, K\, C^{1/(1+\alpha)}\, \varepsilon^{\alpha} \leq q,$$

where $0 < q < 1$ is arbitrary.

Another condition which implies the uniform convergence in (3.11) is given in the following:

Corollary 3.2. Suppose that the Fréchet derivative P is α-Hölder continuous with constant K, i.e., condition (3.16) holds. The inverse $P'(x_o)^{-1}$ exists and

$$||P'(x_o)^{-1}|| \leq B_o. \tag{3.17}$$

The following inequality is satisfied

$$||Px||^{\alpha/(1+\alpha)} + C\, B_oK\, ||x-x_o||^{\alpha} \leq C \quad \text{for all } x \in S(x_o,r), \tag{3.18}$$

where C is a positive constant and $r = (KB_o)^{-1/2}$. Then all hypotheses of Corollary 3.1 are satisfied.

Proof. It follows from (3.16) and (3.17) that the inverse $P'(x)^{-1}$ exists and

$$||P'(x)^{-1}|| \leq B_o/(1-B_oK\, ||x-x_o||^{\alpha}) \tag{3.19}$$

for all x in the open ball $S(x_o,r)$. On the other hand, relation (3.18) implies that

$$||Px|| \cdot ||Px||^{\alpha} \leq [C(1-B_oK\, ||x-x_o||^{\alpha})]^{1+\alpha}\, ||Px||$$

or

$$||Px||/(1-B_oK\, ||x-x_o||^{\alpha}) \leq C\, ||Px||^{1/(1+\alpha)}.$$

Hence, we obtain, by (3.19), that

$$||P'(x)^{-1}Px|| \leq B_o\, ||Px||/(1-B_oK\, ||x-x_o||^{\alpha}) \leq B_oC\, ||Px||^{1/(1+\alpha)}.$$

Thus, conditions b) and (3.15) are satisfied in $S(x_o,r)$ with $B(s) = B_oC\, s^{1/(1+\alpha)}$.

4. THE NEWTON-KANTOROVICH METHOD

One of the most powerful methods widely used for solving operator equations $Px = 0$, where $P\colon\ D(P) \subset X \to Y$, is the Newton-Kantorovich method with iterates defined by the formula

$$(4.1) \qquad x_{n+1} = x_n - P'(x_n)^{-1}\, Px_n, \quad n = 0,1,2,\ldots\ ,$$

where x_0 is a given approximate solution, and assuming that the Fréchet derivative $P'(x)$ exists as well as its inverse $P'(x_n)^{-1}$ for $n = 0,1,\ldots$.

The Newton-Kantorovich method will serve as an illustration to Theorem 3.1 for the general iterative method of contractor directions defined by the formula (3.1), with $\varepsilon_n = 1$, $n = 0,1,\ldots$. Under the assumption that the Fréchet derivative P' is Hölder continuous, it will be shown that P has a (B,g)-strategy δ in the sense of Definition 2.1. But first we prove the following lemma which is crucial in our discussion.

Lemma 4.1. Put

$$(4.2) \qquad t_o = 0,\ t_{n+1} = t_n + (1+\alpha)^{-1}(t_*-t_n^\alpha)^{1/\alpha},\ n = 0,1,\ldots\ ,$$

where $t_* > 0$ and $0 < \alpha \leq 1$. Then

$$t_n \to t_*^{1/\alpha} \text{ as } n \to \infty.$$

Proof. It is clear that if the sequence $\{t_n\}$ has a limit t, then $t = t_*^{1/\alpha}$. Consider the function

$$f(t) = (t_*^{1/\alpha}-t) - (1+\alpha)^{-1}(t_*-t^\alpha),\ 0 \leq t \leq t_*^{1/\alpha}.$$

We have

$$f'(t) = -\alpha(t_*^{1/\alpha}-t)^{\alpha-1} + \alpha(1+\alpha)^{-\alpha}\, t^{\alpha-1}.$$

Hence, it follows that $f'(t) = 0$ if and only if $t = t_c = t_*^{1/\alpha}[1+(1+\alpha)^{\alpha/(1-\alpha)}]^{-1}$, and $f'(t) < 0$ for $t > t_c$ and $f'(t) > 0$ for $t < t_c$. In addition, we have that $f(0) = t_*^{1/\alpha}[1-(1+\alpha)^{-1}] > 0$ and $f(t_*^{1/\alpha}) = 0$. Hence, we infer that $f(t) \geq 0$ for $0 \leq t \leq t_*^{1/\alpha}$. Hence, we conclude that

$$t_*^{1/\alpha} - t - (1+\alpha)^{-1}(t_*-t^{\alpha})^{1/\alpha} \geq 0 \text{ for } 0 \leq t \leq t_*^{1/\alpha}. \tag{4.3}$$

Since

$$t_*^{1/\alpha} - t_{n+1} = t_*^{1/\alpha} - t_n - (1+\alpha)^{-1}(t_*-t_n^{\alpha})^{1/\alpha},$$

it results from (4.3) that $t_n \leq t_*^{1/\alpha}$ for $n = 0,1,\ldots$. But the sequence $\{t_n\}$ is increasing, therefore it has a limit and the lemma is proved.

Kantorovich [6] has introduced the concept of a majorant equation $Q(t) = 0$, where Q is a real valued function of the real variable t, in order to compare Newton's abstract method (4.1) with the same for Q. His method is successful under the assumption that the Fréchet derivative P' is Lipschitz continuous. Kantorovich's argument is not applicable if P' is only Hölder continuous. However, we can use the concept of a majorant function in a broader sense.

Definition 4.1. We say that the real valued function Q is a majorant function for the operator P if for all $n = 0,1,\ldots$ we have

$$||x_n-x_0|| \leq t_n \text{ and } ||Px_n|| \leq Q(t_n), \tag{4.4}$$

where the sequence $\{x_n\}$ is defined by the formula (4.1) and $\{t_n\}$ is a sequence such that

$$||x_{n+1}-x_n|| \leq t_{n+1} - t_n \text{ for all } n = 0,1,\ldots \text{ with } t_o = 0.$$

In our case, the sequence $\{t_n\}$ will be determined by the iterative process (4.2).

Now suppose that the nonlinear operator $P: D(P) \subset X \to Y$ is differentiable in the sense of Fréchet and the inverse of the derivative $P'(x)$ exists at the given point $x = x_o$ with

$$(4.5) \qquad ||P'(x_o)^{-1}|| \leq B_o.$$

We also assume that P' is Hölder continuous, that is, there exist a positive constant K and $0 < \alpha \leq 1$ such that

$$(4.6) \qquad ||P'(x)-P'(\bar{x})|| \leq K \, ||x-\bar{x}||^{\alpha} \text{ for all } x,\bar{x} \in U,$$

where $U = D(P) \cap S(x_o,r) = S(x_o,r)$ being the open ball with center $x_o \in D(P)$ and radius $r = (KB_o)^{-1/\alpha}$. Consider the function

$$(4.7) \qquad Q(t) = K(1+\alpha)^{-1} \, b^{-1}(t_*-t^{\alpha})^m, \; m = (1+\alpha)/\alpha,$$

where $t_* = (KB_o)^{-1}$, and $b \geq 1$ is such that

$$(4.8) \qquad (1+\alpha)^{-1} + (1+\alpha)^{-1/\alpha} \, b^{-\alpha^2/(1+\alpha)} \leq 1.$$

Lemma 4.2. Suppose that $Q(t)$, $0 \leq t \leq t_*^{1/\alpha} = r$, defined by relation (4.7) is such that $||Px_o|| \leq \eta \leq Q(0)$. Then Q is a majorant function for P if t_* is the smallest positive root of the equation

$$(4.9) \quad f(t) = [t_*-(1+\alpha)^{-\alpha}b^{-\alpha^2/(1+\alpha)}(t_*-t)]^{1/\alpha} - t^{1/\alpha} - (1+\alpha)^{-1}(t_*-t)^{1/\alpha} = 0$$

Proof. Since $f(0) \geq 0$, by virtue of (4.8), and $f(t_*) = 0$, it follows from (4.9) that $f(t) \geq 0$ for all $0 \leq t \leq t_*$. Hence, we obtain that

$$t_* - (1+\alpha)^{-\alpha}b^{-\alpha^2/(1+\alpha)}(t_*-t) \geq [t^{1/\alpha}+(1+\alpha)^{-1}(t_*-t)^{1/\alpha}]^{\alpha}$$

for all $0 \leq t \leq t_*$, or replacing t by t^{α}, we have that

$$(4.10) \quad (1+\alpha)^{-\alpha}b^{-\alpha^2/(1+\alpha)}(t_*-t^{\alpha}) \leq t_* - [t+(1+\alpha)^{-1}(t_*-t^{\alpha})^{1/\alpha}]^{\alpha}$$

for all $0 \leq t \leq t_*^{1/\alpha}$.

We can easily show that $P'(x_n)^{-1}$ exists if $||x_n - x_o|| < t_n$ and

(4.11) $$||P'(x_n)^{-1}|| \leq B_o/(1-KB_o\, t_n^{\alpha}), \quad n = 0,1,\dots .$$

In fact, we have

$$P'(x_n) = P'(x_o)\{I-P'(x_o)^{-1}[P'(x_o)-P'(x_n)]\}$$

and (4.11) results from (4.5) and (4.6) since $KB_o\, t_n^{\alpha} < 1$, by virtue of Lemma 4.1.

Now suppose that the inequalities

$$||x_i - x_o|| \leq t_i, \ ||Px_i|| \leq Q(t_i), \text{ and } ||x_i - x_{i-1}|| \leq t_i - t_i$$

are satisfied for $i = 1,2,\dots, n$. Then we can easily show that

$$||x_{n+1}-x_o|| \leq t_{n+1}, \ ||Px_{n+1}|| \leq Q(t_{n+1}),$$

and $||x_{n+1}-x_n|| \leq t_{n+1} - t_n$.

In fact, we have

$$||Px_{n+1}|| = ||Px_{n+1}-Px_n+P'(x_n)[P'(x_n)^{-1}Px_n]|| \leq$$
$$K(1+\alpha)^{-1}(||P'(x_n)^{-1}||\cdot||Px_n||)^{1+\alpha} = K(1+\alpha)^{-1}||P'(x_n)^{-1}||^{1+\alpha}\cdot||Px_n||^{\alpha}\cdot ||Px_n||,$$

by virtue of (4.6). Hence, we obtain

(4.12) $$||Px_{n+1}|| \leq K(1+\alpha)^{-1}||P'(x_n)^{-1}Px_n||^{1+\alpha} \leq (1+\alpha)^{-(1+\alpha)}b^{-\alpha}||Px_n|| = q||Px_n|| \leq (1+\alpha)^{-(1+\alpha)}b^{-\alpha}Q(t_n) \leq Q(t_{n+1}),$$

by virtue of (4.10) and (4.2). Further, we have, by (4.12), that

$$||Px_{n+1}|| \leq K(1+\alpha)^{-1}\ ||P'(x_n)^{-1}Px_n||^{1+\alpha} \leq (1+\alpha)^{-(1+\alpha)}b^{-\alpha}\,||Px_n||.$$

Hence, we get

(4.13) $$||h_n|| = ||\delta(x_n)|| = ||P'(x_n)^{-1}Px_n|| \leq C||Px_n||^{1/(1+\alpha)},$$

where $C = [(1+\alpha)b]^{-\alpha/(1+\alpha)} K^{-1/(1+\alpha)}$. Furthermore, we have

$$||x_{n+1}-x_n|| \leq ||P'(x_n)^{-1}|| \cdot ||Px_n|| \leq (1+\alpha)^{-1}(t_*-t_n^{\alpha})^{1/\alpha} = t_{n+1} - t_n,$$

by virtue of (4.7), (4.11), (4.12) and (4.2), and this completes the proof of the lemma.

Moreover, relations (4.12) and (4.13) show that $h = \delta(x) = -P'(x)^{-1}Px$ is a (B,g)-strategy with $g(x) \equiv 1$ and $B(s) = Cs^{1/(1+\alpha)}$ or $g(x) \equiv [(1+\alpha)b]^{-\alpha} K^{-1}$ and $B(s) = s^{1/(1+\alpha)}$. We have in (4.12) that $q = (1+\alpha)^{-(1+\alpha)}b^{-\alpha} < 1$, since $b \geq 1$. Therefore, the requirements (1.1) and (1.2) of Definition (2.1) are satisfied for $x = x_n$, $y = -Px_n$ with $\varepsilon = 1$ and $h_n = \delta(x_n) = -P'(x_n)Px_n$. In other words, the proof of Lemma 4.2 shows that the (B,g)-strategy δ is defined only for the sequence $\{x_n\}$ determined by the Newton-Kantorovich procedure (4.1). But this is sufficient for the application of the general Theorem 3.1. Let us mention that the following lemma is an immediate consequence of Lemmas 4.1 and 4.2, independently of Theorem 3.1.

Lemma 4.3. Let $P: D(P) \subset X \to Y$ be a nonlinear operator and assume that $x_0 \in D(P)$ is such that $||Px_0|| \leq \eta \leq Q(0)$, where Q is the function defined by formula (4.7). There exists the inverse $P'(x_0)^{-1}$ of the Fréchet derivative $P'(x_0)$ and $||P'(x_0)^{-1}|| \leq B_o$. The derivative P' is Hölder continuous in $S(x_0,r) \subset D(P)$, i.e., condition (4.6) is satisfied with $r = t_*^{1/\alpha} = (KB_o)^{-1/\alpha}$. The smallest positive root of equation (4.9) is t_*. Then the sequence $\{x_n\}$ determined by the Newton-Kantorovich procedure (4.1) is well-defined, remains in $S(x_0,r)$, and converges to a solution $x^* \in \bar{S}(x_0,r)$ of equation $Px = 0$. The rate of convergence is the same as that of a geometric series with ratio $q^{1/(1+\alpha)} = (1+\alpha)^{-1}b^{-\alpha/(1+\alpha)}$, where $b \geq 1$ satisfied condition (4.8).

Proof. It follows from Lemma 4.2 that condition (4.4) is satisfied for all $n = 0,1,\ldots$. Hence $\{x_n\} \subset S(x_0,r)$, by virtue of Lemma 4.1. It results from relation (4.12) that $Px_n \to 0$ as $n \to \infty$. Since $||Px_{n+1}|| \leq q\,||Px_n||$, by virtue of (4.12), it follows from relation (4.13) that the series $\sum_{n=0}^{\infty} ||x_{n+1}-x_n|| = \sum_{n=0}^{\infty} ||h_n||$ is convergent as a series dominated by a geometric series with ratio $q^{1/(1+\alpha)}$. This concludes the proof.

The proof of Lemma 4.3 is based on the assumption that t_* is the smallest positive root of equation (4.9). The verification of such a hypothesis creates a separate problem which can be a difficult one in general case. However, the following lemma shows that in this particular case the hypothesis just mentioned can be deleted.

Lemma 4.4. If b is such that

$$(4.14) \qquad (1+\alpha)^{\alpha}\, b^{\alpha^2/(1+\alpha)} > (1+\alpha)/\alpha^2$$

then t_* is the smallest positive root of equation (4.9).

Proof. We have that $f(t_*) = 0$ and $f(0) \geq 0$, by virtue of (4.14), where f is the function defined by formula (4.9). It remains to show that $f(t) > 0$ for all $0 < t < t_*$. Let us put $t = ct_*$, where c is an arbitrary number such that $0 < c < 1$. Then $f(t) > 0$ if

$$[t_*-[(1+\alpha)\, b^{\alpha^2/(1+\alpha)}]^{-1}(1-c)t_*]^{1/\alpha} - (ct_*)^{1/\alpha} - (1+\alpha)^{-1}[(1-c)t_*]^{1/\alpha}>0$$

or the more so if

$$\{1-[(1+\alpha)^{\alpha}b^{\alpha^2/(1+\alpha)}]^{-1}(1-c)\}^{1/\alpha} > c + (1+\alpha)^{-1}(1-c) = (1+\alpha)^{-1}(1+c\alpha).$$

Hence, we obtain that

$$1 - [(1+\alpha)^{\alpha}b^{\alpha^2/(1+\alpha)}]^{-1}(1-c) > [(1+\alpha)^{-1}(1+c\alpha)]^{\alpha}$$

or, equivalently,

$$(1+\alpha)^{\alpha} b^{\alpha^2/(1+\alpha)} > (1-c)(1+\alpha)^{\alpha}/[(1+\alpha)^{\alpha}-(1+c\alpha)^{\alpha}]. \tag{4.15}$$

Now put $\phi(s) = (1+s\alpha)^{\alpha}$. Then we have

$$(1+\alpha)^{\alpha} - (1+c\alpha)^{\alpha} = \phi(1) - \phi(c) = \phi'(\theta)(1-c) = \alpha^2(1+\theta\alpha)^{\alpha-1}(1-c).$$

Hence, we obtain that

$$(1-c)(1+\alpha)^{\alpha}/[(1+\alpha)^{\alpha}-(1\pm c\alpha)^{\alpha}] \leq (1+\alpha)/\alpha^2. \tag{4.16}$$

We now prove that

$$(1+\alpha)^{\alpha}/[(1+\alpha)^{\alpha}-1] < (1+\alpha)/\alpha^2. \tag{4.17}$$

In fact, put

$$f(c) = \alpha^2 c + (1+\alpha c)^{1-\alpha} - (1+\alpha c), \quad 0 \leq c \leq 1.$$

Then we have

$$f'(c) = \alpha^2 + \alpha(1-\alpha)(1+\alpha c)^{-\alpha} - \alpha < 0 \quad \text{for } 0 < c \leq 1.$$

Since $f(0) = 0$, we infer that $f(c) < 0$ for $0 < c \leq 1$, and consequently, $f(1)(1+\alpha)^{\alpha} < 0$ which implies (4.17). Relations (4.16) and (4.17) imply that the inequalities (4.15) and (4.8) are satisfied if condition (4.14) holds. This completes the proof of the lemma. As a consequence, we obtain the following:

<u>Theorem 4.1</u>. Let $P\colon D(P) \subset X \to Y$ be a nonlinear operator and assume that $x_0 \in D(P)$ is such that $||Px_0|| \leq \eta \leq Q(0)$, where Q is the function defined by formula (4.7). There exists the inverse $P'(x_0)^{-1}$ of the Fréchet derivative $P'(x_0)$ and $||P'(x_0)|| \leq B_0$. The derivative P' is Hölder continuous in the open ball $S(x_0,r) \subset D(P)$, i.e., condition (4.6) is satisfied with $r = t_*^{1/\alpha} = (KB_0)^{-1/\alpha}$. If b satisfies

condition (4.14), then the sequence $\{x_n\}$ determined by the Newton-Kantorovich procedure (4.1) is well-defined and remains in $S(x_o,r)$, and converges to a solution $x^* \varepsilon \bar{S}(x_o,r)$ of equation $Px = 0$. The rate of convergence is the same as that of a geometric series with ratio $q^{1/(1+\alpha)} = (1+\alpha)^{-1} b^{-\alpha/(1+\alpha)}$.

Proof. The proof follows from Lemmas 4.3 and 4.4.

Let us observe that the condition imposed on b in Lemma 4.4, i.e., condition (4.14) depends only upon α and is independent of the operator P and the initial approximation x_o and other constants involved. In the case where $\alpha = 1$, i.e., the derivative P' is Lipschitz continuous, we put $b = 1$ and $2KB_o\eta = 1/2$. Then we have $Q(t) = K2^{-1}(t_*-t)^2$. This is exactly the case where the Newton-Kantorovich method yields linear convergence (see [6]).

Linear convergence is by nature characteristic for iterative methods of contractor directions in general, whereas contractor methods with nonlinear majorant functions (see [3]) can be used in order to investigate fast convergent iterative methods.

The theorem proved above is a good example which shows that the concept of the function $B \varepsilon \mathbb{B}$ which is involved in the (B,g)-strategy, seems to be an essential, natural notion and provides a new insight into the Newton-Kantorovich method, despite the fact that the theory of iterative methods based upon this concept is rather very general and embraces also equations with nonlinear functionals for which the Newton-Kantorovich method is not applicable.

J. Rokne [8] investigated the convergence of the Newton-Kantorovich method for operators with Hölder continuous derivatives. His results are based upon the concept of majorizing sequences introduced by Rheinboldt [7]. He actually examines a more general iterative method that reduces to the Newton-Kantorovich method under appropriate conditions. He also quotes an unpublished result by H. B. Keller [9] concerning the convergence of the Newton-Kantorovich method under similar mild differentiability conditions. The results of both authors require to find the smallest positive root of a "polynomial" of degree $1 + \alpha$. The authors arrive at different polynomials whose smallest positive roots are subject to certain inequalities. The advantage of Theorem 4.2 is that there is no need for finding the smallest positive root of an equation. We have seen above that the assumptions in the case of Hölder continuity are very similar to those under Lipschitz continuity and if $\alpha = 1$, then we obtain the case where the Newton-Kantorovich method is linear convergent.

5. THE NEWTON-KANTOROVICH STRATEGY WITH SMALL STEPS

Suppose that the operator P possesses the Fréchet derivative P' which is nonsingular and Lipschitz continuous with Lipschitz constant K. The well-known Kantorovich theorem says that the iterates (4.1) converge to a solution $x^* \in S(x_o,r)$ of $Px = 0$ if the following hypothesis is satisfied.

(5.1) $$k_o = B_o \eta_o K \leq 1/2,$$

where $||P'(x_o)^{-1}Px_o|| \leq \eta_o$; $||P'(x_o)^{-1}|| \leq B_o$; $r = [1-(1-2k_o)^{1/2}]\eta_o/k_o$. Let us emphasize that the Kantorovich hypothesis that

(5.2) $$||P'(x_o)^{-1}|| \cdot ||P'(x_o)^{-1}Px_o|| \leq 1/2K$$

is rather fundamental in his convergence proof. Now let us make a step further and assume, instead of (5.2), that

(5.3) $$||P'(x)^{-1}|| \cdot ||P'(x)^{-1}Px|| \leq M$$

for all x in the open ball $S(x_o,r)$ with radius r to be defined. It is easy to see that condition (5.3) implies the following one

(5.4) $$||P'(x)^{-1}Px|| \leq M^{1/2}\ ||Px||^{1/2}, \quad x \in S(x_o,r).$$

In other words, the generalized Kantorovich hypothesis (5.3) implies that the Newton-Kantorovich strategy is a (B,g)-strategy with $g(x) \equiv 1$ and $B(s) = M^{1/2}\ s^{1/2}$. This motivation leads to an iterative procedure to be proposed below. Since the operator P possesses a (B,g)-strategy (5.4), the existence of a solution $x^* \in S(x_o,r)$ of the equation $Px = 0$ follows from Theorem 3.2, where

(5.5) $$r = (1-q)^{-1}2(e^{1-q}M||Px_o||)^{1/2},$$

by virtue of condition c) of Theorem 3.2, where q is arbitrary with $0 < q < 1$. The method in question will be defined by formula (3.10). What remains, is to choose there the sequence $\{\varepsilon_n\}$ so as to satisfy condition (3.2). We consider now three cases:

(i) $$k_n = K\ ||P'(x_n)^{-1}||\ \cdot\ ||P'(x_n)^{-1}Px_n|| > 1;$$

(ii) $$1/2 < k_n \leq 1;$$

(iii) $$k_n \leq 1/2,$$

for $n = 0,1,\ldots$. We have, by (3.10), that

$$||Px_{n+1}-(1-\varepsilon_n)Px_n|| < \varepsilon_n^2\ K\ ||P'(x_n)^{-1}Px_n||^2/2 \leq$$
$$\leq 2^{-1}\ \varepsilon_n^2\ K\ ||P'(x_n)^{-1}||\ \cdot\ ||P'(x_n)^{-1}Px_n||\ \cdot\ ||Px_n||.$$

Hence, we obtain that

(5.6) $$||Px_{n+1}|| \leq q(\varepsilon_n)\ ||Px_n||,$$

where $q(\varepsilon_n) = 1 - \varepsilon_n + \varepsilon_n^2\ k_n/2$. We minimize the ratio

$$q(\varepsilon) = 1 - \varepsilon + \varepsilon^2\ k_n/2, \qquad 0 < \varepsilon \leq 1.$$

Its minimum is achieved for $\varepsilon = \varepsilon_n = 1/k_n$ and is

(5.7) $$q(\varepsilon_n) = 1 - \varepsilon_n/2 = 1 - 1/k_n, \quad \varepsilon_n = 1/k_n < 1.$$

Thus, in case (i), we put $\varepsilon_n = 1/k_n$ and we obtain

(5.8) $$||Px_{n+1}-(1-\varepsilon_n)Px_n|| \leq 2^{-1} \varepsilon_n ||Px_n|| = q \varepsilon_n ||Px_n||.$$

In case (ii), if $k_n = 1$, then we put $\varepsilon_n = 1$, and we have $q(\varepsilon_n) = 1/2$ and therefore formula (5.8) remains valid with $q = 1/2$. Now let $k_n < 1$ in case (ii). Then we have $q'(\varepsilon) = -1 + \varepsilon k_n < 0$ for $0 < \varepsilon \leq 1 < 1/k_n$, that is the ratio $q(\varepsilon)$ is a decreasing function. Hence, its minimum is attained at $\varepsilon = \varepsilon_n = 1$, and $q(\varepsilon_n) = k_n/2 < 1/2$. Thus, $q(\varepsilon_n) \leq 1/2$ in case (ii), and formula (5.8) remains valid with $q = 1/2$ and, consequently, we put $\varepsilon_n = 1$ in case (ii). It is clear that we put $\varepsilon_n = 1$ in case (iii), since the Kantorovich hypothesis (5.1) is fulfilled for x_n, and the Kantorovich theorem can be used. Thus we obtain the following:

Theorem 5.1. Suppose that the hypothesis (5.3) is satisfied for all x in the open ball $S(x_o,r)$ with radius defined by (5.5) with $q = 1/2$, i.e., $r = 4(e^{1/2}M||Px_o||)^{1/2}$; in addition, the Fréchet derivative P' is Lipschitz continuous. Then the iterates $\{x_n\}$ defined by (3.10) with $\varepsilon_n = 1$ for $k_n \leq 1$ and $\varepsilon_n = 1/k_n$ for $k_n > 1$- remain in $S(x_o,r)$ and converge to a solution $x^* \in \bar{S}(x_o,r)$ of equation $Px = 0$, and the following error estimate is valid

(5.9) $$||x^*-x_n|| \leq 4(e^{1/2}M||Px_o||e^{-t_n/2})^{1/2}, \quad t_n = \sum_{i=0}^{n-1} \varepsilon_i.$$

Proof. The proof follows from Theorem 3.1, where $B(s) = M^{1/2} s^{1/2}$, since condition (3.2) is satisfied, by virtue of (5.3), i.e., $\varepsilon_n = 1/k_n \geq 1/M$ or $\varepsilon_n = 1$. The error estimate (5.9) results from the formula (3.3), where $q = 1/2$, $C = 1$, and $B(s) = M^{1/2} s^{1/2}$.

Let us observe that if $||P'(x)^{-1}||$ is bounded from above in $S(x_o,r)$, then for some sufficiently large n, the Kantorovich hypothesis (5.1) will be satisfied for $x_n \in S(x_o,r)$, and the convergence will become quadratic.

In the case where the Lipschitz constant K is not available, the tactics of choosing $\{\varepsilon_n\}$ in the Newton-Kantorovich strategy is to start with $\varepsilon_o = 1$. If $||Px_1||$ fails to diminish, then we gradually reduce the value of ε_o, until we obtain that $||Px_1|| < ||Px_0||$. The purpose is to keep a balance between $0 < \varepsilon_n \leq 1$ and the ratio $0 < q(\varepsilon_n) < 1$ in the formula (5.6), so as to minimize $q(\varepsilon_n)$ and to maximize $\varepsilon_n \leq 1$. If the convergence is getting fast enough, for some n, then $\varepsilon_m = 1$ for $m > n$ seems to be the best guess. The following theorem can be proved independently of Theorem 3.1.

Theorem 5.2. Suppose that the hypothesis (5.3) is satisfied for all x in the open ball $S(x_0,r)$ with radius $r = (M||Px_0||)^{1/2}/[1-(1-1/2M)^{1/2}]$ In addition, the Fréchet derivative P' is Lipschitz continuous in $S(x_0,r)$ with Lipschitz constant K. Then the iterates $\{x_n\}$ defined by

$$(5.10)\quad x_{n+1}=x_n-P'(x_n)^{-1}Px_n \text{ if } k_n=K||P'(x_n)^{-1}||\cdot||P'(x_n)^{-1}Px_n||\leq 1$$

and

$$(5.11)\quad x_{n+1} = x_n - k_n^{-1}P'(x_n)^{-1}Px_n \text{ if } k_n > 1,\ n = 0,1,\ldots ,$$

remain in $S(x_0,r)$ and converge to a solution $x^* \in \bar{S}(x_0,r)$ of equation $Px = 0$, and the following error estimate holds

$$(5.12)\quad ||x_n-x^*|| \leq (M||Px_0||^{1/2}(1-1/2M)^{n/2}/[1-(1/2M)^{1/2}].$$

Proof. From the cases (i)-(iii) considered in the proof of Theorem 5.1, we conclude that

$$||Px_{n+1}|| \leq (1-1/2k_n)||Px_n|| \leq (1-1/2M)||Px_n||.$$

Hence, we obtain that

By virtue of (5.4), (5.10) and (5.13) and (5.14), we have that

$$(5.14) \qquad ||x_{n+1}-x_n|| \leq M^{1/2}\ ||Px_n||^{1/2} \leq (M||Px_0||)^{1/2}(1-1/2M)^{n/2}.$$

By using a standard argument, we conclude from (5.14) that the sequence $\{x_n\}$ remains in $S(x_0,r)$ and is a Cauchy sequence. If $x^* \in \bar{S}(x_0,r)$ is the limit of $\{x_n\}$, then $Px^* = 0$, by virtue of relation (5.13). The error estimate (5.12) results from (5.14) by the same standard argument.

Corollary 5.1. Suppose that the hypotheses of Theorem 5.2 are satisfied with $M = 1$. Then the iterates $\{x_n\}$ defined by (5.10) remain in $S(x_0,r)$ with $r = ||Px_0||^{1/2}/[1-(1/2)^{1/2}]$ and converge to a solution $x^* \in \bar{S}(x_0,r)$ of equation $Px = 0$. The error estimate is given by the relation

$$||x_n-x^*|| \leq ||Px_0||^{1/2}(1/2)^{n/2}/[1-(1/2)^{1/2}],\ n = 0,1,\ldots\ .$$

Proof. The proof follows from Theorem 5.2, since $M = 1$ implies that under consideration are only two cases: (ii) and (iii), and in both of them we have $\varepsilon_n = 1$ for $n = 0,1,\ldots$.

Notice that in case of boundedness of $||P'(x)^{-1}||$ in $S(x_0,r)$, the convergence of $\{x_n\}$ in Theorem 5.2 becomes quadratic after a finite number of steps.

Remark 5.1. Theorem 5.1 and 5.2 as well as Corollary 5.1 remain valid if condition (5.3) is replaced by relation (5.4). In this case we put everywhere $k_n = K\ ||P'(x_n)^{-1}Px_n||^2/||Px_n||$ and the argument is exactly the same.

Now let us suppose that the operator P possesses the Fréchet derivative P' which is nonsingular and α-Hölder continuous with Hölder constant K, where $0 < \alpha \leq 1$, i.e.,

(5.15) $$||P'(x)-P'(\bar{x})|| \le K\,||x-\bar{x}||^{\alpha}$$

for all $x,\bar{x} \in S(x_o,r)$, where r is to be defined below. We assume, in addition, that there exists a positive constant M such that

(5.16) $$||P'(x)^{-1}Px|| \le M^{1/(1+\alpha)}||Px||^{1/(1+\alpha)} \text{ for all } x \in S(x_o,r).$$

Consider the following iterative method

(5.17) $$x_{n+1}=x_n-P'(x_n)^{-1}Px_n \text{ if } k_n=K||P'(x_n)^{-1}Px_n||^{1+\alpha}/||Px_n|| \le 1$$

and

(5.18) $$x_{n+1} = x_n - k_n^{-1/\alpha}\,P'(x_n)^{-1}Px_n \quad \text{if } k_n > 1.$$

This is a special case of the method (3.10). We have, by virtue of (5.15), that

$$||Px_{n+1}-(1-\varepsilon_n)Px_n|| \le (1+\alpha)^{-1}\,K\,||P'(x_n)^{-1}Px_n||^{1+\alpha}\,\varepsilon_n^{1+\alpha} =$$
$$= (1+\alpha)^{-1}k_n\,\varepsilon_n^{1+\alpha}\,||Px_n||,$$

where x_{n+1} is defined by (3.10). Hence, we obtain that

$$||Px_{n+1}|| \le q(\varepsilon_n)\,||Px_n||,$$

where $q(\varepsilon_n) = 1 - \varepsilon_n + \varepsilon_n^{1+\alpha}\,k_n/(1+\alpha)$. Now we minimize the ratio

$$q(\varepsilon) = 1 - \varepsilon + \varepsilon^{1+\alpha}\,k_n/(1+\alpha), \quad 0 < \varepsilon \le 1.$$

Its minimum is achieved at $\varepsilon = \varepsilon_n = k_n^{-\alpha} < 1$, and is

(5.19) $$q(\varepsilon_n) = 1 - \varepsilon_n[1-1/(1+\alpha)] = 1 - \alpha/(1+\alpha)k_n^{1/\alpha},$$

provided that $k_n > 1$. Since

$$q'(\varepsilon) = -1 + \varepsilon^{\alpha}\,k_n < 0 \quad \text{for } \varepsilon < 1/k_n^{1/\alpha},$$

that is, the ratio $q(\varepsilon)$ is a decreasing function. Hence, its minimum is attained at $\varepsilon = \varepsilon_n = 1$, and $q(1) = k_n/(1+\alpha) \leq 1/(1+\alpha)$ if $k_n \leq 1$. Thus, in both cases (5.17) and (5.18), we have, by virtue of (5.19),

$$(5.20) \qquad ||Px_{n+1}|| \leq [1-\alpha/(1+\alpha)k_n^{1/\alpha}]\ ||Px_n||,$$

where $k_n \geq 1$. We can now prove the following:

<u>Theorem 5.3</u>. Suppose that conditions (5.15) and (5.16) are satisfied for all $x,\bar{x}$ in the open ball $S(x_o,r)$ with radius

$$(5.21) \qquad r = (M||Px_o||)^{1/(1+\alpha)}/(1-q^{1/(1+\alpha)}), \text{ where } q = 1 - \alpha/(1+\alpha)(MK)^{\alpha}.$$

Then the iterates $\{x_n\}$ defined by (5.17) and (5.18) remain in $S(x_o,r)$ and converge to a solution $x^* \in \bar{S}(x_o,r)$ of equation $Px = 0$, and the following error estimate holds

$$(5.22) \qquad ||x_n-x^*|| \leq (M||Px_o||)^{1/(1+\alpha)}q^{n/(1+\alpha)}/(1-q^{1/(1+\alpha)}),$$

where q is defined by (5.21).

<u>Proof</u>. We conclude from (5.20) and (5.16) that

$$(5.23) \qquad ||Px_n|| \leq ||Px_o||\ q^n \quad \text{for } n = 0,1,\dots ,$$

since $k_n \leq MK$, by (5.16). On the other hand, we have, by virtue of (5.17), (5.18) and (5.16), and (5.23) that

$$(5.24) \qquad ||x_{n+1}-x_n|| \leq M^{1/(1+\alpha)}||Px_n||^{1/(1+\alpha)} \leq (M||Px_o||)^{1/(1+\alpha)}q^{n/(1+\alpha)}.$$

By using a standard argument, we conclude from (5.24) that the sequence $\{x_n\}$ remains in $S(x_o,r)$ and is a Cauchy sequence. If $x^* \in \bar{S}(x_o,r)$ is the limit of $\{x_n\}$, then $Px^* = 0$, by virtue of relation (5.23). The error estimate (5.22) follows from (5.24) by the same standard argument.

Corollary 5.2. Suppose that the hypotheses of Theorem 5.3 are satisfied with MK = 1. Then the iterates $\{x_n\}$ defined by (5.17) remain in $S(x_o,r)$ with r defined by (5.21), where $q = 1 - \alpha/(1+\alpha)$, and converge to a solution $x^* \varepsilon \bar{S}(x_o,r)$. The error estimate is given by relation (5.22) with $q = 1 - \alpha/(1+\alpha)$.

Proof. The proof follows from Theorem 5.3, since MK = 1 implies that $k_n = K \,||P'(x_n)^{-1}Px_n||^{1+\alpha}/||Px_n|| \leq KM \leq 1$, by virtue of (5.16), for all n = 0,1,..., . Thus, the iterative method defined by (5.17) and (5.18) reduces to the Newton-Kantorovich method (5.17).

Remark 5.2. It is easily seen that conditon (5.16) is satisfied if, e.g., the following relation holds

(5.25) $\quad ||P'(x)^{-1}|| \cdot ||P'(x)^{-1}Px||^{\alpha} \leq M$ for all $x \,\varepsilon\, S(x_o,r)$.

If $\alpha = 1$, then condition (5.25) coincides with (5.3).

Now suppose that the operator P possesses the Fréchet derivative P' which is nonsingular and α-Hölder continuous with Hölder constant K, where $0 < \alpha \leq 1$, i.e.,

(5.26) $$||P'(x)-P'(\bar{x})|| \leq K \,||x-\bar{x}||^{\alpha}$$

for all $x,\bar{x} \,\varepsilon\, S(x_o,r)$, where the radius r is to be defined. We assume, in addition, that there exist positive constants M, $\beta < 1$ such that

(5.27) $$||P'(x)^{-1}Px||^{1+\alpha} \leq M \,||Px||^{\beta}$$

for all $x \,\varepsilon\, S(x_o,r)$.

The case where $\beta \geq 1$ actually reduces to that discussed in Theorem 5.3.

Now we consider the following method

$$(5.28)\qquad x_{n+1} = x_n - P'(x_n)^{-1}Px_n \quad \text{if } KM < ||Px_n||^{1-\beta}$$

and

$$(5.29)\qquad x_{n+1} = x_n - \varepsilon_n P'(x_n)^{-1}Px_n \quad \text{if } KM \geq ||Px_n||^{1-\beta},$$

where $\varepsilon_n = (KM)^{-1/\alpha}\,||Px_n||^{(1-\beta)/\alpha}$, $n = 0,1,\ldots$. In order to investigate the method defined by (5.28), (5.29), a different argument will be required. Thus, we first prove the following

Lemma 5.1. Let the positive sequence $\{a_n\}$ be defined as follows

$$(5.30)\qquad a_{n+1} = (1-q\,\varepsilon_n)a_n, \qquad n = 0,1,\ldots ,$$

where $0 < q < 1$ and $0 < \varepsilon_n \leq 1$ for $n = 0,1,\ldots$. Then the series $\sum_{n=0}^{\infty} \varepsilon_n a_n^{\gamma}$ is convergent for arbitrary $0 < \gamma \leq 1$, and we have

$$(5.31)\qquad \sum_{n=0}^{\infty} \varepsilon_n a_n^{\gamma} \leq (q\gamma)^{-1} e^{q\gamma} a_0^{\gamma}$$

and the remainder then satisfies

$$(5.32)\qquad \sum_{i=n}^{\infty} \varepsilon_i a_i^{\gamma} \leq (q\gamma)^{-1} e^{q\gamma} a_0^{\gamma}(e^{-q\gamma t_n} - e^{-q\gamma T}),$$

where $t_o = 0$, $t_n = \sum_{i=0}^{n-1} \varepsilon_i$, $T = \sum_{i=0}^{\infty} \varepsilon_i$.

Proof. First we prove by induction that

$$(5.33)\qquad a_n \leq e^{-qt_n} a_o; \ (t_o=0,\ t_n=\sum_{i=0}^{n-1} \varepsilon_i),\ n = 0,1,\ldots .$$

In fact, we have, by virtue of (5.30) and (5.33), that

$$a_{n+1} \leq (1-q\varepsilon_n)a_n \leq e^{-q\varepsilon_n} e^{-qt_n} a_0 = e^{-qt_{n+1}} a_0.$$

Furthermore, it follows from (5.33) that

$$\varepsilon_n a_n^{\gamma} \leq \varepsilon_n e^{-q\gamma t_n} a_0^{\gamma} = \varepsilon_n e^{q\gamma(t_{n+1}-t_n)} e^{-q\gamma t_{n+1}} a_0^{\gamma} \leq$$

$$\leq e^{q\gamma} a_0^{\gamma}(t_{n+1}-t_n)e^{-q\gamma t_{n+1}} \leq e^{q\gamma} a_0^{\gamma} \int_{t_n}^{t_{n+1}} e^{-q\gamma t}dt.$$

Hence, we obtain that

$$\sum_{i=n}^{m} \varepsilon_i \, a_i^\gamma \leq e^{q\gamma} \, a_o^\gamma \int_{t_n}^{t_{m+1}} e^{-q\gamma t} \, dt,$$

and, consequently, all assertions of the lemma follow immediately.

We can now prove the following

Theorem 5.4. Suppose that conditions (5.26) and (5.27) are satisfied for all $x, \bar{x}$ in the open ball $S(x_o, r)$ with radius

$$r = M^{1/(1+\alpha)}(1+\alpha)^2(\alpha\beta)^{-1} \, e^{\alpha\beta(1+\alpha)^{-2}} \, ||Px_o||^{\beta/(1+\alpha)}.$$

Then the iterates $\{x_n\}$ defined by (5.28) and (5.29) remain in $S(x_o, r)$ and converge to a solution x^* of equation $Px = 0$ and the following error estimate holds.

$$(5.34) \qquad ||x_n - x^*|| \leq M^{1/(1+\alpha)}(q\gamma)^{-1} \, e^{q\gamma} \, ||Px_o||^\gamma \, (e^{-q\gamma t_n} - e^{-q\gamma T}),$$

where $t_n = \sum_{i=0}^{n-1} \varepsilon_i$, $t_o = 0$, $q = \alpha/(1+\alpha)$, $\gamma = \beta/(1+\alpha)$, and $T = \sum_{i=0}^{\infty} \varepsilon_i$.

Proof. We have, by (5.26) and (5.27), that

$$||P(x_n - \varepsilon_n P'(x_n)^{-1}Px_n) - (1-\varepsilon_n)Px_n|| \leq (1+\alpha)^{-1}K \, \varepsilon_n^{1+\alpha} ||P'(x_n)^{-1}Px_n||^{1+\alpha} \leq$$
$$\leq (1+\alpha)^{-1} \, \varepsilon_n \, ||Px_n||,$$

where $\{\varepsilon_n\}$ is the same as in (5.28) and (5.29). Hence, we obtain that

$$(5.35) \qquad ||Px_{n+1}|| \leq [1 - \alpha(1+\alpha)^{-1}\varepsilon_n] \, ||Px_n||,$$

by virtue of (5.28) and (5.29). Furthermore, it follows from (5.28), (5.29) and (5.27) that

$$(5.36) \qquad ||x_{n+1} - x_n|| = \varepsilon_n \, ||P'(x_n)^{-1}Px_n|| \leq M^{1/(1+\alpha)}\varepsilon_n \, ||Px_n||^{\beta/(1+\alpha)}.$$

It results from (5.35) that the sequence $\{a_n = ||Px_n||\}$ satisfies the hypotheses of Lemma 5.1 with $q = \alpha/(1+\alpha)$ and $\gamma = \beta/(1+\alpha)$. Hence, by virtue of (5.36), we have that $\sum_{n=0}^{\infty} ||x_{n+1}-x_n|| \leq r$, and, therefore, $\{x_n\}$ remains in $S(x_o,r)$ and is a Cauchy sequence. Denote its limit by x^*. Then we can show that $Px^* = 0$. In fact, we have that

$$(5.37) \qquad \varepsilon_n ||Px_n||^{\beta/(1+\alpha)} = (KM)^{-1/\alpha} ||Px_n||^{\delta}, \ n = 0,1,\dots ,$$

where $\delta = (1-\beta+\alpha)/\alpha(1+\alpha)$, and $\{\varepsilon_n\}$ is the same as in (5.29) except for a finite number of n. Since the series of positive terms of (5.37) is convergent, by virtue of Lemma 5.1, it follows that the sequence $\{||Px_n||\}$ converges to 0, and this implies that $Px^* = 0$. Now suppose that $\varepsilon_n = 1$ in formula (5.28) for an infinite number of n. Then the iterates $\{x_n\}$ defined by (5.28) and (5.29) satisfy condition (3.2). Moreover, it follows from (5.27) that the operator P has a (B,g)-strategy with $g(x) \equiv 1$ and $B(s) = M^{1/(1+\alpha)}s^{\beta/(1+\alpha)}$. Hence, we conclude that the hypotheses of Theorem 3.3 are satisfied and as a consequence, we obtain that $Px^* = 0$, indeed, in both cases. The error estimate (5.34) follows immediately from (5.36) and (5.32).

Notice that Lemma 5.1 can be generalized as follows.

Lemma 5.2. Let the positive sequence $\{a_n\}$ be defined by relation (5.30) and let B be some function from class $\mathbb{B}$. Then the series $\sum_{n=0}^{\infty} \varepsilon_n B(a_n)$ is convergent, and we have

$$(5.38) \qquad \sum_{n=0}^{\infty} \varepsilon_n B(a_n) \leq q^{-1} \int_0^a s^{-1} B(s)ds, \quad a = e^q a_o.$$

and the remainder then satisfies

$$(5.39) \qquad \sum_{i=n}^{\infty} \varepsilon_i B(a_i) \leq q^{-1} \int_b^{b_n} s^{-1} B(s)ds,$$

where $b_n = e^q a_o e^{-qt_n}$, $b = e^q a_o e^{-qT}$, $t_n = \sum_{i=0}^{n-1} \varepsilon_i$, and $T = \sum_{i=0}^{\infty} \varepsilon_i$, $t_o = 0$.

Proof. The proof is actually the same as that of Lemma 5.1, where the integrals should be replaced by $\int s^{-1} B(s)ds$ with appropriate integration limits.

Based on Lemma 5.2 is the following

general convergence theorem

for the iterative method defined by relations (5.17) and (5.18).

Theorem 5.5. Suppose that the operator P possesses the Fréchet derivative P' which is nonsingular and α-Hölder continuous with $0 < \alpha \leq 1$ and Hölder constant K in the open ball $S(x_0,r)$ with radius

$$r = q^{-1} \int_0^a s^{-1} B(s)ds, \quad a = e^q \, ||Px_0||, \quad q = \alpha/(1+\alpha).$$

Moreover, there exists a function $B \in \mathbb{B}$ such that

$$(5.40) \qquad ||P'(x)^{-1}Px|| \leq B(||Px||)$$

for all $x \in S(x_0,r)$. Then the iterates

$$(5.41) \qquad x_{n+1} = x_n - \varepsilon_n \, P'(x_n)^{-1}Px_n, \quad n = 0,1,\ldots ,$$

where $\varepsilon_n = 1$ if $k_n = K \, ||P'(x_n)^{-1}Px_n||^{1+\alpha}/||Px_n|| \leq 1$ and $\varepsilon_n = k_n^{-\alpha}$ if $k_n > 1$, - remain in $S(x_0,r)$ and converge to a solution x^* of equation $Px = 0$. The following error estimate holds.

$$(5.42) \qquad ||x_n - x^*|| \leq q^{-1} \int_b^{b_n} s^{-1} B(s)ds,$$

where $q = \alpha/(1+\alpha)$, $b_n = e^q \, ||Px_0|| \, e^{-qt_n}$, $b = e^q \, ||Px_0|| \, e^{-qT}$, $t_o = 0$, $t_n = \sum_{i=0}^{n-1} \varepsilon_i$, and $T = \sum_{i=0}^{\infty} \varepsilon_i$.

Proof. In the same way as in the proof of Theorem 5.2, we prove by induction that

$$(5.43) \qquad ||Px_{n+1}|| \leq (1-q\varepsilon_n) \, ||Px_n||, \quad n = 0,1,\ldots .$$

Since we have that

$$(5.44)\qquad ||x_{n+1}-x_n|| = \varepsilon_n\, ||P'(x_n)^{-1}Px_n|| \leq \varepsilon_n\, B(||Px_n||),$$

we can apply the Lemma 5.2 to the sequence $\{a_n=||Px_n||\}$ which satisfies relation (5.43). Hence, by virtue of (5.44), we have that $\sum_{n=0}^{\infty} ||x_{n+1}-x_n|| \leq r$, and, therefore, $\{x_n\}$ remains in $S(x_o,r)$ and is a Cauchy sequence. Denote by x^* its limit. We have to show that $Px^* = 0$. If $\varepsilon_n = 1$ holds for an infinite number of n, then we can apply Theorem 3.3, since condition (3.2) as well as the other hypotheses are satisfied, and as a consequence, we obtain that $Px^* = 0$. Suppose now that $\varepsilon_n = k_n^{-\alpha}$ for all n, except for a finite number of them. Then we obtain that

$$(5.45)\qquad \varepsilon_n\, ||P'(x_n)^{-1}Px_n|| = ||Px_n||^{1/\alpha}\, K^{-1/\alpha}\, ||P'(x_n)^{-1}Px_n||^{-1/\alpha}.$$

Since $\{||P'(x_n)^{-1}Px_n||\}$ is bounded, by virtue of (5.40), it follows from (5.44) and (5.45) that $\{||Px_n||\}$ converges to 0. The error estimate (5.42) results from (5.39) with $\{a_i=||Px_i||\}$.

Remark 5.3. Suppose that the Fréchet derivative P' is Lipschitz continuous. In this case, Theorem 5.5 has the following advantage. While starting with x_o which is not good enough for the application of the Kantorovich theorem, we may arrive at some $x_n = \bar{x}_o$ which does satisfy condition (5.2), and then the convergence becomes quadratic.

6. THE NEWTON-KANTOROVICH METHOD WITH SMALL STEPS GENERATED BY ITERATES OF A POSITIVE FUNCTION

The methods discussed in Section 5 are of the type (3.10) and satisfy condition (3.2). However, except for Theorem 5.4, the last property is mainly due to the fact that the step sequence $\{\varepsilon_n\}$ has a positive lower bound. We can show that it is possible to construct a general method of the type (3.10) with the property (3.2) and such that $\{\varepsilon_n\}$ converges to zero. For this purpose we need a certain method of generating convergent series of positive terms as iterates of a positive function.

Let Q be a real valued function with the properties:

a) $\qquad 0 < Q(s) < s \quad \text{for } 0 < s \leq s_o$

b) $\qquad g(s) = x/(s-Q(s))$ be nonincreasing

<u>Proposition A. [3].</u> Put

$$s_{n+1} = Q(s_n) \qquad \text{for } n = 0,1,\dots .$$

Then the series $S = \sum_{n=0}^{\infty} s_n$ generated by Q is convergent if

c) $$\int_o^{s_o} g(s)ds < \infty,$$

and the remainder then satisfies

$$\sum_{i=n}^{\infty} s_i \leq \int_o^{s_n} g(s)ds, \quad n = 0,1,\dots .$$

As a particular case of Proposition A, let us put

(6.1) $\qquad Q(s) = (1-q\ s^{\beta})s, \text{ where } 0 < g < 1,\ 0 < \beta < 1.$

Consider the series $S = \sum_{n=0}^{\infty} s_n$ generated by Q, i.e.,

(6.2) $\qquad s_{n+1} = (1-q\ s_n^{\beta})s_n, \text{ where } q\ s_o^{\beta} < 1,\ n = 0,1,\dots .$

It is clear that all hypotheses of Proposition A are satisfied and we have

(6.3) $$\sum_{n=0}^{\infty} s_n \leq \int_o^{s_o} g(s)ds = q^{-1}\int_o^{s_o} s^{-\beta}ds = s_o^{1-\beta}/q(1-\beta)$$

(6.4) $$\sum_{i=n}^{\infty} s_i \leq \int_o^{s_n} g(s)ds = q^{-1}\int_o^{s_n} s^{-\beta}ds = s_n^{1-\beta}/q(1-\beta),$$

where $g(s) = s/(s-Q(s)) = s/(1-q\ s^{\beta})s = s^{-\beta}/q$.

Now put in (6.1) $q = \beta = \alpha/(1+\alpha)$, where $0 < \alpha \leq 1$. Then the series $\sum_{n=0}^{\infty} s_n$ generated by Q is defined as follows

(6.5) $$s_{n+1} = [1-\alpha(1+\alpha)^{-1}s_n^{\alpha/(1+\alpha)}]s_n, \quad s_o \leq 1, \; n = 0,1,\ldots .$$

Suppose now that the operator P possesses the Fréchet derivative P' which is **nonsingular** and α-Hölder continuous with Hölder constant K in the open ball $S(x_o,r)$ with radius r to be defined, i.e.,

(6.6) $$||P'(x)-P'(\bar{x})|| \leq K||x-\bar{x}||^{\alpha} \text{ for all } x,\bar{x} \in S(x_o,r).$$

Consider the following iterative method

(6.7) $$x_{n+1} = x_n - \varepsilon_n P'(x_n)^{-1}Px_n, \quad n = 0,1,\ldots ,$$

where $\varepsilon_n = s_n^{\alpha/(1+\alpha)}$, $n = 0,1,\ldots$, the sequence $\{s_n\}$ being defined by the formula (6.5).

Theorem 6.1. Suppose that the Fréchet derivative is nonsingular and satisfies condition (6.6) with $r = (s_o/K)^{1/(1+\alpha)}(1+\alpha)^2/\alpha$. In addition: $||Px_0|| \leq s_o \leq 1$ and

(6.8) $$K\,||P'(x_n)^{-1}Px_n||^{1+\alpha} \leq s_n \text{ for } n = 0,1,\ldots .$$

Then the iterates $\{x_n\}$ defined by (6.7) remain in $S(x_o,r)$ and converge to a solution $x^* \in \bar{S}(x_o,r)$ of equation $Px = 0$, and the following error estimate holds.

(6.9) $$||x_n-x^*|| \leq (s_n/K)^{1/(1+\alpha)}(1+\alpha)^2/\alpha, \quad n = 0,1,\ldots .$$

Proof. We prove by induction that

(6.10) $$||Px_n|| \leq s_n \text{ for } n = 0,1,\ldots .$$

In fact, we have, by virtue of (6.7), that

$$||Px_{n+1}-(1-\varepsilon_n)Px_n|| \leq (1+\alpha)^{-1}K\ s_n^{\alpha}\ ||P'(x_n)^{-1}Px_n||^{1+\alpha} \leq$$
$$\leq s_n^{\alpha}(1+\alpha)^{-1}\ s_n,$$

by virtue of (6.8). Hence, we obtain that

$$||Px_{n+1}|| \leq (1-s_n^{\beta}+(1+\alpha)^{-1}s_n^{\alpha})s_n \leq (1-q\ s_n^{\beta})s_n = s_{n+1},$$

where $q = \beta = \alpha/(1+\alpha)$, and provided that $||Px_n|| \leq s_n$.

It follows from (6.8) and (6.7) that

$$(6.11)\qquad ||x_{n+1}-x_n|| = \varepsilon_n||P'(x_n)^{-1}Px_n|| \leq K^{-1/(1+\alpha)}s_n,\ n = 0,1,\ldots .$$

It is easily seen from (6.11) that $\{x_n\}$ remains in $S(x_0,r)$, by virtue of (6.3), and is a Cauchy sequence. Denote by x^* its limit. Then we have $Px^* = 0$, by virtue of (6.10). Finally, the error estimate (6.9) results from (6.11) and (6.4).

Remark 6.1. If condition (6.8) is replaced by the stronger one

$$K||P'(x_n)^{-1}Px_n||^{1+\alpha} \leq ||Px_n|| \quad \text{for } n = 0,1,\ldots ,$$

then we have the same situation as in the Corollary 5.2.

It is easy to verify that $\sum_{n=0}^{\infty} \varepsilon_n = \infty$. In fact, we have

$$q\ s_n^{\beta} = (s_n-s_{n+1})/s_n \geq (1-q\ s_n^{\beta})\int_{s_{n+1}}^{s_n} s^{-1}ds,$$

by virtue of (6.2), where the sequence $\{s_n\}$ is decreasing and converges to zero.

Now Theorem 6.1 can be generalized as follows.

Theorem 6.2. Suppose that there exist positive constants M, q, β with

$$(6.12)\qquad \beta/q(1+\beta) < 1,\ 0 < q < 1 \text{ and } 0 < \beta < 1$$

which have the following property:

For each $x \in S(x_0,r)$ with $r \geq M\, s_o^{1-\beta}/(1-q)(1-\beta)$ and s with

(6.13) $$||Px|| \leq s \leq s_o = ||Px_0|| < (1-q)^{-1/\beta},$$

there exist ε and $h \in X$ such that

(6.14) $$||P(x+\varepsilon h)-(1-\varepsilon)Px|| \leq q\, \varepsilon^{(1+\beta)/\beta}$$

(6.15) $$[\beta||Px||/q(1+\beta)]^{\beta} < \varepsilon \leq s^{\beta}$$

(6.16) $$||h|| \leq M\, s^{1-\beta}.$$

Put

(6.17) $$s_{n+1} = [1-(1-q)s_n^{\beta}]s_n \quad \text{for } n = 0,1,\ldots .$$

Then for s_n $(n=0,1,\ldots)$ satisfying

(6.18) $$||Px_n|| \leq s_n \leq s_o < (1-q)^{-1/\beta}$$

choose ε_n satisfying

(6.19) $$[\beta||Px_n||/q(1+\beta)]^{\beta} < \varepsilon_n < s_n^{\beta}, \quad n = 0,1,\ldots ,$$

and $h_n \in X$ satisfying

(6.20) $$||h_n|| \leq M\, s_n^{1-\beta}, \quad n = 0,1,\ldots .$$

and

(6.21) $$||P(x_n+\varepsilon_n h_n)-(1-\varepsilon_n)Px_n|| \leq q\, \varepsilon_n^{(1+\beta)/\beta}, \quad n = 0,1,\ldots ,$$

i.e., conditions (6.13)-(6.16) are satisfied for $x = x_n$, $s = s_n$, $h = h_n$ and $\varepsilon = \varepsilon_n$. Now put

(6.22) $$x_{n+1} = x_n + \varepsilon_n\, h_n, \quad n = 0,1,\ldots .$$

Then the iterates $\{x_n\}$ defined by (6.22) remain in $S(x_0,r)$ and converge to a solution $x^* \in \bar{S}(x_0,r)$ of equation $Px = 0$, and the following error estimate holds.

(6.23) $$||x_n-x^*|| \leq M\, s_n^{1-\beta}/(1-q)(1-\beta) \quad \text{for } n = 0,1,\ldots .$$

Proof. First we show that the iterates (6.22) are well-defined by showing that condition (6.18) holds for all n = 0,1,... . We prove this by induction. Put $\eta_n = \varepsilon_n^{1/\beta}$. Then it follows from (6.21) that

$$||Px_{n+1}|| \leq (1-\eta_n^{\beta})||Px_n|| + q\,\eta_n^{1+\beta}.$$

Put

$$f(\eta) = (1-\eta^{\beta})||Px_n|| + q\,\eta^{1+\beta}.$$

Then

$$f'(\eta) = -\beta||Px_n||\eta^{\beta-1} + q(1+\beta)\eta^{\beta} > 0$$

if $\eta > \beta||Px_n||/q(1+\beta)$ or $\varepsilon_n = \eta_n^{\beta}$ satisfies the first inequality in (6.19). Then the function $f(\eta)$ is increasing and we obtain that $f(\eta_n) \leq f(s_n)$, and, consequently, we have that

$$||Px_{n+1}|| \leq f(s_n) \leq [1-(1-q)s_n^{\beta}]s_n = s_{n+1},$$

by virtue of (6.18). Furthermore, it results from (6.22) and (6.20) that

(6.24) $$||x_{n+1}-x_n|| = \varepsilon_n||h_n|| \leq s_n^{\beta}||h_n|| \leq M\, s_n, \quad n = 0,1,\ldots ,$$

by virtue of the second inequality in (6.19). We conclude from (6.24) and (6.3) with q replaced by 1 - q, that $\{x_n\}$ remains in $S(x_o,r)$ and is a Cauchy sequence. Denote by x^* its limit. Then $Px^* = 0$, by virtue of (6.18), since the sequence $\{s_n\}$ defined by (6.17) is decreasing and converges to zero. The error estimate (6.23) results from (6.4) with q replaced by 1 - q.

Remark 6.2. Suppose that (6.6) holds and

$$(6.25)\qquad ||P'(x)^{-1}Px|| \le K^{-1/(1+\alpha)}||Px||^{(1+2\alpha)/(1+\alpha)^2} \text{ for } ||Px|| \le 1$$

Then relations (6.13)-(6.16) are satisfied with $\varepsilon = s^{\beta}$, $\beta = \alpha/(1+\alpha)$, $q = 1/(1+\alpha)$, $h = -P'(x)^{-1}Px$, $M = K^{-1/(1+\alpha)}$, and $s_o = ||Px_o|| \le 1$. In fact, we have that

$$||P(x-s^{\beta}P'(x)^{-1}Px)-(1-s^{\beta})Px|| \le (1+\alpha)^{-1}K||P'(x)^{-1}Px||^{1+\alpha}\ s^{\beta(1+\alpha)} \le$$
$$\le q\ s^{\alpha}||Px||^{(1+2\alpha)/(1+\alpha)} \le q\ s^{1+\beta} = q\ \varepsilon^{(1+\beta)/\beta},$$

and

$$||h|| \le M\ s^{(1+2\alpha)/(1+\alpha)^2} \le M\ s^{1-\beta},$$

by virtue of (6.25), provided that $||Px|| \le s \le 1$.

Notice that the convergence of the methods with small steps discussed above occurs in general under rather weak conditions. However, because of the rather slow convergence, these methods can be used as a starting procedure in order to get a good starting point, e.g., in the case of the Newton-Kantorovich method if applicable. The advantage of the Newton-Kantorovich method with small steps generated by iterates of a positive function over the method (5.17)-(5.18) is that $\{\varepsilon_n\}$ in (6.7) is defined by (6.5) and, therefore, independent of the constant K in (6.6).

A parallel theory will be developed elsewhere for Newton's method proposed by Altman [5] for nonlinear functionals on Banach spaces.

Open Problems

1. In order to solve the equation $Px = 0$, the following general iterative method of contractor directions is used

(i) $$x_{n+1} = x_n + \varepsilon_n h_n, \quad n = 0,1,\dots ,$$

where

a) $$||P(x_n+\varepsilon_n h_n)-(1-\varepsilon_n)Px_n|| \leq q\, \varepsilon_n ||Px_n||,$$

b) $$||h_n|| \leq B(||Px_n||),\ B \in \mathbb{B},\ 0 < q < 1,\ 0 < \varepsilon \leq 1.$$

Find general conditions which imply that

c) $$\sum_{i=0}^{\infty} \varepsilon_i = \infty$$ (Then, by Theorem 3.1 [1], $x_n \to x^*$ and $Px^* = 0$).

2. If $||P'(x)^{-1}Px|| \leq B(||Px||)$ for $x \in S(x_0,r)$, $B \in \mathbb{B}$, what is the best practical tactics of choosing $\{\varepsilon_n\}$ in (i) so as to satisfy a) with $h_n = -P'(x_n)^{-1}Px_n$, provided the Lipschitz constant for P' exists but is unknown? (see Theorem 5.5 [1]).

3. The method of contractor directions has been applied to nonlinear differential and integral equations. In order to widen the scope of applicability of this theory, extensive studies of a priori estimates of solutions of linearized equations (which depend on a variable vector of a Banach space) are needed in both fields.

REFERENCES

[1] M. Altman, Contractor directions, directional contractors and directional contractions for solving equations, Pacific J. Math., 62 (1976), 1-18.

[2] ________________, General solvability theorems, to appear.

[3] ________________, Contractors and contractor directions - Theory and applications, to appear.

[4] ________________, An existence principle in nonlinear functional analysis, to appear.

[5] ________________, Concerning approximate solutions of nonlinear functional equations, Bull. Acad. Polon. Sci., Cl. III, vol. 5 (1957), 461-465.

[6] L. V. Kantorovich, G. P. Aiklov, Functional analysis in normed spaces, Oxford, Pergamon Press, 1964.

[7] W. C. Rheinboldt, A unified convergence theory for a class of iterative processes, SIAM J. Numer. Anal. 5 (1968), 42-63.

[8] J. Rokne, Newton's method under mild differentiability conditions with error analysis, Numer. Math. 18 (1972), 401-412.

[9] H. B. Keller, Newton's method under mild differentiability conditions, J. Comput. System Sci. 4 (1970), 15-28.

A Unified Approach to the Approximate Solution of Linear Integral Equations

by

P. M. Anselone and J. W. Lee

Department of Mathematics
Oregon State University
Corvallis, Oregon 97331

1. Introduction

Methods of approximate solution of Fredholm integral equations or compact operator equations in Banach spaces usually are formulated directly or indirectly in terms of matrix equations in finite dimensional spaces. A unified framework for such methods is presented here in terms of maps which connect the spaces involved. With considerable latitude in the determination of these maps, most of the approximation schemes in current use, including projection and numerical integration methods, are encompassed. In addition, new alternatives are revealed. The unified framework facilitates theoretical and practical comparison of alternative methods and provides a basis for a choice.

Similar techniques have been used in related settings and in various degrees of generality by Linz [5], Noble [6], Spence [9], and Thomas [10]. For recent contributions to projection and numerical integration methods see Anselone and Lee [2], Atkinson [3], Ikebe [4], Phillips [7], and Prenter [8].

Of the preceding references those by Noble and Linz were most influential for the present work. In [6] Noble uses the approximation scheme

$$\begin{cases}(I - K)x = y \sim (\hat{I}_n - \hat{K}_n)\hat{x}_n = \hat{y}_n, \\ x \sim x_n = p_n\hat{x}_n,\ \hat{x}_n = r_n x_n,\ \hat{y}_n = r_n y,\end{cases}$$

and the convergence relation $\|\hat{K}_n - r_n K p_n\| \to 0$. Here r_n and p_n are suitable maps used to relate the original problem $(I - K)x = y$ to an approximate matrix problem $(\hat{I}_n - \hat{K}_n)\hat{x}_n = \hat{y}_n$. Many desirable features of this approximation scheme are developed in [6]; however, error and convergence analysis is awkward because the problems compared are formulated in different spaces. Our approach is to augment and extend the approximation scheme so that both the original and approximate problems are formulated in the same space, and so that the desirable features of [5] and [6] are maintained. As already mentioned this augmented formulation has the additional advantage that it includes projection methods, prolongation-restriction methods, and numerical integration methods within a single general setting.

The following notation is adopted. Let X be a Banach space and $\hat{X}_n = R^n$ or C^n according as X is real or complex. The spaces $\hat{X}_n$ will be given appropriate norms. Spaces of bounded linear operators are denoted by $L(X)$, $L(\hat{X}_n)$, $L(X,\hat{X}_n)$ and $L(\hat{X}_n,X)$. The identity operators on X and $\hat{X}_n$ are I and $\hat{I}_n$. Let K be a compact linear operator in $L(X)$.

A general scheme for the determination of approximate solutions $x_n \in X$ of $(I - K)x = y$ is represented by

$$(1.1)\quad (I - K)x = y \sim (\hat{I}_n - \hat{K}_n)\hat{x}_n = \hat{y}_n,\ y \mapsto \hat{y}_n,\ \hat{x}_n \mapsto x_n \sim x.$$

The choices of the operators $\hat{K}_n \in L(\hat{X}_n)$ and the maps $y \mapsto \hat{y}_n$ and $\hat{x}_n \mapsto x_n$ determine particular cases.

An effective way to relate problems in X and $\hat{X}_n$ is by means of so-called restriction and prolongation maps:

$$r_n \in L(X,\hat{X}_n),\ p_n \in L(\hat{X}_n,X),\ r_np_n = \hat{I}_n,\ p_nr_n = E_n.$$

Then $E_n^2 = E_n \in L(X)$, so E_n is a projection. We assume $E_n \to I$ pointwise on X as $n \to \infty$. (Pointwise convergence on a suitable subspace of X often suffices.) In typical applications such as collocation, r_n restricts a function to a finite set, p_n interpolates a polynomial or spline function, and E_n approximates a function by a polynomial or spline function. Other applications include series truncation.

The scheme (1.1) is appropriate for the numerical determination of approximate solutions x_n. But it is not very convenient for convergence and error analysis. For these purposes it is advantageous to be given or to derive linear problems $(I - K_n)x_n = y_n$ for the x_n, where $K_n \in L(X)$. Then we have an augmented reformulation of the approximation scheme:

$$(1.2)\quad \begin{cases}(I - K)x = y \sim (I - K_n)x_n = y_n \iff (\hat{I}_n - \hat{K}_n)\hat{x}_n = \hat{y}_n, \\ x \sim x_n \leftrightarrow \hat{x}_n,\ \hat{x}_n = r_nx_n,\ y \sim y_n \leftrightarrow \hat{y}_n.\end{cases}$$

The operators K_n and $\hat{K}_n$ may be given in terms of K or in terms of each other. For example, if $K_n = E_nK$, $\hat{K}_n = r_nKp_n$, $\hat{y}_n = r_ny$ and $x_n = p_n\hat{x}_n$, then we obtain the standard projection method with

$$(I - K)x = y \sim (I - E_nK)x_n = E_ny.$$

In every case to be considered, $\dim K_nX \leq n$, so K_n is compact. The usual reduction of $(I - K_n)x_n = y_n$ to a matrix problem can be put into the setting (1.2) with $\hat{K}_n = r_nK_np_n$ and

$$x_n = p_n\hat{x}_n + (I - E_n)y_n, \; y_n = r_n[I + K_n(I - E_n)]y_n.$$

This framework also includes numerical integration methods for integral equations, generally with $\hat{K}_n = p_nK_nr_n$ and

$$x_n = y_n + K_np_n\hat{x}_n, \; \hat{y}_n = r_ny_n.$$

In all, eight subcases of (1.2) will be examined. Various applications to integral equations will be described.

From (1.2) consider

$$(1.3) \qquad (I - K)x = y \sim (I - K_n)x_n = y_n.$$

Typically, $y_n \to y$ and either $K_n \to K$ or $\|K_n - K\| \to 0$, sometimes by assumption and sometimes by derivation. Of course, $\|K_n - K\| \to 0$ implies $K_n \to K$. Available operator approximation theories can be brought to bear on the questions of the existence and convergence of $(I - K_n)^{-1}$. These theories generally give better theoretical results and error bounds than do methods expressed directly in terms of (1.1). Since the general operator approximation theories are framed in terms of minimal hypotheses they avoid extraneous hypotheses sometimes imposed.

Suppose that $(I - K)^{-1}$ and $(I - K_n)^{-1}$ exist at least for n sufficiently large. Then they belong to $L(X)$ by the Fredholm alternative. It follows from the Banach-Steinhaus theorem and the resolvent identity,

$$(I - K_n)^{-1} - (I - K)^{-1} = (I - K_n)^{-1}(K_n - K)(I - K)^{-1},$$

that $(I - K_n)^{-1} \to (I - K)^{-1}$ iff the operators $(I - K_n)^{-1}$ are uniformly bounded. From (1.3), by standard manipulations,

$$x = (I - K)^{-1}y, \; x_n = (I - K_n)^{-1}y_n,$$

$$x_n - x = [(I-K_n)^{-1} - (I-K)^{-1}]y + (I-K_n)^{-1}(y_n-y) = (I_n-K_n)^{-1}[(K_n-K)x + y_n-y]$$

$$\|x_n-x\| \leq \|(I-K_n)^{-1}\| \{\|(K_n-K)x\| + \|y_n-y\|\}.$$

If $(I - K_n)^{-1} \to (I - K)^{-1}$ in addition to $K_n \to K$ and $y_n \to y$, then $x_n \to x$ and the approximation scheme is well-posed. The bound for $\|x_n - x\|$ is of limited practical value since it involves the unknown solution x.

Suppose now that $\|K_n - K\| \to 0$. For example, $K_n = E_nK$. Then there exists $(I - K)^{-1}$ iff there exist uniformly bounded $(I - K_n)^{-1}$ for n sufficiently large, in which case $\|(I - K_n)^{-1} - (I - K)^{-1}\| \to 0$ and, hence, $x_n \to x$. Moreover, if $\Delta_n = \|(I - K_n)^{-1}\| \|K_n - K\| < 1$, then $(I - K)^{-1}$ exists and

$$\|(I - K_n)^{-1} - (I - K)^{-1}\| \leq \frac{\Delta_n \|(I - K_n)^{-1}\|}{1 - \Delta_n}.$$

This yields a more practical bound for $\|x_n - x\|$.

Often $K_n \to K$ but $\|K_n - K\| \not\to 0$, so the foregoing theory cannot be applied. This happens, for example, if numerical integration is used by choice or necessity, in the approximate solution of an integral equation. An alternative theory, presented in [1], is based on the hypotheses that $K_n \to K$ and $\{K_n : n = 1,2,\ldots\}$ is collectively compact, i.e., the set $\{K_n x : \|x\| \leq 1,\ n = 1,2,\ldots\}$ is relatively compact. Then there exists $(I - K)^{-1}$ iff there exist uniformly bounded $(I - K_n)^{-1}$ for n sufficiently large, in which case $(I - K_n)^{-1} \to (I - K)^{-1}$ and, hence, $x_n \to x$. Moreover, if $\Delta_n = \|(I - K_n)^{-1}\| \, \|(K_n - K)K\| <$ then $(I - K)^{-1}$ exists and

$$\|(I - K_n)^{-1}y - (I - K)^{-1}y\| = \frac{\|(I - K_n)^{-1}\| \ \|(K_n - K)y\| + \Delta_n\|x_n\|}{1 - \Delta_n},$$

which yields a practical bound for $\|x_n - x\|$.

The operator norm and collectively compact operator approximation theories have desirable features:

(a) the existence of $(I - K)^{-1}$ follows from a criterion based on $(I - K_n)^{-1}$ for a single value of n;

(b) $(I - K_n)^{-1}$ converges to $(I - K)^{-1}$ in norm or pointwise;

(c) error bounds are available for $\|x_n - x\|$ in terms of $(I - K_n)^{-1}$ and other available quantities.

These are minimal requirements for a satisfactory operator approximation theory. A fully practical scheme should have the additional feature that the abstract operator approximation theory pertains to approximations K_n, y_n and x_n which are determined numerically. In terms of the scheme (1.2) this means that K_n should be available in matrix form, the matrix elements $\hat{K}_{nij}$ and the vectors $\hat{y}_n$ are known explicitly or can be computed with errors which are demonstrably insignicant, and the maps $\hat{x}_n \mapsto x_n$ are computationally realizable. Otherwise, the abstract operator approximation theory would be applied to operators and elements which, in fact, are not available.

These are serious concerns, often overlooked, in the numerical solution of integral equations. Several common methods ostensibly yield $\|K_n - K\| \to 0$. However, $\hat{K}_{nij}$ involves single or double integrals which must be approximated numerically. When $\hat{K}_{nij}$ is redefined so as to include the numerical integration, then (cf. [2]), $\|K_n - K\| \to 0$ reduces to $K_n \to K$, and $\{K_n\}$ is collectively compact.

The organization of the paper is as follows. In Section 2, general properties of p_n, r_n and E_n are developed and examples in terms of particular spaces are given. The general approximation scheme (1.2) is explored in Section 3 along with a variety of special cases. In Section 4 applications to integral equations are presented. Alternative methods are compared and problems of practical implementation are discussed. For numerical results which support and complement our findings, see Atkinson [3].

The general framework presented here can be applied directly to eigenvalue problems. To do this simply set $y, y_n, \hat{y}_n = 0$ and replace K, K_n, etc. by $\lambda^{-1}K, \lambda_n^{-1}K_n$, etc. Also various features of our analysis extend in a straightforward way to nonlinear problems. This extension will not be discussed further.

2. Restriction, Prolongation, and Projection Maps

In this section we examine the basic connection between approximation methods defined by means of projections and those defined in terms of prolongation and restriction maps.

Consider any maps $r_n \in L(X,\hat{X}_n)$ and $p_n \in L(\hat{X}_n,X)$ such that

$$r_n p_n = \hat{I}_n. \tag{2.1}$$

Let X_n denote the range of p_n and Y_n the null space of r_n. Thus,

$$X_n = p_n\hat{X}_n \subset X, \qquad Y_n = \mathcal{N}(r_n) \subset X. \tag{2.2}$$

Then p_n is an isomorphism from $\hat{X}_n$ onto X_n, $\dim X_n = \dim \hat{X}_n = n$, X_n and Y_n are closed subspaces of X, and

$$r_n X = r_n X_n = \hat{X}_n, \qquad p_n^{-1} = r_n\big|_{X_n}. \tag{2.3}$$

Define

$$E_n = p_n r_n \in L(X). \tag{2.4}$$

Then $E_n^2 = E_n$, so E_n is a projection. Moreover,

$$E_n X = X_n, \qquad \mathcal{N}(E_n) = Y_n, \tag{2.5}$$

$$X = X_n \oplus Y_n, \tag{2.6}$$

$$r_n E_n = r_n = p_n^{-1}E_n, \qquad E_n p_n = p_n. \tag{2.7}$$

For the moment, let E_n be any projection in $L(X)$ with $\dim E_nX = n$. Define $X_n = E_nX$ and $Y_n = \mathcal{N}(E_n)$. Choose any $p_n \in L(\hat{X}_n, X)$ such that $p_n\hat{X}_n = X_n$. Then there exists $p_n^{-1} \in L(X_n, \hat{X}_n)$. Define $r_n = p_n^{-1}E_n \in L(X, \hat{X}_n)$. Then $E_np_n = p_n$, $r_np_n = p_n^{-1}E_np_n = p_n^{-1}p_n = \hat{I}_n$, and (2.1) - (2.7) are satisfied. Thus, either the pair (r_n, p_n) or E_n can be used as the basis for the formulation and analysis of an approximation method. As a consequence, the usual types of restriction-prolongation methods can be recast as projection methods and vice versa.

We assume the pointwise convergence on X,

(2.8) $$E_n \to I,$$

which is necessarily uniform on compact sets. The norm on $\hat{X}_n$, which is at our disposal, is specified by

(2.9) $$\|\hat{x}_n\|_n = \|p_n\hat{x}_n\| .$$

This choice is natural and convenient for theoretical and practical convergence and error analysis. Now p_n is an isometry and

(2.10) $$\|p_n\| = 1, \qquad \|r_n\| = \|E_n\| ,$$

(2.11) $$\|r_nx\|_n = \|E_nx\| \to \|x\|, \qquad x \in X.$$

From (2.8), (2.10) and the Banach-Steinhaus theorem, the operators E_n, r_n and p_n are uniformly bounded. Often, in contrast to the present situation, $\|r_n x\|_n \to \|x\|$ and the uniform boundedness of r_n and p_n are assumed. It follows from (2.4), (2.8) and the uniform boundedness of the maps r_n that

$$x_n \to x \iff r_n x_n - r_n x \to 0, \qquad x \in X,\ x_n \in X_n. \tag{2.12}$$

The maps r_n and p_n have unique representations

$$r_n x = [f_{n1}(x), \ldots, f_{nn}(x)], \qquad f_{ni} \in X^*, \tag{2.13}$$

$$p_n \hat{x}_n = \sum_{j=1}^{n} \hat{x}_{nj}\, \phi_{nj}, \qquad \phi_{nj} \in X_n. \tag{2.14}$$

Practical considerations often dictate that $X_n \subset X_{n+1}$ for all n, in which case it is convenient to choose $\phi_{nj} = \phi_j$ independent of n, with $\{\phi_j\}$ a basis for X. The hypothesis $r_n p_n = \hat{I}_n$ is equivalent to

$$f_{ni}(\phi_{nj}) = \delta_{ij}. \tag{2.15}$$

So $\{\phi_{nj} : j = 1, \ldots, n\}$ is a basis for X_n and $\{f_{ni}\big|_{X_n} : i = 1, \ldots, n\}$ is the dual basis for X_n^*. Since $E_n = p_n r_n$,

$$E_n x = \sum_{j=1}^{n} f_{nj}(x)\phi_{nj}. \tag{2.16}$$

As an example, let $X = C$, the space of continuous functions on $[0,1]$ with the max norm. Two choices for $r_n = [f_{n1}, \ldots, f_{nn}]$, both with practical significance, are described in the following cases.

Case A.

(2.17) $f_{ni}(x) = \langle x, \theta_{ni}\rangle = \int_0^1 x(s)\theta_{ni}(s)ds, \qquad \theta_{ni} \in C,$

(2.18) $\langle \phi_{nj}, \theta_{ni}\rangle = \delta_{ij}, \qquad \phi_{ni} \in C,$

(2.19) $r_n x = [\langle x, \theta_{n1}\rangle, \ldots, \langle x, \theta_{nn}\rangle],$

(2.20) $E_n x = \sum_{j=1}^{n} \langle x, \theta_{nj}\rangle \phi_{nj}.$

The special case $\theta_{nj} = \phi_{nj} = \phi_j$ where $\{\phi_j\}$ is orthonormal, perhaps relative to a weight function, is used in the Galerkin method. Then (2.20) is a truncated orthogonal series.

Case B.

(2.21) $f_{ni}(x) = x(t_{ni}), \qquad 0 \leq t_{n1} < \cdots < t_{nn} \leq 1,$

(2.22) $\phi_{nj}(t_{ni}) = \delta_{ij},$

(2.23) $r_n x = [x(t_{n1}), \ldots, x(t_{nn})],$

(2.24) $E_n x = \sum_{j=1}^{n} x(t_{nj})\phi_{nj}.$

For example, $E_n x$ may be a spline function which interpolates x at knots t_{ni}. In this situation, $\{\phi_{nj} : j = 1, \ldots, n\}$ is an appropriate spline basis for X_n.

3. Specializations of the Approximation Scheme

Consider the general approximation scheme introduced in Section 1:

$$(3.1)\quad \begin{cases} (I - K)x = y \sim (I - K_n)x_n = y_n \iff (\hat{I}_n - \hat{K}_n)\hat{x}_n = \hat{y}_n, \\ x \sim x_n \leftrightarrow \hat{x}_n, \quad \hat{x}_n = r_n x_n, \quad y \sim y_n \leftrightarrow \hat{y}_n. \end{cases}$$

We shall examine and compare eight cases of (3.1) determined by specifying relations among the operators and vectors. These include the most commonly used methods and some new alternatives. The cases are related as follows:

Cases 2, 3, 5, 7 specialize Case 1,

Cases 5 - 8 specialize Case 4,

Case 7 specializes Case 5.

Typically by assumption or derivation, $y_n \to y$ and either $K_n \to K$ or $\|K_n - K\| \to 0$. In particular, we may have $y_n = y$ for all n. Since K is compact and $E_n \to I$ uniformly on compact sets,

$$(3.2)\qquad \|E_n K - K\| \to 0.$$

Thus, for a projection method with $K_n = E_n K$, we have $\|K_n - K\| \to 0$.

Case 1. Let

$$(3.3)\qquad \hat{K}_n = r_n K_n p_n, \quad E_n K_n = K_n.$$

The relations in Section 2 yield

$$(3.4) \qquad K_nX \subset X_n, \qquad p_n\hat{K}_n = K_np_n,$$

$$(3.5) \qquad \hat{K}_n = [\hat{K}_{nij}], \qquad \hat{K}_{nij} = f_{ni}(K_n\phi_{nj}).$$

If K_n is any operator in $L(X)$ with $\dim K_nX \leq n$, then X_n can be selected such that $K_nX \subset X_n$ and $\dim X_n = n$, after which E_n, p_n and r_n can be specified. Then $E_nK_n = K_n$ and we can define $\hat{K}_n = r_nK_np_n$.

It is well-known that $(I - K_n)x_n = y_n$ with $\dim K_nX < \infty$ reduces to a matrix problem. Such a reduction is expressed in the context of (3.1) by

$$(3.6) \quad \begin{cases} (I - K)x = y \sim (I - K_n)x_n = y_n \iff (\hat{I}_n - \hat{K}_n)\hat{x}_n = \hat{y}_n, \\ x_n = p_n\hat{x}_n + (I - E_n)y_n, \quad \hat{x}_n = r_nx_n, \; \hat{y}_n = r_n[I+K_n(I-E_n)]y_n. \end{cases}$$

To verify this, first define $x_n = p_n\hat{x}_n + (I - E_n)y_n$. Then $r_nx_n = \hat{x}_n$ and, by routine steps,

$$(I - K_n)x_n = p_n(\hat{I}_n - \hat{K}_n)\hat{x}_n + (I - K_n)(I - E_n)y_n,$$

$$(\hat{I}_n - \hat{K}_n)\hat{x}_n = r_n(I - K_n)x_n + r_nK_n(I - E_n)y_n,$$

from which (3.6) follows.

In Case 1 no relation between K_n and K is specified in (3.6). To make (3.6) practical K_n must be related to K. The following specializations of Case 1 include such relations which yield norm or pointwise convergence of K_n to K. The following is pertinent to this convergence. Since

$$K_n - K = E_n(K_n - K) + E_nK - K,$$

(3.2) and $E_n = p_nr_n$ yield

$$(3.7)\quad \|K_n - K\| \to 0 \iff \|E_n(K_n - K)\| \to 0 \iff \|r_n(K_n - K)\| \to 0,$$

$$(3.8)\quad K_n \to K \iff E_n(K_n - K) \to 0 \iff r_n(K_n - K) \to 0.$$

Case 2. Let

$$(3.9)\quad \hat{K}_n = r_n K p_n, \qquad K_n = E_n K, \qquad y_n = y.$$

This specializes Case 1. Now

$$(3.10)\quad \hat{K}_n = [\hat{K}_{nij}], \qquad \hat{K}_{nij} = f_{ni}(K\phi_{nj}),$$

and (3.6) pertains to the projection method

$$(3.11)\quad (I - K)x = y \sim (I - E_n K)x_n = y.$$

Since $\|E_n K - K\| \to 0$, the operator norm convergence theory applies.

Case 3. Let

$$(3.12)\quad \hat{K}_n = r_n K p_n, \qquad K_n = E_n K, \qquad y_n = E_n y.$$

This is another specialization of Case 1. Again (3.10) holds. Now (3.6) becomes the widely used projection method:

$$(3.13)\quad \begin{cases} (I - K)x = y \sim (I - E_n K)x_n = E_n y \iff (\hat{I}_n - \hat{K}_n)\hat{x}_n = \hat{y}_n, \\ x_n = p_n \hat{x}_n, \qquad \hat{x}_n = r_n x_n, \qquad \hat{y}_n = r_n y. \end{cases}$$

In this case, $x_n \in X_n$ for all n. As before, $\|E_n K - K\| \to 0$, so the operator norm approximation theory applies. Case 2 should be superior to Case 3 since y is not approximated in the former, whereas y is replaced by $E_n y$ in the latter.

Case 4. Let

$$r_n K_n = \hat{K}_n r_n, \qquad K_n E_n = K_n. \tag{3.14}$$

Specializations will include certain projection methods and numerical integration approximations of integral operators. Consequences of (3.14) are

$$\hat{K}_n = r_n K_n p_n, \qquad p_n \hat{K}_n r_n = E_n K_n, \tag{3.15}$$

$$\mathcal{N}(K_n) \supset Y_n, \qquad \dim K_n X = \operatorname{codim} \mathcal{N}(K_n) \leq n, \tag{3.16}$$

where $Y_n = \mathcal{N}(E_n) = \mathcal{N}(r_n)$. Define $\tilde{K}_n \in L(\hat{X}_n, X)$ by

$$\tilde{K}_n = K_n p_n. \tag{3.17}$$

From (3.14) and (3.17),

$$K_n = \tilde{K}_n r_n, \qquad \hat{K}_n = r_n \tilde{K}_n. \tag{3.18}$$

In Case 4, (3.1) becomes

$$\begin{cases} (I - K)x = y \sim (I - K_n)x_n = y_n \iff (\hat{I}_n - \hat{K}_n)\hat{x}_n = \hat{y}_n, \\ x_n = y_n + \tilde{K}_n \hat{x}_n, \qquad \hat{x}_n = r_n x_n, \qquad \hat{y}_n = r_n y_n. \end{cases} \tag{3.19}$$

To verify this, first let $(I - K_n)x_n = y_n$ and $\hat{x}_n = r_n x_n$. Then

$$(\hat{I}_n - \hat{K}_n)\hat{x}_n = r_n(I - K_n)x_n = r_n y_n = \hat{y}_n.$$

Now let $(\hat{I}_n - \hat{K}_n)\hat{x}_n = \hat{y}_n$ and $x_n = y_n + \tilde{K}_n\hat{x}_n$. Then

$$r_n x_n = r_n y_n + r_n \tilde{K}_n \hat{x}_n = \hat{y}_n + \hat{K}_n \hat{x}_n = \hat{x}_n,$$

$$K_n x_n = \tilde{K}_n r_n x_n = \tilde{K}_n \hat{x}_n, \qquad (I - K_n) x_n = x_n - \tilde{K}_n \hat{x}_n = y_n.$$

Thus, (3.19) is established.

<u>Case 5</u>. Let

$$K_n = p_n \hat{K}_n r_n. \tag{3.20}$$

This is included in Cases 1 and 4. We now have

$$\begin{cases} (I - K)x = y \sim (I - K_n)x_n = y_n \iff (\hat{I}_n - \hat{K}_n)\hat{x}_n = \hat{y}_n, \\ x_n = p_n \hat{x}_n + (I - E_n) y_n, \quad \hat{x}_n = r_n x_n, \quad \hat{y}_n = r_n y_n. \end{cases} \tag{3.21}$$

It follows from (3.7), (3.8) and $r_n K_n = \hat{K}_n r_n$ that

$$\|K_n - K\| \to 0 \iff \|\hat{K}_n r_n - r_n K\| \to 0, \tag{3.22}$$

$$K_n \to K \iff \hat{K}_n r_n - r_n K \to 0. \tag{3.23}$$

The right members of (3.22) and (3.23) are forms of discrete convergence. Equations (3.22) and (3.23) clarify the role of the discrete convergence hypothesis which usually is made in the context of (1.1) where K is "approximated" directly by the matrix $\hat{K}_n$ without reference to K_n.

By (3.17) and (3.20), $\tilde{K}_n = K_n p_n = p_n \hat{K}_n$ and $K_n = \tilde{K}_n r_n$ so,

(3.24) $\{K_n\}$ collectively compact $\iff$ $\{\tilde{K}_n\}$ collectively compact, where the right member means that the set

$$\{p_n \hat{K}_n \hat{x}_n : \hat{x}_n \in \hat{X}_n, \quad \|\hat{x}_n\|_n \leq 1, \quad n = 1, 2, \ldots\}$$

is relatively compact. This version of collective compactness, with the $\hat{x}_n$ in different spaces, is due to Linz [5]. From (3.23) and (3.24),

$$(3.25) \quad \begin{cases} K_n \to K, \quad \{K_n\} \text{ collectively compact } \iff \\ \hat{K}_n r_n - r_n K \to 0, \quad \{p_n \hat{K}_n\} \text{ collectively compact.} \end{cases}$$

When these conditions are satisfied, the collectively compact operator approximation theory gives convergence results and error bounds for $\|x_n - x\|$.

The work of Noble in [6] can be put into the context of Case 5. He studies the basic approximation scheme

$$(3.26) \quad \begin{cases} (I - K)x = y \sim (\hat{I}_n - \hat{K}_n)\hat{x}_n = \hat{y}_n, \\ x_n = p_n \hat{x}_n, \quad \hat{x}_n = r_n x_n, \quad \hat{y}_n = r_n y, \end{cases}$$

under the hypothesis that

$$(3.27) \qquad \|\hat{K}_n - r_n K p_n\| \to 0.$$

He does not define operators K_n, but rather deals directly with $\hat{K}_n$ and K to obtain convergence theorems and error bounds for approximate solutions. However, the bounds are rather complicated and, because of several triangle inequalities, they seem to be not very sharp. Thus, their practical value is limited.

Define $K_n = p_n\hat{K}_n r_n$ as in Case 5. Also let $y_n = E_n y$. Then (3.21) and (3.26) are equivalent. Note that

$$\hat{K}_n r_n - r_n K = (\hat{K}_n - r_n K p_n) r_n + r_n K(E_n - I),$$

$$p_n \hat{K}_n = p_n(\hat{K}_n - r_n K p_n) + E_n K p_n.$$

It follows by means of (3.27) that $\hat{K}_n r_n - r_n K \to 0$ and $\{\hat{K}_n\}$ is collectively compact, as Linz [5] observed. Hence, by (3.25), $K_n \to K$ and $\{K_n\}$ is collectively compact, and we have convergence results and practical error bounds for $\|x_n - x\|$.

The connection between Case 4 and the narrower Case 5 is closer than one might suspect. If $\hat{K}_n$ is nonsingular, which is likely, then there are unique choices of X_n, E_n and p_n in Case 4 such that $K_n = p_n\hat{K}_n r_n$, which defines Case 5. This assertion is verified as follows. From (3.14) - (3.16),

$$n \geq \dim K_n X \geq \dim r_n K_n X = \dim \hat{K}_n r_n X = \dim \hat{K}_n \hat{X}_n = n,$$

$$\dim K_n X = n \geq \operatorname{codim} \mathcal{N}(K_n), \qquad \mathcal{N}(K_n) = Y_n,$$

$$K_n = p_n \hat{K}_n r_n \iff K_n = E_n K_n \iff X_n = K_n X.$$

Define $X_n = K_n X$. Then (3.14) and the nonsingularity of $\hat{K}_n$ yield

$$K_n^2 x = 0 \Rightarrow \hat{K}_n^2 r_n x = r_n K_n^2 x = 0 \Rightarrow r_n x = 0 \Rightarrow K_n x = 0,$$

$$X_n \cap Y_n = K_n X \cap \mathcal{N}(K_n) = \{0\}, \qquad X = X_n \oplus Y_n.$$

Define E_n as the projection with $E_n X = X_n$ and $\mathcal{N}(E_n) = Y_n$. Define $p_n = \left(r_n \middle| X_n\right)^{-1}$. Then $r_n p_n = \hat{I}_n$, $p_n r_n = E_n$, and $K_n = E_n K_n = p_n \hat{K}_n r_n$. Thus, Case 5 is obtained.

Case 6. Let

$$\hat{K}_n = r_n K p_n, \qquad K_n = KE_n. \tag{3.28}$$

This specializes Case 4 to the projection method:

$$\begin{cases} (I - K)x = y \sim (I - KE_n)x_n = y_n \iff (\hat{I}_n - \hat{K}_n)\hat{x}_n = \hat{y}_n, \\ x_n = y_n + Kp_n\hat{x}_n, \quad \hat{x}_n = r_n x_n, \quad \hat{y}_n = r_n y_n. \end{cases} \tag{3.29}$$

As before, greater accuracy is to be expected if $y_n = y$ than if $y_n = E_n y$ or $y_n \to y$ otherwise. Since $E_n \to I$ we have $KE_n \to K$. Since $\{E_n\}$ is bounded and K is compact, $\{KE_n\}$ is collectively compact. Therefore the collectively compact theory compares $(I - K)^{-1}$ and $(I - KE_n)^{-1}$ and gives computable error bounds. Alternatively, we can take advantage of the fact that

$$\exists(I - KE_n)^{-1} \iff \exists(I - E_nK)^{-1} \implies$$

$$(I - KE_n)^{-1} = I + K(I - E_nK)^{-1}E_n.$$

Since $\|E_nK - K\| \to 0$, the operator norm approximation theory is applicable. It seems likely, however, that the collectively compact theory will yield sharper error bounds because fewer triangle inequality applications are involved.

Case 7. Let

$$\hat{K}_n = r_n K p_n, \qquad K_n = p_n \hat{K}_n r_n = E_n K E_n. \tag{3.30}$$

This is a specialization of Cases 1, 4 and 5. We now have another familiar projection method:

$$(3.31)\quad \begin{cases} (I - K)x = y \sim (I - E_nKE_n)x_n = y_n \iff (\hat{I}_n - \hat{K}_n)\hat{x}_n = \hat{y}_n , \\ x_n = p_n\hat{x}_n + (I - E_n)y_n, \quad \hat{x}_n = r_nx_n, \quad \hat{y}_n = r_ny. \end{cases}$$

It should be expected that Cases 3 and 6 are superior because they involve fewer approximations.

By straightforward arguments, $E_nKE_n \to K$ and $\{E_nKE_n\}$ is collectively compact. So the corresponding operator approximation theory applies. Alternatively, we can use

$$\exists(I - E_nKE_n)^{-1} \iff \exists(I - E_nK)^{-1} \implies$$

$$(I - E_nKE_n)^{-1} = I + E_nK(I - E_nK)^{-1}E_n,$$

and apply the operator norm theory.

<u>Case 8</u>. This is a reformulation of Case 4 which is particularly appropriate for numerical integration approximations of integral operators. Assume given any $\tilde{K}_n \in L(\hat{X}_n, X)$ and any $r_n \in L(X, \hat{X}_n)$ with $r_nX = \hat{X}_n$. Define K_n and $\hat{K}_n$ by (3.18). Then (3.19) follows. Although p_n and E_n are not needed in this version of Case 4, they can be introduced as follows. Choose any $p_n \in L(\hat{X}_n, X)$ such that $r_np_n = \hat{I}_n$ and define $E_n = p_nr_n$ as usual. Then (3.14)ff are satisfied. If $\hat{K}_n$ is nonsingular, which is likely, then Case 8 can be put under Case 5 and, therefore, it becomes related to discrete convergence.

4. Integral Equations

Let $X = C[0,1]$. Let $K \in L(X)$ be a Fredholm integral operator,

$$(4.1) \qquad (Kx)(s) = \int_0^1 k(s,t)x(t)dt,$$

with a continuous kernel $k(s,t)$, $0 \leq s,t \leq 1$. Then K is compact and $||K|| \leq ||k|| = \max_{s,t}|k(s,t)|$. We shall apply various cases of the general approximation scheme to the Fredholm integral equation $(I - K)x = y$.

First consider the familiar projection method, Case 3. Thus, let

$$(4.2) \qquad \hat{K}_n = r_n K p_n, \qquad K_n = E_n K, \qquad y_n = E_n y.$$

Since $||E_n K - K|| \to 0$, the operator norm convergence theory applies. Now $E_n K$ is a Fredholm integral operator with a continuous kernel:

$$(4.3) \qquad (E_n Kx)(s) = \int_0^1 (E_n k)(s,t)x(t)dt,$$

where E_n acts on k as a function of s. A simple proof of this is based on the uniform approximation of k by kernels of finite rank. The proof is even easier when E_n is given by (2.20) or (2.24). It follows from $E_n \to I$ and the fact that $k(s,t)$ can be regarded as a compact set of functions of s parameterized by t that

(4.4) $\qquad (E_n k)(s,t) \to k(s,t)$ uniformly in s,t,

(4.5) $$\|E_n K - K\| \leqq \|E_n k - k\| \to 0.$$

For further elaboration of (4.3) - (4.5) and related results see [2].

Now let us specialize E_n and r_n to Cases A and B of Section 2. The notation, Case 3-A, will mean that Case 3 is used with E_n and r_n given by Case A, and so forth.

<u>Case 3-A</u>. Define r_n by (2.19). It follows from (3.10) and (3.13) that

(4.6) $$\hat{K}_{nij} = \int_0^1\int_0^1 k(s,t)\phi_{nj}(t)\theta_{ni}(s)\,dtds, \qquad i, j = 1,\ldots, n,$$

(4.7) $$\hat{y}_{ni} = \int_0^1 y(t)\theta_{ni}(t)\,dt, \qquad i = 1,\ldots, n,$$

(4.8) $$x_n = \sum_{j=1}^{n} \hat{x}_{nj}\,\phi_{nj},$$

(4.9) $$(E_n k)(s,t) = \sum_{j=1}^{n} \phi_{nj}(s) \int_0^1 k(s,t)\theta_{nj}(s)\,ds.$$

<u>Case 3-B</u>. Define r_n by (2.23). Then

(4.10) $$\hat{K}_{nij} = \int_0^1 k(t_{ni},t)\phi_{nj}(t)\,dt, \qquad i, j = 1,\ldots, n,$$

(4.11) $$\hat{y}_{ni} = y(t_{ni}), \qquad i = 1,\ldots, n,$$

(4.12) $x_n = \sum_{j=1}^{n} \hat{x}_{nj} \phi_{nj}$,

(4.13) $(E_n k)(s,t) = \sum_{j=1}^{n} k(t_{ni},t)\phi_{nj}(s)$.

The practical application of these approximation schemes requires that $\hat{K}_n = [\hat{K}_{nij}]$ is explicitly available, hence that the integrals in (4.6) and (4.10) can be evaluated in closed form. This limits the methods to rather simple kernels. We also need $r_n y$. This presents no problem in Case 3-B, in view of (4.11). In Case 3-A, however, the integrals in (4.7) must be evaluated. If they must be done numerically, this leads to the approximation of y by a convergent sequence $\{y_n\}$. Then a simple triangle inequality argument accounts for the additional approximation.

If the integrals in (4.6) or (4.10) must be evaluated numerically, then the corresponding approximation method of Section 3 does not apply to the numerically available solution. Rather, as mentioned in the introduction, the computed solution corresponds to a related double approximation scheme whose convergence and error analysis depends on collective compactness and pointwise convergence, not the operator norm convergence in (4.5). This type of double approximation and its consequences are treated in [2].

It is clear from the preceding examples that the methods of Section 3 using Case B are easier to implement than those using Case A. For this reason and for simplicity of exposition, we examine below only methods depending on Case B.

Case 2-B. For this projection method $\hat{K}_{nij}$ and $E_n k$ are given by (4.10) and (4.13), while

$$\hat{y}_{ni} = y(t_{ni}) + Ky(t_{ni}) - \sum_{j=1}^{n} y(t_{nj})\hat{K}_{nij}, \tag{4.14}$$

$$x_n = y + \sum_{j=1}^{n} [\hat{x}_{nj} - y(t_{nj})]\phi_{nj}. \tag{4.15}$$

For this case, the explicit availability of the discrete system $(\hat{I}_n - \hat{K}_n)\hat{x}_n = \hat{y}_n$ requires that all the integrals appearing in $\hat{K}_n$ and $Ky(t_{ni})$ can be found analytically. As pointed out in Section 3, this method should be superior, in general, to Case 3-B. On the other hand, Case 3-B may yield better numerical results if the integrals $K\phi_{nj}(t_{ni})$ can be evaluated in closed form but the integrals $Ky(t_{ni})$ cannot. In this case $\hat{K}_n$ and $\hat{y}_n$ are known explicitly for Case 3-B, but $\hat{y}_n$ is not for Case 2-B. If the integrals in both $K\phi_{nj}(t_{ni})$ and $Ky(t_{ni})$ must be done numerically, then Case 2-B should be more accurate.

The projection methods above are based on kernel approximation. Methods based on numerical quadrature fall within the scope of Case 4 and its specializations. Assume below that a convergent quadrature formula,

$$\sum_{j=1}^{n} w_{nj}\, x(t_{nj}) \to \int_0^1 x(t)\,dt, \qquad x \in X, \tag{4.16}$$

is given.

Case 4 - B. In the Nyström method, the integral operator K is approximated by numerical-integral operators $K_n \in L(X)$ of the form

(4.17) $$(K_n x)(s) = \sum_{j=1}^{n} w_{nj}\, k(s, t_{nj}) x(t_{nj}).$$

Also define $\tilde{K}_n \in L(\hat{X}_n, X)$ and $\hat{K}_n \in L(\hat{X}_n)$ by

(4.18) $$(\tilde{K}_n \hat{x}_n)(s) = \sum_{j=1}^{n} w_{nj}\, k(s, t_{nj}) \hat{x}_{nj},$$

(4.19) $$\hat{K}_{nij} = w_{nj}\, k(t_{ni}, t_{nj}).$$

Then

(4.20) $$K_n = \tilde{K}_n r_n, \qquad \hat{K}_n = r_n K_n,$$

so (3.18) is satisfied and we obtain Case 8. Since p_n and E_n can be introduced so that (3.14) holds, we also have a specialization of Case 4. The well-known reduction of $(I - K_n)x_n = y_n$ to an algebraic system $(\hat{I}_n - \hat{K}_n)\hat{x}_n = \hat{y}_n$ is formulated in (3.19). If the matrix $\hat{K}_n$ is nonsingular, then we also have a specialization of Case 5.

Case 5 - B. In (3.20) choose

(4.21) $$\hat{K}_{nij} = w_{nj}\, k(t_{ni}, t_{nj}).$$

Then

(4.22) $$K_n x = \sum_{i=1}^{n} \{ \sum_{j=1}^{n} w_{nj}\, k(t_{ni}, t_{nj}) x(t_{nj}) \} \phi_{nj},$$

the approximation procedure is specified by (3.21), and

$$x_n = p_n\hat{x}_n + (I - E_n)y_n. \tag{4.23}$$

The sequence of operators $\{K_n\}$ is collectively compact and $K_n \to K$. This can be verified directly or by means of [2, Th. 5.1]. Thus, convergence results and error bounds follow from the collectively compact theory.

The foregoing case complements the procedure studied by Noble in [6]. When $y_n = E_n y$, the discrete system $(\hat{I}_n - \hat{K}_n)\hat{x}_n = \hat{y}_n$ and the approximate solution $x_n = p_n\hat{x}_n$ are the same as those in [6]. The two approaches differ in that the error bounds in [6] depend on the operator norm convergence in (3.27) rather than on collective compactness, and also in that the choice of the convergent quadrature rule is further restricted in [6].

<u>Case 6-B</u>. In this situation,

$$\hat{K}_{nij} = (r_n K p_n)_{ij} = K\phi_{nj}(t_{ni}), \qquad i, j = 1,\ldots, n, \tag{4.24}$$

$$\hat{y}_n = y_n(t_{ni}), \qquad i = 1,\ldots, n, \tag{4.25}$$

$$x_n(s) = y_n(s) + \sum_{j=1}^{n} \hat{x}_{nj}\, K\phi_{nj}(s), \tag{4.26}$$

$$(K_n x)(s) = (KE_n x)(s) = \sum_{j=1}^{n} x(t_{nj})K\phi_{nj}(s). \tag{4.27}$$

When $y_n = y$ this case should be more accurate than the standard projection method, Case 3-B. Notice also that the quantites $\hat{K}_n$ and $\hat{y}_n$ are explicitly available for Case 6-B whenever they are available for Case 3-B.

When $y_n = E_n y$ there is a close connection between the standard (left) projection method, Case 3-B, and the (right) projection method, Case 6-B. In this case, $\hat{K}_n$ and $\hat{y}_n$ are identical for the two methods and hence (assuming unique solvability) $\hat{x}_n$ is the same for both methods. Reference to (3.13) and (3.29) yields

$$(4.28) \qquad \begin{cases} x_{n,6} = E_n y + K x_{n,3}, \\ x_{n,3} = E_n x_{n,6}, \end{cases}$$

where $x_{n,3}$ and $x_{n,6}$ are the approximate solutions in Case 3-B and Case 6-B.

The reader may have noticed that very little has been said about the choice of p_n. Indeed, when r_n is specified by (2.23) in Case B, the functions ϕ_{nj} defining p_n by (2.14) are subject only to two conditions; namely, they must be continuous and satisfy $\phi_{nj}(t_{ni}) = \delta_{ij}$. Thus, there is considerable flexibility in the choice of the prolongation maps p_n. This flexibility has important practical significance. For example, in Case 5-B, (4.21) implies that $\hat{x}_n$ is independent of the choice of $\{\phi_{nj}\}$ while the solution x_n depends on $\{\phi_{nj}\}$ through (4.23). Thus, after $\hat{x}_n$ is found, we can determine $x_n = p_n \hat{x}_n + (I - E_n) y_n$ in terms of any set of basis splines $\{\phi_{nj}\}$ adequate for our purposes. For instance, if $y_n = y$ and y has only two continuous derivatives, an appropriate choice for $\{\phi_{nj}\}$ would be the piecewise linear basis splines with knots $\{t_{nj}\}$. Use of higher degree splines would increase computation time with no probable increase in accuracy. Thus, the choice of basic splines $\{\phi_{nj}\}$ can be tailored to the smoothness of k and y.

For numerical evidence of some of the qualitative comparisons made above, see Atkinson [3].

References

1. P. M. Anselone, Collectively Compact Operator Approximation Theory, Prentice-Hall, Englewood Cliffs, New Jersey (1971).

2. P. M. Anselone, and J. W. Lee, Double approximation methods for the solution of Fredholm integral equations, in Numerische Methoden der Approximationstheorie, Bd. 3, Internationale Schriftenreihe zur numerische Mathematik, 30 (1976), p. 9-34.

3. K. E. Atkinson, A Survey of Numerical Methods for the Solution of Fredholm Integral Equations of the Second Kind, SIAM Publications, Philadelphia, 1976.

4. Y. Ikebe, The Galerkin method for the numerical solution of Fredholm integral equations of the second kind, SIAM Review 14, (1972), p. 465-491.

5. P. Linz, A general theory for the approximate solution of operator equations of the second kind, SIAM J. Num. An., to appear.

6. B. Noble, Error analysis of collocation methods for solving Fredholm integral equations, in Topic in Numerical Analysis, J. J. H. Miller, ed., Academic Press, 1973, p. 211-232.

7. J. L. Phillips, The use of collocation as a projection method for solving linear operator equations, SIAM J. Numer. Anal. 9, (1972), p. 14-28.

8. P. M. Prenter, A collocation method for the numerical solution of integral equations, SIAM J. Numer. Anal. 10, (1973), p. 570-581.

9. A. Spence, The numerical solution of the integral equation eigenvalue problem, Ph.D. thesis, University of Oxford, 1974.

10. K. S. Thomas, On the Approximate solution of operator equations, Numer. Math 23 (1975), p. 292-301.

The Topological Degree Applied to Some Problems in Approximation Theory

David L. Barrow*

Department of Mathematics
Texas A&M University
College Station, Texas 77843 USA

1. Introduction

The problems to be described herein can be reduced ultimately to the following question: does a certain system of n equations in n unknowns have a solution; if so, is the solution unique? The primary tool used in answering this question will be the finite dimensional topological degree of a mapping (cf. Nirenberg [11] or Schwartz [14]). In this section we give a brief discussion of the degree, including the properties which will be needed in the subsequent analysis.

Let D be a bounded open subset of the Euclidean space R^n, and let $F:\bar{D} \to R^n$ be continuous. Then if $q \in R^n$ and $q \neq F(x)$ for $x \in \partial D$ (the boundary of D), the degree of F with respect to D and q is defined, is an integer, and will be denoted deg(F,D,q). The following are some properties of the degree:

(i) Let $F \in C^1(D) \cap C(\bar{D})$ satisfy $x \notin \partial D$ and $\det F'(x) \neq 0$ whenever $F(x) = q$. Then there are a finite number of points $x_i \in D$ where $F(x_i) = q$, and $\deg(F,D,q) = \sum_i \operatorname{sgn}(\det F'(x_i))$.

(ii) If $\deg(F,D,q) \neq 0$, there is at least one solution in D to the equation $F(x) = q$.

(iii) If $F:\bar{D} \times [a,b] \to R^n$ is continuous and $F(x,\lambda) \neq q$ for $x \in \partial D$, $a \leq \lambda \leq b$, then $\deg(F(\cdot,\lambda),D,q)$ is constant, independent of λ.

(iv) If D is a symmetric bounded neighborhood of the origin and F is an odd continuous mapping ($F(-x) = -F(x)$) with $0 \notin F(\partial D)$, then deg(F,D,0) is an odd number (Borsuk's Theorem).

*This work was partially supported by National Science Foundation Grants MCS 77-02464 and DC R75-04545.

The problems in the following sections will be reduced to examining solutions to an equation $F(x) = q$. In §§2-4, the existence of solutions (in D) will be accomplished by showing that $\deg(F,D,q) = 1$; this will be done by constructing a homotopy, as in (iii) above, deforming F to a map F_0 for which $\deg(F_0,D,q) = 1$ can be established directly. In §§2 and 3 we also show that these solutions are unique via property (i), by showing that $\det F'(x) > 0$ whenever $F(x) = q$. In §5 a corollary to Borsuk's Theorem is used to prove the existence of a solution in ∂D to $F(x) = 0$.

2. Gaussian quadrature formulae with multiplicities

A quadrature formula is an approximation to a definite integral by a linear combination of values of the integrand and its derivatives at selected points. It is a classical result that the formula

$$\int_a^b f(t)\phi(t)dt \simeq \sum_{i=1}^{k} a_i f(t_i) , \quad \phi(t) \geq 0, \tag{2.1}$$

can, by an appropriate choice of the points $a < t_1 < \ldots < t_k < b$ and weights a_i, be made exact for all integrands f which are polynomials of degree at most 2k-1. Furthermore, this choice of points and weights is unique. An equivalent statement is that a certain system of 2k equations in 2k unknowns has a unique solution. Formulas of this type, and their various generalizations, are usually known as Gaussian quadrature formulas.

One such generalization is Theorem 2.1 (below), which is proven in [4]. This theorem concerns "multiple node" formulas, meaning those which include derivatives of the integrand, of order 0 through μ_i-1 at the points t_i. Such formulas have been studied by various authors; [7], [13], [16] and [18] deal with the case when the basis functions are polynomials, while [9] treats the more general case of Tchebycheff systems (see below).

Turán [18] proved existence and uniqueness of the formulae when all the multiplicities μ_i are equal and odd, using a variational argument. Popoviciu [13] proved existence when the μ_i are merely odd, and Ghizzetti and Ossicini [7] proved uniqueness for this case. Stroud and Stancu [17] discussed how these formulae may be computed and presented some numerical results. Karlin and Pinkus [9] proved a slightly weaker form of Theorem 2.1 using an argument based on induction and the implicit function theorem.

Before stating Theorem 2.1, we first give some definitions we will need concerning <u>Tchebycheff systems</u> (cf. [10]).

Let $\{u_i\}_{i=1}^n$ be functions of class $C^{N-1}[a,b]$. Then the $\{u_i\}$ are said to be an <u>extended Tchebycheff system</u> (ET-system) <u>of order N</u> provided $\det U(t_1,t_2,\dots,t_n) > 0$, where $a \le t_1 \le \dots \le t_n \le b$ and U is the $n \times n$ matrix whose (i,j)-th entry is $u_i(t_j)$ (or $\partial^k u_i(t_j)/\partial t^k$, in case $t_j = t_{j-k}$ and for no larger k is this true). At most N consecutive t_j's are allowed to coincide. An ET-system or order n is called simply an extended Tchebycheff system (ET-system); if the functions $\{u_i\}_{i=1}^n$ satisfy the property that for any k, $1 \le k \le n$, the system $\{u_i\}_{i=1}^k$ is an ET-system, then we say the $\{u_i\}_{i=1}^n$ form an <u>extended complete Tchebycheff</u> (ECT) system.

Let $\mu_1,\dots,\mu_k$ be odd positive integers and let $n = \sum_{i=1}^k (\mu_i+1)$. Let $\{u_i\}_{i=1}^n$ be an ET-system on [a,b], and let L be a positive linear functional on $U \equiv \mathrm{span}(\{u_i\})$ (i.e., if $p \in U$ is non-trivial and non-negative, then $L(p) > 0$). Finally, let $\Delta_k = \{\underline{t} = (t_1,\dots,t_k)\colon\ a < t_1 < \dots < t_k < b\}$.

<u>Theorem 2.1</u> There is a unique $\underline{t} \in \Delta_k$ and coefficient array $\{a_{ij}\}_{i=1}^{k}{}_{j=0}^{\mu_i-1}$ such that

$$L(p) = \sum_{i=1}^{k} \sum_{j=0}^{\mu_i-1} a_{ij} p^{(j)}(t_i) \tag{2.2}$$

for all $p \in U$. This is sharp in the sense that no formula of the form (2.2) can hold for an ET-system having $\overline{n} > n$ functions.

The details of the proof can be found in [4]. Here we describe the mapping F, the set D, and the point q, as in the Introduction, to which degree theory can be applied.

For any $\underline{t} \in \Delta_k$ and $i = 1,\dots,k$, let $p_i(\underline{t};\cdot) \in U$ satisfy

$$p_i^{(j)}(\underline{t};t_\ell) = 0, \text{ for } j = 0,\ldots,\mu_\ell - 1, \quad \ell = 1,\ldots,k,$$

and

$$p_i^{(\mu_j)}(\underline{t};t_j) = \delta_{ij} \quad \text{(Kronecker } \delta) \quad .$$

Then the mapping F is defined by

$$(2.3) \qquad F_i(t) = -L(p_i(\underline{t};\cdot)),\ i = 1,\ldots,k.$$

The domain D will be the subset of Δ_k

$$D = \{\underline{t} = (t_1,\ldots,t_k):\ a = t_0 < t_1 < \ldots < t_k < t_{k+1} = b,\ t_i - t_{i-1} > \varepsilon,\ i = 1,\ldots,k+1\}$$

for an appropriately chosen $\varepsilon > 0$. The theorem follows by showing that the equation $F(\underline{t}) = 0$ has a unique solution. That $\deg(F,D,0) = 1$ is shown by using a homotopy based on the linear functional

$$L_\lambda = \lambda L + (1-\lambda)L_0,\ 0 \le \lambda \le 1,$$

and then defining $F(\underline{t},\lambda)$ as in (2.3), using L_λ. L_0 is the linear functional

$$L_0(p) = \sum_{i=1}^{k} p(r_i),$$

where $\underline{r} = (r_1,\ldots,r_k)$ is an arbitrary element of Δ_k. The proof that $\deg(F(\cdot,0),D,0) = 1$ is facilitated by the change of variables $s_i = (t_i - r_i)^{\mu_i}$, which is invertible since the μ_i are odd.

The proof is completed by showing unicity of solutions to the equation $F(\underline{t}) = 0$. This is done by showing $\det F'(\underline{t}) > 0$ whenever $F(\underline{t}) = 0$; in fact, when $F(\underline{t}) = 0$ we have $F'(\underline{t}) = \operatorname{diag}(a_{1\mu_1-1},\ldots,a_{k\mu_k-1})$, and all $a_{i\mu_i-1} > 0$.

The paper [4] also includes versions of the above theorem when the $\{u_i\}_{i=1}^n$ form a periodic Tchebycheff system (e.g., $\{1,\cos t,\sin t,\ldots,\sin mt\}$), and when L is a non-negative linear functional. We conjecture that Theorem 2.1 is also true when the $\{u_i\}$ are merely an ET-system of order $\mu = \max\{\mu_i\}$.

3. Unicity of best mean approximants from a spline manifold.

Let $S_N \subset C[0,1]$ denote the class of all piecewise linear functions with at most N+1 linear segments. In this section we consider the problem of finding a "best approximant" in S_N to a given function f, with $f'' > 0$. The proofs of the following results can be found in [5].

For p = 1 or 2, let $\|\cdot\|_p$ denote the $L_p[0,1]$ norm $\|g\|_p = \int_0^1 |g(t)|^p dt^{1/p}$. A function $\ell \in S_N$ will be called a best $L_2[0,1]$ approximant to f provided

$$\|f-\ell\|_2 = \inf\{\|f-s\|_2 : s \in S_N\}.$$

Similar definitions apply for a best $L_1[0,1]$ approximant and a best one-sided $L_1[0,1]$ approximant to f, where the latter infimum is over all $s \in S_N$ such that $s \geq f$.

The existence of such best approximants has been established in [6] and [15]; we remark that since S_N is both a nonlinear manifold and a non-closed set (in the above norms), arguments regarding existence of best approximants are nontrivial.

In the present discussion we are interested in unicity of best approximants for the above three approximation problems. For ease of exposition we restrict to the $L_2[0,1]$ problem, but all results apply to the other two cases also.

In [5] it is proven that for any integer $N \geq 1$, there is a $C^\infty[0,1]$ function f, with $f'' > 0$, having at least two best $L_2[0,1]$ approximants. Hence, we would like to determine conditions on f and/or N which ensure unicity.

Let $\Sigma^N = \{\underline{t} = (t_1,\ldots,t_N): \ 0 = t_0 < t_1 < \ldots < t_N < t_{N+1} = 1\}$ and for $\underline{t} \in \Sigma^N$, let $h_i = t_{i+1} - t_i$, $i = 0,\ldots,N$. It is shown in [5] that if $\underline{t} \in \Sigma^N$ represents the "knots" in a best approximant to f, then the following equations must hold:

$$F_i(\underline{t}) \equiv h_{i-1}^2 \int_0^1 w(\tau)f''(t_i - \tau h_{i-1})d\tau - h_i^2 \int_0^1 w(\tau)f''(t_i+\tau h_i)d\tau \tag{3.1}$$

$$= 0, \quad i = 1,\ldots,N.$$

Here $w(\tau) = \tau(1-\tau)^2 \geq 0$ on $0 \leq \tau \leq 1$. We define the map $F: \Sigma^N \to R^N$ by (3.1). Letting $f_\lambda(t) = \lambda f(t) + (1-\lambda)f_0(t)$, $0 < \lambda < 1$, where $f_0(t) = t^2/2$, we use (3.1) to define a homotopy $F(\underline{t},\lambda)$. It is clear that $F(\underline{t},\lambda) \neq 0$ for $\underline{t} \in \partial\Sigma^N$, for if $t \in \partial\Sigma^N$ some $h_i = 0$, and so $F(\underline{t},\lambda) = 0$ implies that all $h_i = 0$, an impossibility. Furthermore, it is easy to show that $\deg(F(\cdot,0), \Sigma^N, 0) = 1$ since the only solution to $F(\underline{t},0) = 0$ occurs when all $h_i = 1/N+1$.

The only problem that remains is to show that

(3.2) $$\det F'(\underline{t}) > 0 \text{ if } F(\underline{t}) = 0.$$

Gerschgorin's Theorem on the location of the eigenvalues of a matrix is used to prove (3.2) under the hypotheses of the following theorem:

Theorem 3.1 Let $f \in C^2[0,1]$ with $f'' > 0$ on $[0,1]$ and $\log f''$ concave on $(0,1)$. Then for every positive integer N, f has a unique best $L^2[0,1]$ approximant from S_N.

The proof of the following "eventual uniqueness" result is somewhat more delicate.

Theorem 3.2 Let $f \in C^5[0,1]$ with $f'' > 0$ on $[0,1]$. There is a positive integer N_0 such that for any integer $N \geq N_0$, f has a unique best $L_2[0,1]$ approximant from S_N.

The proof of Theorem 3.2 depends on showing that (3.2) holds for all N sufficiently large. This in turn is proven using the following algebraic result:

Proposition 3.3 Let $A = [a_{i,j}]$ be a tridiagonal N x N real matrix with positive diagonal entries. Then if

$$a_{n,n-1}\, a_{n-1,n} \leq a_{n,n}\, a_{n-1,n-1}\, (1 + \pi^2/4N^2)/4$$

for $n = 2,\dots,N$, it follows that $\det A > 0$.

The above theorems immediately suggest some open problems. What can be said about unicity of best approximants from manifolds of spline functions of order higher than two? One difficulty is that the mapping analogous to the F used above no longer has a tridiagonal (or even banded) derivative. Also, what can be said about best approximants by second order splines to functions f which are no longer convex? Examples suggest that finding such approximants may be an ill-posed problem.

4. Interpolation formulae for partial differential equations.

A. H. Stroud posed the following problem. Let Ω be an open, bounded domain in the plane whose boundary Γ is a Jordan curve. We are interested in a particular kind of approximation to the function $u \in C(\bar{\Omega})$, where u satisfies the Dirichlet problem

$$\frac{\partial^2 u}{\partial x^2} + \frac{\partial^2 u}{\partial y^2} = 0 \qquad \text{in } \Omega, \quad u = f \text{ on } \Gamma. \tag{4.1}$$

Let (x_*,y_*) be a given point in Ω. We seek an approximation to $u(x_*,y_*)$ of the form

$$u(x_*,y_*) \simeq \sum_{k=1}^{N} A_k u(x_k,y_k) , \tag{4.2}$$

where the weights A_k are positive and the points (x_k,y_k) are points on Γ. The points and weights are to be determined, in analogy with Gaussian quadrature formulae, by requiring (4.2) to be exact for as many harmonic polynomials as possible. (A harmonic polynomial is any linear combination of the linearly-independent polynomials 1, $\mathrm{Re}(x + iy)^m$, $\mathrm{Im}(x + iy)^m$, $m = 1,2,\ldots$). Since there are 2N free parameters in (4.2) and there are 2N-1 harmonic polynomials of degree at most N-1, one might expect that by fixing one of the parameters in advance, say (x_N,y_N), the other parameters could be uniquely determined so as to make (4.2) exact for all harmonic polynomials of degree at most N-1. Stroud has called such a formula a <u>Gauss harmonic interpolation formula of degree N-1</u> ([17]), and has asked whether such a formula exists for an arbitrary Ω and (x_*,y_*). This question is answered affirmatively (except for unicity) in [2], using degree theory. An outline of the proof will be presented below.

Let $(x(s),y(s))$ be a parametrization for the curve Γ. With $n = 2N-1$, let $D \subset R^n$ be defined as

$$D = \{x = (A_1,\ldots,A_N,s_1,\ldots,s_{N-1}):\ 0 < A_i < 1,\ i=1,\ldots,N \text{ and } 0 = s_0 < s_1 < \ldots < s_N = 1\}.$$

Define the functions $\{u_i\}_{i=1}^n$ as

$$u_1(x,y) = 1$$
$$u_{2m}(x,y) = \mathrm{Re}(x+iy)^m$$
$$u_{2m+1}(x,y) = \mathrm{Im}(x+iy)^m,\ m = 1,\ldots,N-1\ .$$

Define the mapping $F:D \to R^n$ as

(4.3) $$F_i(x) = \sum_{k=1}^{N} A_k u_i(x(s_k),y(s_k)) - u_i(x_*,y_*),\ i = 1,\ldots,n\ .$$

The existence of the desired formula is thus equivalent to the existence of a solution to the equation $F(x) = 0$, and this will follow immediately from

(4.4) $$\deg(F,D,0) = \pm 1.$$

The details of proving (4.4) are given in [2]. The basic idea is to first establish that $F(x) \neq 0$ if $x \in \partial D$, and then to construct a homotopy from the mapping F_0, where F_0 corresponds to the F defined in (4.3) when $\Gamma = \Gamma_0$ is the unit circle and $(x_*^0,y_*^0) = (0,0)$. The fact that $\deg(F_0,D,0) = 1$ follows from the theory of periodic Tchebycheff systems (cf. [10], chapter 6).

The result just described has been extended in [1] to the one dimensional heat equation, and in [3] to more general one dimensional parabolic equations.

5. A result on L_1 approximation using Borsuk's Theorem.

The following theorem which arises in the study of L_1 approximation was first proved in [8].

Theorem 5.1 Let $\{u_i\}_{i=1}^n$ be real functions in $L_1(d\mu;[0,1])$, where d is a finite, nonatomic measure on $[0,1]$. Then there exist points $\{t_i\}_{i=1}^r$, $0 = t_0 < t_1 < \ldots < t_r < t_{r+1} = 1$, with $r \leq n$, such that

$$(5.1) \qquad \sum_{j=0}^{r} (-1)^j \int_{t_j}^{t_j+1} u_i(x)d\mu(x) = 0, \quad i = 1,\ldots,n.$$

Pinkus [12] has given a short proof of this theorem using the following corollary to Borsuk's Theorem ([14], p. 81):

Theorem 5.2 Let $D \subset R^{n+1}$ be a symmetric, bounded neighborhood of the origin, and let $F: \partial D \to R^k$, $k \leq n+1$, be an odd continuous mapping. Then there is a point $x \in \partial D$ such that $F(x) = 0$.

To apply this to Theorem 5.1, Pinkus constructs the following mapping. Let $D = \{x = (x_1,\ldots,x_{n+1}): \sum_{j=1}^{n+1} |x_j| < 1\}$, so that $x \in \partial D$ satisfies $\sum_{j=1}^{n+1} |x_j| = 1$. For $x \in \partial D$, define $y_0(x) = 0$, $y_j(x) = \sum_{k=1}^{j} |x_k|$, $j = 1,\ldots,n+1$. Then define $F: \partial D \to R^n$ by

$$F_i(x) = \sum_{j=1}^{n+1} (\operatorname{sgn} x_j) \int_{y_{j-1}(x)}^{y_j(x)} u_i(t)d\mu(t), \quad i = 1,\ldots,n.$$

It is clear that F is odd and continuous, and hence has a zero in ∂D. This proves Theorem 5.1.

References

1. D. L. Barrow, "Existence of Gauss interpolation formulas for the one-dimensional heat equation," Math. Comp., v. 30 (1976), pp. 24-34.

2. D. L. Barrow and A. H. Stroud, "Existence of Gauss harmonic interpolation formulas," SIAM J. Numer. Anal., v. 13 (1976), pp. 18-26.

3. D. L. Barrow, "Gauss interpolation formulas and totally positive kernels," Math. Comp. (to appear).

4. D. L. Barrow, "On multiple node Gaussian quadrature formulae," submitted to Math. Comp.

5. D. L. Barrow, C. K. Chui, P. W. Smith, and J. D. Ward, "Unicity of best mean approximation by second order splines with variable knots," submitted to J. Analyse Math.

6. C. K. Chui, P. W. Smith, and J. D. Ward, "On the smoothness of best L_2-norm approximants from nonlinear spline manifolds,"Math Comp., v. 31 (1977), pp. 17-23.

7. A. Ghizzetti and A. Ossicini, "Sull'esistenza e unicitá delle formule di

quadratura Gaussione," Rendiconti de Matematica, v. 8 (1975), pp. 1-15.

8. C. R. Hobby and J. R. Rice, " A moment problem in L_1 approximation," Proc. Amer. Math. Soc., v. 16 (1965), pp. 665-670.

9. S. Karlin and A. Pinkus, "Gaussian quadrature formulae with multiple nodes," in Studies in Spline Functions and Approximation Theory, Academic Press, New York (1976), pp. 113-141.

10. S. Karlin and W. J. Studden, Tchebycheff Systems: With Applications in Analysis and Statistics, Interscience, New York (1966).

11. L. Nirenberg, Topics in Nonlinear Functional Analysis, Courant Inst. of Math. Sciences, New York (1974).

12. A. Pinkus, "A simple proof of the Hobby-Rice Theorem," Proc. Amer. Math. Soc., v. 60 (1976), pp. 82-84.

13. T. Popoviciu, "Sur une généralisation de la formulae d'intégration numérique de Gauss, "Acad. R. P. Romine Fil. Iasi. Stud. Cerc. Sti. v. 6 (1955), pp. 29-57.

14. J. T. Schwartz, Nonlinear Functional Analysis, Gordon and Breach, New York (1969).

15. P. W. Smith, "On the smoothness of local best Lp spline approximations," in Approximation Theory II; G. G. Lorentz, C. K. Chui, L. L. Schumaker, Eds., Academic Press, New York (1976), pp. 563-566.

16. A. H. Stroud and D. D. Stancu, "Quadrature formulas with multiple Gaussian nodes," SIAM J. Numer. Anal., Ser. B., v. 2 (1965), pp. 129-143.

17. A. H. Stroud, "Gauss harmonic interpolation formulas," Comm. ACM, v. 17 (1974), pp. 471-475.

18. P. Turán, "On the theory of mechanical quadrature," Acta. Sci Math. (Szeged), v. 12 (1950), pp. 30-37.

Numerical Solution of Eigentuple-eigenvector Problems in Hilbert Spaces

E.K. Blum*
Mathematics Department, University of Southern California

1. Introduction.

Like ordinary eigenvalue problems of the form $Ax=\lambda x$, the generalized "eigentuple-eigenvector" problems (also known as "mutiparameter eigenvalue problems") which we consider in this paper have their origin in classical analysis. In particular, they arise in various boundary-value problems for ordinary differential equations analogous to the Sturm-Liouville problems which are one of the main sources of ordinary eigenvalue problems.

Perhaps, the best known of these equations is Mathieu's equation, which, like many Sturm-Liouville problems, arises in the solution of a partial differential equation by separation of variables. In this case, the partial differential equation is the reduced wave equation, $\nabla^2\varphi+\lambda\varphi=0$, which governs the vibration of a homogeneous membrane. For elliptic membranes, it is natural to introduce elliptic coordinates [1, 16]. Separation of variables introduces a second parameter (the separation constant) and yields Mathieu's equation,

$$d^2y/dv^2+(\lambda-2q\cos 2v)y=0, \tag{1}$$

and the modified Mathieu's equation

$$d^2z/du^2-(\lambda-2q\cosh 2u)z=0.$$

The variable v is the polar angle of the ellipse and, therefore, solutions y(v) which are periodic of periods π and 2π are of physical importance. These are the Mathieu functions which have been extensively studied. (See [21, 28] and bibliography in [16].) For the functions of period π, we have the periodic boundary conditions,

$$y(0)-y(\pi)=0,\ \ y'(0)-y'(\pi)=0. \tag{2}$$

Now, the general Sturm-Liouville system [2] has the form,

$$L(y)=d/dv(p(v)\,dy/dv)+(\lambda f(v)-g(v))y=0,$$

$$\alpha_1 y(a)+\alpha_2 y(b)+\alpha_3 y'(a)+\alpha_4 y'(b)=0,$$

$$\beta_1 y(a)+\beta_2 y(b)+\beta_3 y'(a)+\beta_4 y'(b)=0.$$

For an arbitrary real value of the parameter q in (1), we obtain a Sturm-Liouville system, $L_q(y)=0$, with $p=1$, $f=1$, $g(v)=2q\cos 2v$ and periodic

*This research was partially supported by NSF Grant MPS74-13332 and NSF Grant MCS76-09172

boundary conditions (2). According to classical Sturm-Liouville theory [17, p. 214], there exists a sequence of eigenvalues, $(\lambda_i(q))$, such that $\lambda_o(q) < \lambda_1(q) \leq \lambda_2(q) < \lambda_3(q) \leq \lambda_4(q) < \cdots$, with corresponding eigenfunctions $\varphi_{iq}(v)$. Furthermore, φ_{oq} has no zeros in $[0, \pi]$ while $\varphi_{2i+1,q}$ and $\varphi_{2i+2,q}$, $i \geq 0$, each have exactly $2i+2$ zeros in $[0, \pi)$.

Another two-parameter problem arises in the solution of Laplace's equation, $\nabla^2\varphi = 0$, in elliptic coordinates [1]. Separation of variables leads to the Lamé equation,

$$\psi(s)\, d^2y/ds^2 + (\psi'(s)/2)\, dy/ds + (\lambda s + \mu)y = 0$$

where $\psi(s) = 4(s-e_1)(s-e_2)(s-e_3)$, $e_1 < e_2 < e_3$, and (λ, μ) is an eigenvalue-pair. The boundary-value problem which is associated with the Lamé equation is of the following kind. Let $(a_1, b_1) \subset (e_1, e_2)$ and $(a_2, b_2) \subset (e_2, e_3)$. Determine (λ, μ) for which there exists a solution y_1 in (a_1, b_1) which vanishes at a_1 and b_1 and a solution y_2 in (a_2, b_2) which vanishes at a_2 and b_2. The existence of solutions is proven for a generalization of this problem which can be formulated as follows [2]. Consider the n-parameter differential equation,

$$d/ds(p(s)\, dy/ds) + (\lambda_1 f_1(s) + \cdots + \lambda_n f_n(s) - g(s))y = 0,$$

with n pairs of boundary conditions, $1 \leq i \leq n$,

$$c_{1i}\, y_i(a_i) + c_{2i}\, y_i'(a_i) = 0,$$

$$d_{1i}\, y_i(b_i) + d_{2i}\, y_i'(b_i) = 0,$$

where $a_1 < b_1 \leq a_2 < b_2 \leq \cdots \leq a_n < b_n$ and $\det(f_j(s_i))$ has the same sign for $s_i \in (a_i, b_i)$. It is required to determine an "eigentuple", $(\lambda_1, \ldots, \lambda_n)$, such that there exist n solutions of the above equation, say $y_1, \ldots, y_n$, where y_i satisfies the ith pair of boundary conditions. Klein's oscillation theorem [2] provides a solution to this problem. It asserts that there exists an infinite set of eigentuples $(\lambda_1, \ldots, \lambda_n)$ for which n eigenfunctions, $(y_1, \ldots, y_n)$, exist satisfying the respective boundary conditions and for each n-tuple of non-negative integers $(m_1, \ldots, m_n)$ there is precisely one eigentuple for which the corresponding eigenfunctions y_i have precisely m_i zeros in (a_i, b_i), $1 \leq i \leq n$. See also [4], [5], [6], and [7] for further theoretical results on multiparameter problems.

As a special case, we have the following two-parameter, three-point boundary-value problem

$$d^2y/ds^2 + (\lambda + \mu f(s) + g(s))y = 0, \qquad y(a) = y(b) = y(c) = 0. \tag{3}$$

The numerical solution of this problem is treated in [9]. For analytic treatments of special instances, see also [3], [8], [18], [19]. Problem (3) has an interesting aspect not encountered in the Mathieu problem (1), (2) which has important implications for numerical solution methods. In (1), (2), for any choice of q there are corresponding values of λ which yield solutions. Therefore, methods of computing eigenpairs (λ, q) and their corresponding eigenfunctions can be based on standard numerical methods for computing single eigenvalues; e.g. for a given value of q, the discretization of (1) leads to the standard matrix problem, $Ax=\lambda x$. (Different techniques based on continued fractions and Fourier expansions of y are used to compile tables of $\lambda_i(q)$ and corresponding Mathieu eigenfunctions y_i in [16, 22].) However, in (3), under certain conditions of f, there is a denumerable number of eigenpairs (λ, μ). Hence, only for certain values of μ are there values of λ which yield solutions of (3). Therefore, (3) is not easily reducible to a conventional eigenvalue problem. This can be understood by referring back to Klein's oscillation theorem. We remark that any solution, y, of (3) gives rise to a pair of functions, y_1, the restriction of y to (a, b) and y_2, the restriction of y to (b, c), such that (λ, μ) and y_1 are a solution of the 2-parameter boundary-value problem,

$$y'' + (\lambda + \mu f + g)y = 0, \qquad y_1(a) = y_1(b) = 0, \tag{4}$$

while (λ, μ) and y_2 are a solution of

$$y'' + (\lambda + \mu f + g)y = 0, \qquad y_2(b) = y_2(c) = 0. \tag{5}$$

Applying Klein's theorem to (4), (5) we see that if $\det(f_j(s_i)) = f(s_2) - f(s_1)$ has the same sign for any $s_1 \in (a, b)$ and $s_2 \in (b, c)$, then there are a denumerable number of eigenpairs (λ_i, μ_j) with corresponding solutions y_{1i}, y_{2j} having respectively i zeros in (a, b) and j zeros in (b, c). In [9], the following particular instance of (3) is solved by a finite-difference shooting method:

$$y'' + (\lambda + \mu \cos s + e^s) y = 0,$$
$$y(0) = y(2) = y(4) = 0. \tag{6}$$

Clearly, the condition $\cos s_2 - \cos s_1 < 0$ holds for all $s_1 \in (0, 2)$ and $s_2 \in (2, 4)$. Hence, (6) has at most a denumerable number of solutions. As pointed out in [9], these solutions can be visualized as the intersections of two families of curves, $\lambda = \alpha_i(\mu)$ and $\lambda = \beta_j(\mu)$, obtained by solving two families of Sturm-Liouville problems, one with boundary conditions $y(0)=y(2)=0$ and the other with $y(2)=y(4)=0$. The families are parametrized by μ in the differential equation (6). For each μ, there exists a denumerable set of eigenvalues, $\alpha_i(\mu)$, with eigenfunctions y_{1i} satisfying $y_{1i}(0) = y_{1i}(2) = 0$ and another set, $\beta_j(\mu)$, with eigenfunctions y_{2j} and $y_{2j}(2) = y_{2j}(4) = 0$. At an intersection of two curves $\alpha_i(\mu)$ and $\beta_j(\mu)$ we obtain a possible solution (λ_i, μ_j) with y_{1i} having i zeros in (0, 2) and y_{2j} having j zeros in (2, 4). (Note that the derivatives of y_{1i} and y_{2j} must agree at s=2 in order to constitue a solution of (6).)

To solve (3) numerically, it seems reasonable to proceed as in [9] to replace the differential equation by a finite-difference approximation. We assume that we can shoose a step-size $h=(c-a)/n = (b-a)/k$. Replacing the second derivative by a second difference we obtain the system of equations,

$$y_{r-1} - (2 - h^2(\lambda + \mu f(s_r) + g(s_r)) y_r + y_{r+1} = 0, 1 \leq r \leq n-1, \; y_0 = y_k = y_n = 0.$$

If we rewrite this as

$$y_{r-1} - (2 - h^2 g(s_r)) y_r + y_{r+1} + \lambda h^2 y_r + \mu h^2 f(s_r) y_r = 0,$$

then we obtain the matrix form of the problem,

$$Ax = \lambda B_1 x + \mu B_2 x, \tag{7}$$

where $x^T = (y_1, \ldots, y_{k-1}, 0, y_{k+1}, \ldots, y_{n-1})$, A is a tri-diagonal symmetric matrix of order n-1 and B_1, B_2 are diagonal matrices.

2. Eigentuple-eigenvector problems.

Equation (7) is a special case of the general class of problems which we wish to discuss. Let H and H′ be real Hilbert spaces. (Extension to complex spaces is possible, but will not be considered here.) Let $A, B_1, \ldots, B_n$ be bounded linear operators from H to H′. The eigentuple-eigenvector problem associated with these operators is to find an eigentuple

$\Lambda = (\lambda_1, \ldots, \lambda_n) \in R^n$ and an eigenvector $x \in H$, $x \neq 0$, such that

$$Ax = \sum_{i=1}^{n} \lambda_i B_i x. \tag{8}$$

The case n=1 has been rather extensively studied in numerical analysis for A and B_1 square matrices [10, 14, 15, 24, 25, 26, 27]. In this case, one writes (8) as

$$Ax = \lambda Bx, \tag{9}$$

and calls (9) the "generalized" eigenvalue problem. Note that (9) need not have any solutions. The case n=2 has already been introduced in equation (7). We can rewrite (7) as

$$(A - \mu B_2)x = \lambda B_1 x. \tag{7'}$$

Again, for various values of μ, there may be no solutions λ. However, if B_1^{-1} exists, then for each value of μ, we obtain a standard eigenvalue problem $B_1^{-1}(A-\mu B_2)x = \lambda x$ and there exists a solution λ (possibly complex-valued). If A, B_1 and B_2 are odd-order matrices, then there is a real solution, $\lambda_1(\mu)$, for each μ. If A, B_1 and B_2 are symmetric operators with B_1 positive definite, then all solutions, $\lambda_i(\mu)$, are real. In the latter two cases, there is a continuum of solutions parametrized by μ. We have already seen this phenomenon in the Mathieu equation. A simpler example of this behavior is the following [11].

<u>Example 1.</u> Take $B_2 = I$, the identity operator, and take A and B_1 to be the matrices of order 3,

$$A = \begin{bmatrix} 1 & 3 & 0 \\ -2 & -2 & 3 \\ 0 & -3 & 2 \end{bmatrix}, \quad B_1 = \begin{bmatrix} -5 & 5 & -15 \\ -15 & -20 & 10 \\ -10 & -5 & 5 \end{bmatrix}.$$

Thus, $H = H' = R^3$ and $(\lambda, \mu) \in R^2$. Let $T = A - \lambda B_1 - \mu B_2$ and $p(\lambda, \mu) = \det T$. A direct calculation yields

$p(\lambda, \mu) = -2000\ \lambda^3 - 25(23-2\mu)\lambda^2 + 5(14+7\mu+4\mu^2)\lambda + (17-11\mu+\mu^2-\mu^3)$. The graph of $p(\lambda, \mu) = 0$ in the real (λ, μ) plane has the usual shape of a bicubic. (See [11].) If we parametrize it, with parameter θ say, then as θ varies from $-\infty$ to ∞, $(\lambda(\theta), \mu(\theta))$ traces the curve of real eigenpairs of this example. The corresponding real unit eigenvector, $x(\theta)$, traces a continuous nonlinear curve on the unit 3-sphere; i.e. the family of eigenvectors lies in a nonlinear manifold. We note that $\lim_{\theta \to \infty} x(\theta) = \lim_{\theta \to -\infty} x(\theta)$. (See [11] for details.)

The existence of a continuum of real solutions of (7) when A, B_1 and B_2 are nth order matrices may not be surprising, since there are n+2 unknowns (λ, μ and the components of x) and only n+1 equations (including a normalization equation such as $\|x\| = 1$). In any application in which (7) arises, one would expect to have another condition prescribed. This could make it feasible to determine μ and thereby reduce the two-parameter problem (7) to the one-parameter generalized eigenproblem (9). Thus, the existence of a continuum of real solutions could make (7) numerically tractable by permitting its reduction to a sequence of solutions of (9). This technique has been used to solve the Mathieu problem (1), (2) with some success. Equation (1) is replaced by a finite-difference approximation, as explained above, giving rise to (7). A value of μ is chosen, and (7′) is solved by one of the numerical methods available for the solution of $Ax = \lambda x$. In this way, one could compile tables of $\lambda_i(\mu)$ and the associated Mathieu functions like those in [16, 22]. (Our tables of the functions would not have the Fourier coefficients, but rather the function values at the mesh points.) The computational efficiency and accuracy of the finite-difference technique remains to be evaluated and compared to the projectional methods used in [16, 22, 29] and the variational methods in [30, 31].

The naive argument that (7) can be expected to have a continuum of real solutions because there is one more unknown that there are equations in the finite-dimensional case ignores the deeper questions of existence and multiplicity of real solutions of a system of algebraic equations. In the next example, we show that it can happen that (7) has only a finite number of real solutions.

<u>Example 2.</u> Consider the generalized eigenproblem (9), where A and B are arbitrary complex square matrices of order n. In general, standard procedures for solving (9) in the real case must be modified to deal with complex A, B. However, we can remain in the real domain by reformulating (9) as follows. Let $A = A_R + i A_I$, $B = B_R + i B_I$, $\lambda = \lambda_1 + i\lambda_2$ and $x = u + iv$, where $i^2 = -1$ and all other quantities are real. It is easy to see that (9) is equivalent to the real two-parameter eigenproblem,

$$A^{(2)} z = \lambda_1 B_1 z + \lambda_2 B_2 z,$$

where

$$A^{(2)} = \begin{bmatrix} A_R & -A_I \\ A_I & A_R \end{bmatrix}, \quad B_1 = \begin{bmatrix} B_R & -B_I \\ B_I & B_R \end{bmatrix}, \quad B_2 = \begin{bmatrix} B_I & -B_R \\ B_R, & -B_I \end{bmatrix}$$

and $z^T = (u, v)$. For example, take

$$A = \begin{bmatrix} 8 & -1 & -5 \\ -4 & 4 & -2 \\ 18 & -5 & -7 \end{bmatrix} \quad \text{and } B=I.$$

This matrix is taken from [32] and its eigenvalues are $1, 2 \pm 4i$. They are computed in [11] by using the device just explained. Thus, $Ax = \lambda x$ is converted to a two-parameter problem of dimension 6. Letting $T = A^{(2)} - \lambda_1 B_1 - \lambda_2 B_2$ as before, we calculate
$p(\lambda_1, \lambda_2) = \det T = [(\lambda_1 - 1)^2 + \lambda_2^2][(\lambda_1 - 2)^4 + 2(\lambda_1 - 2)^2(\lambda_2^2 + 16) + (\lambda_2^2 - 16)^2]$. Although $p(\lambda_1, \lambda_2)$ is a sixth degree polynomial in λ_1, λ_2, we see that it factors in a manner which limits the number of real solutions to three, namely, (1, 0), (2, 4) and (2, -4). Hence, the numerical real solution of $A^{(2)} z = \lambda_1 B_1 z + \lambda_2 B_2 z$ for this example cannot be effected by choosing arbitrary values of λ_2 and solving the resulting one-parameter problems for λ_1. Instead, what is required is a method which treats (λ_1, λ_2) as a 2-vector and iterates on both variables simultaneously. We shall describe two such methods, but first it is of interest to consider some theoretical aspects of the eigentuple-eigenvector problem (8).

3. Collatz's classification

Let H be an inner-product space, with inner-product denoted by $\langle u, v \rangle$. Let A, $B_1, \dots, B_n$ be linear operators on H and define $T = T_\Lambda = A - \sum_1^n \lambda_i B_i$, for any n-tuple of scalars $\Lambda = (\lambda_1, \dots, \lambda_n)$. For $x \in H$, define $\varphi_x(\Lambda) = \langle T_\Lambda x, x \rangle$ and let $K_x = \{\Lambda : \varphi_x(\Lambda) = 0\}$ and $K = \bigcup K_x$, where the union is over all $x \neq 0$.

For example, for n=1 and $B_1 = I$, we have $T = A - \lambda I$ and $\varphi_x(\Lambda) = \langle Ax, x \rangle - \lambda \langle x, x \rangle$. Hence, λ is in K_x if and only if $\lambda = \langle Ax, x \rangle / \langle x, x \rangle$, that is, K is just the numerical range of A.

For n=2, we have $T = A - \lambda_1 B_1 - \lambda_2 B_2$ and $\varphi_x(\Lambda) = \langle Ax, x \rangle - \lambda_1 \langle B_1 x, x \rangle - \lambda_2 \langle B_2 x, x \rangle$. Hence, $K_x = \{(\lambda_1, \lambda_2): \langle B_1 x, x \rangle \lambda_1 + \langle B_2 x, x \rangle \lambda_2 = \langle Ax, x \rangle\}$. Now, let E_Λ be the subspace of eigenvectors belonging to an eigentuple Λ. Then $x \in E_\Lambda$ implies that $T_\Lambda x = 0$, which implies that $\varphi_x(\Lambda) = 0$. Hence, $\Lambda \in K_x \subset K$, that is, K contains the set of eigentuples. With Collatz [5], we define, $K_+ = \{\Lambda : \varphi_x(\Lambda) > 0 \text{ for all } x \neq 0\}$, $K_- = \{\Lambda : \varphi_x(\Lambda) < 0 \text{ for all } x \neq 0\}$. We note that $K = \{\Lambda : \varphi_x(\Lambda) = 0 \text{ for some nonzero } x\}$.

We state some properties of K_+ and K_- without proof. (See [5].)

Lemma. $R^n = K_+ \cup K_- \cup K$.

Lemma. K_+ and K_- are convex sets.

Lemma. (Hadeler) If either K_+ or K_- is a bounded nonempty set, then the other set is empty.

For example, for n=1, if $\lambda \in K_+$, then $\lambda < \langle Ax, x\rangle / \langle x, x\rangle$ for all $x \neq 0$. Thus, λ lies to the left of the numerical range.

Collatz [5] classifies eigentuple-eigenvector problems as follows.

K_+, K_-		Classification
Both sets unbounded		Hyperbolic
One set empty	The other unbounded	Parabolic
One set empty	The other bounded nonempty	Elliptic

To illustrate, we refer to Example 1 above. Letting $x = (\xi, \eta, \zeta)$, we have $\langle Tx, x\rangle = 5(\xi^2 + 2\xi\eta + 5\xi\zeta + 4\eta^2 - \zeta^2)\lambda - (\xi^2 + \eta^2 + \zeta^2)\mu + \xi^2 + \xi\eta - 2\eta^2 + 2\zeta^2$.

For $(\lambda, \mu) = (0, 3)$, $\langle Tx, x\rangle = -2(\xi - \frac{\eta}{4})^2 - \frac{39\eta^2}{8} - \zeta^2 \leq 0$.

Thus, $\langle Tx, x\rangle < 0$ for $x \neq 0$. Hence, $(0, 3) \in K_-$. For $(\lambda, \mu) = (0, -3)$,

$$\langle Tx, x\rangle = 4(\xi + \frac{\eta}{8})^2 + \frac{15}{16}\eta^2 + 5\zeta^2 \geq 0.$$

Hence, $(0, -3) \in K_+$ and we see that Example 1 is hyperbolic.

4. Eigenvectors as stationary points

Let $\hat{x}$ be an eigenvector belonging to the eigentuple $\Lambda = (\lambda_1, \ldots, \lambda_n)$. Thus,

$$A\hat{x} = \lambda_1 B_1 \hat{x}_1 + \cdots + \lambda_n B_n \hat{x}. \tag{10}$$

Now, for any x, let $\alpha_i(x) = \langle Ax, B_i x\rangle$, $1 \leq i \leq n$, and $\alpha(x) = (\alpha_1(x), \ldots, \alpha_n(x))$. Further, let $\beta_{ij}(x) = \langle B_i x, B_j x\rangle$, $1 \leq i, j \leq n$, and $\beta(x) = (\beta_{ij}(x))$. Taking inner-products of both sides of (10) with $B_i\hat{x}$, $1 \leq i \leq n$, yields the system of linear equations for Λ,

$$\alpha_i(\hat{x}) = \sum_{j=1}^{n} \beta_{ij}(\hat{x})\lambda_j, \quad 1 \leq i \leq n.$$

Assumption: We assume that $\beta^{-1}(\hat{x})$ exists for any eigenvector $\hat{x}$, or equivalently, that the set of vectors $\{B_1\hat{x}, \ldots, B_n\hat{x}\}$ is linearly independent.

This is a reasonable assumption to make, since otherwise there can be a degenerate case in which $A\hat{x} = 0$ and any scalar multiple of Λ is an eigentuple. If we further assume that the B_i are bounded operators, then $\beta^{-1}(x)$ exists for all x in a neighborhood of $\hat{x}$.

Henceforth, we assume that all operators are bounded. With this assumption, we can define $\Lambda(x) = (\lambda_1(x), \ldots, \lambda_n(x))$ to be the solution of $\alpha(x) = \beta(x) \cdot \Lambda(x)$ for x in a neighborhood of $\hat{x}$. This allows us to define the functional F by

$$F(x) = (\tfrac{1}{2}) \| Ax - \sum_1^n \lambda_i(x) B_i x \|^2. \tag{11}$$

Lemma. Any eigenvector is a stationary point of F and, conversely, any stationary point is an eigenvector.

Proof. Let $T_\Lambda(x) = A - \sum_1^n \lambda_i(x) B_i$, so that $F(x) = (\tfrac{1}{2}) \| T_{\Lambda(x)} x \|^2$. The differential of F is given by $dF(x;h) = \langle T_{\Lambda(x)} h - \sum_1^n d\lambda_i(x;h) B_i x, T_{\Lambda(x)} x \rangle$. Since $\langle B_j x, T_{\Lambda(x)} x \rangle = \alpha_j(x) - \sum_{i=1}^n \lambda_i(x) \beta_{ji}(x) = 0$, it follows that $dF(x;h) = \langle T_{\Lambda(x)} h, T_{\Lambda(x)} x \rangle$ and, therefore, the gradient is given by $\nabla F(x) = T^*_{\Lambda(x)} T_{\Lambda(x)} x$. ($T^*$ is the adjoint of T.) Hence, we have $\langle \nabla F(x), x \rangle = 2F(x)$, so that $\nabla F(x) = 0$ if and only if $F(x) = 0$.

This result suggest the use of a gradient procedure as a method of numerical solution. Although gradient methods are slow, there is always the possibility of accelerating them. Furthermore, for certain special cases, for example when A and the B_i are sparse matrices, suitably accelerated gradient procedures may be feasible computationally. We mention one such gradient method for which a convergence proof has been obtained for the eigentuple-eigenvector problem. This is the procedure given in [10, 14] and defined as follows.

$$y_{k+1} = x_k + h(x_k), \quad k = 0, 1, 2, \ldots, \tag{12}$$

$$x_{k+1} = y_{k+1} / \| y_{k+1} \|, \tag{13}$$

$$h(x) = \begin{cases} \dfrac{2F(x)}{\| \nabla F(x) \|^2} \nabla F(x), & \text{if } \nabla F(x) \neq 0, \\ 0 & \text{if } \nabla F(x) = 0. \end{cases} \tag{14}$$

In [10], the following theorem is proved for an arbitrary real Hilbert space H, F any real functional and the sequence (x_k) defined as above.

Theorem 1. Let E be a stationary set of F, with $E \subset V$ and $V = x_0 + V_0$, where V_0 is a closed subspace of H. Suppose there exists a neighborhood, N, of E such that (i) for all $x \in N$ there exists a unique $y \in E$ with $\|x-V\| = \|x-y\|$; (ii) for all $x \in N$, the second Fréchet derivative, $F''(x)$, exists and is continuous in x; (iii) there exists a constant $c > 0$ for which $\langle F''(x)h, h\rangle \geq c\|h\|^2$ for all $x \in E$ and $h \in V_0^{\perp}$; (iv) there exists a constant $M > 0$ such that $\|F''(x)h\| \leq M\|h\|$ for all $x \in E$ and $h \in H$; (v) for all $x \in N - E$ the condition $\nabla F(x) \neq 0$ holds. Then there is a neighborhood, $N_0(E)$, of E and a constant $L < 1$ such that

$$\|x_k - E\| \leq L^k \|x_0 - E\|.$$

This is the usual linear rate of convergence result for gradient methods. However, it holds for non-isolated solutions; i.e. the sequence (x_k) converges to a set of stationary points. In applying this result to the generalized eigenvalue problem (9), Rodrigue [14] obtains a sufficient condition on A and B for Theorem 1 to be applicable. In [12], we generalize this condition as follows.

Definition. Let $\Lambda = (\lambda_1, \ldots, \lambda_n)$ be an eigentuple of (8) and let E_Λ be its eigensubspace. Let $M_\Lambda = B_1 E_\Lambda + \cdots + B_n E_\Lambda$. Let $R(T_\Lambda)$ be the range of $T_\Lambda = A - \lambda_1 B_1 - \cdots - \lambda_n B_n$. If $R(T_\Lambda) \cap M_\Lambda = \{0\}$, then we call Λ a non-defective eigentuple.

This definition reduces to the usual one in the standard problem where $n = 1$ and $B_1 = I$, for then $M_\Lambda = E_\Lambda$ = null space of $A - \lambda I$. If λ is non-defective, then there can be no vectors of grade $r \geq 2$, since $(A-\lambda I)^r x = 0$ implies $(A-\lambda I)^{r-1} x \in R(T_\lambda) \cap M_\lambda = \{0\}$.

The gradient procedure defined above converges for the functional in (11) in a neighborhood of an eigenspace belonging to a non-defective isolated eigentuple. This result is proven in [12]. We state it here as Theorem 2 and sketch the proof.

Theorem 2. Let Λ be a non-defective eigentuple of (8), with eigensubspace E_Λ. Then conditions (i) and (ii) of Theorem 1 hold for $E = E_\Lambda$ and F given by (11). If Λ is an isolated eigentuple, then condition (v) holds. If $H = H' = R^n$, then (iii) and (iv) hold and the gradient method (12)-(14) converges to E_Λ.

Proof. Since E_Λ is a closed subspace, (i) holds. To establish (ii), we compute $F''(x)h$ as follows. Let

$$G(s) = F'(x+sh) = T^*_{\Lambda(x+sh)} T_{\Lambda(x+sh)}(x+sh).$$

Then

$$F''(x)h = G'(s)\Big|_{s=0} = \frac{d}{ds} T^*_{\Lambda(x+sh)}\Big|_{s=0} T_{\Lambda(x)}x +$$

$$T^*_{\Lambda(x)} \frac{d}{ds} T_{\Lambda(x+sh)}\Big|_{s=0} x + T^*_{\Lambda(x)} T_{\Lambda(x)} h.$$

Hence,

$$(+)\quad F''(x)h = -\Sigma d\lambda_i(x,h)\,(B_i^* T_{\Lambda(x)}x + T^*_{\Lambda(x)}B_i x) + T^*_{\Lambda(x)} T_{\Lambda(x)} h.$$

For $x \varepsilon E_\Lambda$ we have

$$d\lambda_i(x;h) = \sum_{j=1}^{n} \gamma_{ij}(x) < T_{\Lambda(x)}h, B_j x > , \quad 1 \leq i \leq n,$$

where $(\gamma_{ij}(x)) = \beta^{-1}(x)$. (See above for definition of $\beta(x)$). Hence, $F''(x)$ is continuous in x.

To establish (iii), we use (+) with $T_{\Lambda(x)}x = 0$ and $h \varepsilon E_\Lambda^\perp - \{0\}$. This gives us

$$< F''(x)h, h> = \|T_{\Lambda(x)}h\|^2 - \sum_1^n d\lambda_i(x;h) < T_{\Lambda(x)}h, B_i x>$$

$$= \| T_{\Lambda(x)}h - \sum_1^n d\lambda_i(x;h) B_i x\|^2$$

$$\geq \|T_{\Lambda(x)}h - P_M T_{\Lambda(x)}h\|^2 = \|P_{M^\perp} T_{\Lambda(x)}h\|^2$$

Since $R(T_\Lambda) \cap M = \{0\}$, there exists $c_1 > 0$ with

$$\|P_{M^\perp} T_{\Lambda(x)}h\| \geq c_1 \|T_{\Lambda(x)}h\|.$$

Since $h \varepsilon E^\perp$ there exists $c_2 > 0$ with

$$\|T_{\Lambda(x)}h\| \geq c_2 \|h\| .$$

Hence, $< F''(x)h, h > \geq c\|h\|^2$. □

For H infinite-dimensional we require that the subspaces $M, R(T_\Lambda)$ and $P_{M^\perp} R(T_\Lambda)$ be closed.

5. Iterative Methods

Consider the case n=1 again. Let $\rho(x)$ be any functional and define

$$G(x) = (1/2)\| (A-\rho(x)B)x\|^2 = (1/2)\| T_{\rho(x)}x\|^2. \tag{15}$$

Any mininum of G is an eigenvector. For any x, it is easy to show that the functional ρ which minimizes G(x) is given by

$$\rho(x) = \frac{<Ax, Bx>}{<Bx, Bx>} = \Lambda(x), \qquad (16)$$

rather than by the usual generalization of the Rayleigh quotient, $<Ax, x>/<Bx, x>$. Note that $<Bx, x>$ can be zero even when $Bx \neq 0$. The minimized residual property of $\Lambda(x)$ for $\rho(x)$ has another advantage. If we use a gradient method such as

$$x_{i+1} = x_i - s_i \nabla G(x_i),$$

where s_i is the "step-size", we get

$$\nabla G(x) = T^*_{\rho(x)} T_{\rho(x)} x - < Bx, T_{\rho(x)} x > \nabla\rho(x).$$

When $\rho(x) = \Lambda(x)$, we have the simplification,

$$<Bx, T_{\rho(x)} x> = < Bx, Ax - \frac{<Ax, Bx>}{<Bx, Bx>} Bx > = 0.$$

Thus,

$$x_{i+1} = (I - s_i T^*_{\rho(x_i)} T_{\rho(x_i)}) x_i = C x_i,$$

where C is a polynomial in A, A^*, B and B^*. Hence, the gradient method in this case is a generalized power method. The method with $s_i = 2F(x_i)/\|\nabla F(x_i)\|^2$ has already been discussed in section (4). Numerical results for the two-parameter eigenproblem are given in [11]. As Theorem 1 above indicates, convergence is quite slow, although when A, B_1 and B_2 are band matrices or otherwise sparse, the slow convergence rate is somewhat compensated for by the small number of operations per iteration. Some acceleration procedures, such as conjugage gradient techniques, are currently being investigated.

Another numerical procedure for solving (8) is a generalization of the Rayleigh quotient iteration (RQI) used for the ordinary eigenvalue problem. RQI can be derived from Newton's method, but not applied to the function $(A - \rho(x) I)x$, where $\rho(x)$ is homogeneous in x, as when $\rho(x) = R(x)$, where $R(x) = < Ax, x >/ <x, x>$, the Rayleigh quotient. This function, being homogeneous in x, causes Newton's method to converge to the solution $x = 0$ in one iteration for all starting values. To motivate the RQI, we apply Newton's method to the function $f(x, \mu) = (A - \mu I)x$, with the normalizing equation $\|x\|^2 = 1$. As shown in [13], this leads to the method of "inverse iteration with shifts", the shifts being those of Newton's method. This type of iteration with arbitrary shifts, μ_i, takes the form

$$y_{i+1} = (A - \mu_i I)^{-1} y_i, \qquad x_{i+1} = y_{i+1} / \|y_{i+1}\|. \tag{17}$$

If the shifts, μ_i, are taken to be the Rayleigh quotients ar each step, that is, if

$$\mu_i = R(x_i) = \langle Ax_i, x_i \rangle / \langle x_i, x_i \rangle, \tag{18}$$

then (17), (18) define RQI.

The Rayleigh quotion, $R(x)$, is the value of $\rho(x)$ which minimizes the residual $G(x)$ in (12) when $B = I$. For arbitrary B, the minimizing functional is given by $\Lambda(x)$ in (16). For the eigentuple-eigenvector problem, we consider the corresponding functional,

$$G(x) = (1/2) \|Ax - \sum_1^n \mu_i B_i x\|^2,$$

and by a straightforward computation, we find that $G(x)$ is minimized with respect to the $\mu_i = \lambda_i(x)$, where the $\lambda_i(x)$ are as in (11). This leads to the generalization of RQI called minimum residual quotient iteration (MRQI) introduced in [13] for the double eigenvalue three-point boundary-value problem (3).

The discretized form of (3) is given by (7) together with the condition $x_k = 0$. We now set $x_i = y_i$, $1 \le i \le n-1$, and regard (7) as a system of n-1 equations in the n+1 unknowns $\lambda, \mu, x_1, \ldots, x_{n-1}$. Two additional equations are given by the normalization, $\|x\|^2 = 1$, and the middle boundary condition, $x_k = 0$. By Klein's oscillation theorem, under certain conditions on f in (3), the continuous system has a denumerable number of solutions. Hence, for a sufficiently small step h, we can expect the corresponding discretized system (7) with the condition $x_k = 0$ and $\|x\| = 1$ to have a finite number of real solutions. This must be proved, of course. It should be possible to carry out such a proof by the methods in [34]. This presents us with the computational problem referred to in section 2, namely, to reduce (7) to a one-parameter problem as in (7′) and then solve a sequence of problems (7′) for various values of μ, thereby obtaining a sequence of unit eigenvectors $(x^{(m)})$ which converge to an eigenvector satisfying the middle boundary condition. As pointed out in section z, this procedure may fail if there are no real solution λ for certain values of μ. Although this does not happen when A, B_1 and B_2 are symmetric with B_1 positive definite, even in this case it may be preferable computationally to use a method which computes (λ, μ) simultaneously. Such a method should be applied to (7) with additional equations to restrict the set of solutions to be finite, since convergence would be quite slow when there are a continuum of solutions. Thus, we seek a method to solve the augmented system

$$\begin{aligned} Ax - \lambda B_1 x - \mu B_2 x &= 0, \\ g(x) &= 0, \\ \|x\|^2 - 1 &= 0, \end{aligned} \tag{19}$$

where $g(x)$ is some differentiable function of x. For the boundary-value problem (3), $g(x) = x(b)$. In the discretized form (7), we have $g(x) = x(k)$, the kth component of x. In [9], some initial-value shooting methods are developed for the numerical solution of (19) and applied to the particular problem in (6) above. In [13], the same problem is solved by the MRQI iteration.

As we have said, MRQI is a generalization of RQI. Like RQI, it can be regarded as a special case of inverse iteration. Inverse iteration for (19) is derived from Newton's method. See [13] for details. MRQI is obtained from inverse iteration by replacing the arbitrary "shifts" by the functionals $\lambda_i(x)$ given above. (See (11)). For the two-parameter case in (19), we have

$$\begin{aligned} \langle B_1x, B_1x\rangle \lambda_1(x) + \langle B_2x, B_1x\rangle \lambda_2(x) &= \langle Ax, B_1x\rangle, \\ \langle B_1x, B_2x\rangle \lambda_1(x) + \langle B_2x, B_2x\rangle \lambda_2(x) &= \langle Ax, B_2x\rangle. \end{aligned} \tag{20}$$

Under the assumption of linear independence of $\{B_1x, B_2x\}$ made in section 4, there is a unique solution, $(\lambda_1(x), \lambda_2(x))$, of (20). This is used in the iteration defined by

$$\begin{aligned} z_{i+1} &= (A - \lambda_1(x_i) B_1 - \lambda_2(x_i) B_2)^{-1} (B_1 x_i + c_i B_2 x_i), \\ x_{i+1} &= z_{i+1} / \|z_{i+1}\|, \end{aligned} \tag{21}$$

to generate a sequence of unit vectors (x_i). The constants c_i are chosen to make $g(x_i) = 0$ in (19). When $g(x)$ is a homogeneous linear function of x, then it is easy to determine c_1, since the first equation in (21) together with $g(z_{i+1}) = 0$ is a system of n+1 linear equations for the n+1 unknowns(z_{i+1}, c_i) and x_{i+1} given by the second equation in (21) also satisfies $g(x_{i+1}) = 0$.

Numerical computations on the discretization of problem (6) indicate that the MRQI sequence (x_i) converges to a solution for various starting vectors [13]. The convergence rate appears to be at least quadratic, although it may be faster, as in RQI. (See [35, 36].) The proof of convergence of MRQI for the generalized eigenvalue problem (9) as well as the

two parameter problem (19) remains as an open problem. For local convergence, it seems possible to relate this to the local convergence of inverse iteration and Newton's method. However, the use of the minimum residual functionals, $\lambda_1(x_i)$, $\lambda_2(x_i)$, at each step increases the rate of convergence. There appears to be a general principle at work here which remains to be analyzed and, perhaps, generalized to Newton's method in other contexts.

Other open problems concern the general spectral theory of problems (8) and (19); i.e. existence and multiplicity of solutions.

References

1. Courant-Hilbert, Mathematische Physik I Springer 1931.
2. E. L. Ince, Ordinary Differential Equations Dover, 1956, 197-213.
3. F. M. Arscott, Two-parameter eigenvalue problems in differential equations, Proc. London Math. Soc. (3), 14 (1964), 459-470.
4. K. P. Hadeler, Mehrparametrige and nicht lineare eigenwertaufgaben, Archive Rat. Mech. Anal. (1967), 306-328.
5. L. Collatz, Multiparameter eigenvalue problems in inner-product spaces, J. Computer and System Sciences 2(1968), 333-341.
6. ____________, Mehrparametrige Eigenwertaufgaben in unitaren Raumen, Vortrag Bucaresti, 1967.
7. M. Faierman, The completeness and expansion theorems associated with the multiparameter eigenvalue problems in ODE, J. Diff. Eqs. 5(1969), 197-213.
8. M. Gregus, F. Neuman and F. M. Arscott, Three-point boundary value problems in differential equations, J. London Math. Soc. (2) 3(1971), 429-436.
9. L. Fox, L. Hayes and D. F. Mayers, The double eigenvalue problem.
10. E. K. Blum and G. H. Rodrigue, Solution of eigenvalue problems in Hilbert spaces by a gradient method, J. Computer and System Sciences 2(1974), 220-237.
11. E. K. Blum and P. B. Geltner, Numerical solution of eigentuple-eigenvector problems in Hilbert spaces by a gradient method, USC Math. Dept. Report, July 1975, Submitted for publication.

12. E.K. Blum and A.R. Curtis, A convergent gradient method for matrix eigenvector-eigentuple problems, Math. Dept. Report, U.S.C., March 1976. (Submitted.)
13. E.K. Blum and Albert F. Chang, A new algorithm for the double eigenvalue problem, Math. Dept. Report, U.S.C., June 1976 (Submitted.)
14. Garry Rodrigue, A gradient method for the matrix eigenvalue problem $Ax = \lambda Bx$, Numer. Math. 22, 1-16 (1973).
15. S.F. McCormick, A general approach to one-step iterative methods with applications to eigenvalue problems, J. Comp. Sys. Sci. 6, 354-372 (1972).
16. Gertrude Blanch, Mathieu functions, in Handbook of Mathematical Functions (M. Abromowitz and I.A. Stegun, Eds.) Nat. Bur. Standards Appl. Math. Ser. 55, U.S. Government Printing Office, Washington, D.C. 1964.
17. E.A. Coddington and N. Levinson, Theory of Ordinary Differential Equations, McGraw-Hill, 1955.
18. R.S. Taylor and F.M. Arscott, Lame's equation and its use in elliptic cone problems, J. Inst. Math. Applics. 7(1971), 331-336.
19. R.S. Taylor, A new approach to the delta wing problem, J. Inst. Maths. Applics. 7(1971), 337-347.
20. B.D. Sleeman, The two-parameter Sturm-Liouville problem for ordinary differential equations, Proc. Roy. Soc. Edin., ALXIX Part II, No, 10, (1971), 139-148.
21. N.W. McLachlan, Theory and Application of Mathieu Functions, Dover, New York (1964).
22. Tables Relating to Mathieu Functions, Nat. Bur. Standards, App. Math. Ser. vol. 59, Washington, D.C. (1967).
23. B.A. Troesch and H.R. Troesch, Eigenfrequencies of an elliptic membrane, Math. of Comp., vol. 27, No. 124, (1973), 755-765.
24. C.B. Moler and G.W. Stewart, An algorithm for the generalized matrix eigenvalue problems, SIAM J. Numer. Anal. 10(1973).
25. G. Peters and J.H. Wilkinson, $Ax = \lambda Bx$ and the generalized eigenproblem, SIAM J. Numer. Anal., 7(1970), 479-492.
26. S.F. McCormick, An outline of the rectangular eigenvalue problem $Ax = \lambda Bx$: Theory, Iterative solution and deflation, Personal communication.
27. J.B. Campbell and B.A. Chartres, An eigenvalue problem for rectangular matrices, Report NSF-E-782 App. Math. Div., Brown University, Jan. 1964.
28. F.M. Arscott, Periodic Differential Equations, MacMillan, New York, 1964.
29. Y. Ikebe, Matrix theoretic approach for the numerical computation of the eigenvalues of Mathieu's equation, Report CNA-100, Center Num. Anal., U. Texas, May 1975.
30. A. Weinstein and W. Stenger, Methods of Intermediate Problems for Eigenvalues, Academic Press, N.Y. 1972.
31. H.F. Weinberger, Variational Methods for Eigenvalue Approximation, SIAM, Pa. 1974.
32. R.T. Gregory and D.L. Karney, A Collection of Matrices for Testing Computational Algorithms, Wiley-Interscience 1969.

33. L. Collatz, Eigenwertaufgaben mit technischen Anwendungen, Leipzig, 1963.

34. H. B. Keller, Numerical Methods for Two-point Boundary-Value Problems, Blaisdell Publishing Co., 1968.

35. B. N. Parlett, The Rayleigh quotient iteration and some generalizations for non-normal matrices, Math. of Comp. 28, No. 127, (1974), 679-693.

36. Nai-Fu Chen, The Rayleigh quotient iteration for non-normal matrices, Ph. D. Dissertation, U. of Calif. Berkeley, 1975.

Improved convergence for linear systems using three-part splittings*

John de Pillis

University of California, Santa Cruz 95064
Univeristy of California, Riverside 92502

Introduction

We seek the solution vector x in Hilbert space H for the linear system $Ax = y_o$, where $y_o \in H$ is given in advance and A is a bounded linear invertible operator on H. We shall compare convergence of two-part splittings (i.e., linear stationary methods of first degree) with convergence of three-part splittings (i.e., linear stationary methods of second degree). In fact, we write the two-part and three-part splittings of A as follows:

$$A = A_1 + A_2' \tag{1.1a}$$

and $$A = A_1 + A_2 + A_3, \tag{1.1b}$$

where A_1, A_2', A_2 and A_3 are all bounded linear operators on H and A_1^{-1} is presumed easy to compute (for example, if H is finite-dimensional, then A_1 might be represented by a diagonal matrix). We are now ready to define the respectively induced two-part and three-part sequences, $\{x_k'\}$ and $\{x_k\}$.

<u>Definition 1.1</u>. Given $A = A_1 + A_2'$ where $A_2' = A_2 + A_3$, with fixed $y_o \in H$. Then for any $x_o = x_o' \in H$ the <u>two-part sequence</u>

$$\{x_o, x_1', x_2', \ldots, x_k', \ldots\}$$

is defined iteratively by

* This research was partially supported by AFOSR 75-2858.

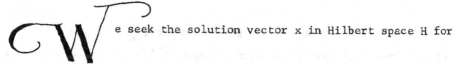

$$A_1x'_{k+1} + A_2'x'_k = y_o \ , \ k = 0,1,2,\ldots \tag{1.2a}$$

The three-part sequence $\{x_k\}$ is defined iteratively by

$$A_1x_{k+2} + A_2x_{k+1} + A_3x_k = y_o \ , \ k = 0,1,2,\ldots \tag{1.2b}$$

Note that if either the two-part sequence $\{x_k'\}$ or the three-part sequence $\{x_k\}$ converges, then since A is invertible, convergence is necessarily to the solution vector x where $Ax = y_o$. We shall comment on the case where A is singular, at the end of this paper. Now if A is a non-singular matrix on finite-dimensional H, then consider A may be decomposed as

$$A = D + E + F,$$

where D is the diagonal part of A, E is the (lower triangular) matrix whose entries match those of A below the main diagonal and F is the corresponding upper triangular matrix taken from the entries of A above the main diagonal. The two-part splitting (1.1a) then yields the well-known schemes

(a) Choosing $A_1 = D$, we have the Jacobi iteration scheme,

(b) Choosing $A_1 = D + E$, we have the Gauss-Seidel iteration scheme, (1.2c)

(c) Choosing $A_1 = \frac{1}{\omega}D + E$, with $0 < \omega < 2$, we have the successive overrelaxation (SOR) method.

Three-part splittings are given mention in [7] and [9]. Also, in a paper of J. D. P. Donnelly, a theorem on chaotic relaxation is proven [5, Theorem 2.1]. This result is recaptured (and generalized) in the setting of three-part splittings, cf. [2, Theorem 5.4].

How are the two-part and three-part splittings related? Observe that (1.2b) is equivalent to

$$\underbrace{\begin{bmatrix} x_{k+2} \\ x_{k+1} \end{bmatrix}}_{Z_{k+1}} = \underbrace{\begin{bmatrix} -A_1^{-1}A_2 & -A_1^{-1}A_3 \\ I & 0 \end{bmatrix}}_{B_2} \underbrace{\begin{bmatrix} x_{k+1} \\ x_k \end{bmatrix}}_{Z_k} + \underbrace{\begin{bmatrix} A_1^{-1}y_o \\ 0 \end{bmatrix}}_{b}, \qquad (1.3)$$

where $x_k \to x$ in H if and only if $Z_k \to Z = \begin{bmatrix} x \\ x \end{bmatrix}$ in $H \oplus H$. Moreover, since (1.3) is the result of a two-part splitting on $H \oplus H$,

$$(Z_k - Z) = B_2^k(Z_o - Z). \qquad (1.4)$$

Compare (1.4) to the condition resulting from the two-part splitting (1.1a), viz.,

$$(x_k' - x) = B^k(x_o - x),\ B = -A^{-1}A_2'. \qquad (1.5)$$

In other words, as $B = -A_1'A_2$ is the transition (or iteration) operator for the two-part splitting (1.1a), so is B_2 (given in (1.3)) the transition matrix operator induced by the three-part splitting (1.1b). The importance of these remarks is that a sufficient condition for the convergence of sequences $\{x_k'\}$ of (1.2a) or $\{x_k\}$ of (1.2b) is that the spectral radius of the respective transition operators $\rho(B)$, and $\rho(B_2)$, be less one (this condition is necessary for finite-dimensional H). We are ready for the following definition:

Definition 1.2. Given the sequences $\{x_k'\}$ of (1.2a) and $\{x_k\}$ of (1.2b), then their respective asymptotic convergence rates, R' and R, are given by

$$R' = -\log_{10}\rho(B),$$
$$R = -\log_{10}\rho(B_2),$$

where $B = -A_1^{-1}A_2'$ from (1.1a) and B_2 is given in (1.3).

Remark. We indicate base ten in the definition above only as a convenience since 1/R will then indicate, roughly, the number of iterations or "hits" which will produce one more decimal place of accuracy (cf. [8, page 63]).

We conclude this section with an explicit statement of the goals of this paper. Given $A_1x_{k+2} + A_2x_{k+1} + A_3x_k = y_o$ with A_1 fixed (and presumed easy to invert). Then we seek to

1) Explicitly construct A_3 (and hence, A_2). (Theorem 2.1, display (2.2), Hypothesis 3).
2) Explicitly compute $\rho(B_2)$, and to
3) Compare with $\rho(B)$ (noting that $A = A_1 + A_2' = A_1(I-B)$). (Theorem 3.1, display (3.4) if $\rho(B)$ is real; Theorem 3.3 if $\rho(B)$ is imaginary.)

Thus, if for a fixed two-part splitting $A = A_1 + A_2'$, we can define that A_3 in the three-part splitting $A = A_1 + A_2 + A_3$ where $\rho(B_2) < 1$ and $\rho(B_2) < \rho(B)$, then we will have established the greater convergence rate of $\{x_k\}$ in (1.2b) over that of $\{x_k'\}$ in (1.2a).

2. Relating $\sigma(B)$ and $\sigma(B_2)$ via analytic ϕ.

Recall that we are attempting to compare the spectral radius $\rho(B)$ of iteration operator B (see 1.5)) with the spectral radius $\rho(B_2)$ of the iteration matrix operator B_2 (see (1.3), (1.4)). In the following theorem, we assume that some bounds on $\sigma(B)$, the spectrum of B, are known (Theorem 2.1, Hypothesis 1). Then, A_3 is explicitly constructed as a certain analytic operator function of B (Hypothesis 3). This automatically fixes A_2 and hence, defines B_2 in (1.3). The theorem characterizes $\sigma(B_2)$ in terms of $\sigma(B)$, depending on the analytic $\phi(\cdot)$ chosen (cf. (2.3)).

Theorem 2.1 ([4, Th. 3.1]). Given invertible $A = A_1 + (A_2 + A_3)$ where $-B = A_1^{-1}(A_2 + A_3)$. Suppose

1. $\sigma(-B)$ lies inside the cardioid (cf. Figure 1)

$$\mathcal{C} = \{2Z(1 + \mathrm{Re}(Z)) - 1 : Z = e^{i\theta}\} . \tag{2.1}$$

2. $\phi(\cdot)$ is a complex analtyic function whose domain of definition contains $\sigma(-B)$, and does not assume the value -1 on $\sigma(-B)$.

3. $$A_3 = -A_1\phi(-B)\frac{(\phi(-B) + B)}{(\phi(-B) + I)}, \tag{2.2}$$

where $\phi(-B)$ is the corresponding analytic operator-valued function of -B, and I is the identity operator. Then with B_2 given in (1.3),

$$-\sigma(B_2) = \sigma(\phi(B)) \cup \sigma\left(\frac{\phi(-B) + B}{\phi(-B) + I}\right) . \tag{2.3}$$

Proof. Use $\phi(-B)$ to define U,V and A_3 as follows:

$$U = \phi(-B)$$

$$V = -\left(\frac{\phi(-B) + B}{\phi(-B) + I}\right)$$

$$A_3 = A_1UV .$$

Then verify the identity

$$-B = U + V + UV .$$

Note that from (1.3),

$$B_2 = \begin{bmatrix} U+V & UV \\ -I & 0 \end{bmatrix} .$$

Let $W = \begin{bmatrix} I & -V \\ 0 & I \end{bmatrix}$ so that $W^{-1}B_2W$ (which has the same spectrum as B_2) has form

$$W^{-1}B_2W = \begin{bmatrix} U & 0 \\ -I & V \end{bmatrix}$$

$$\Rightarrow \sigma(B_2) = \sigma(W^{-1}B_2W) = \sigma(U) \cup \sigma(V) .$$

The conclusion now follows, and the proof is done. ■

Remark. Why does the cardiod $\mathcal{C}$[equation (2.1)] enter into the Theorem (2.1)? The reason is that we require that $\rho(B_2) < 1$ so as to guarantee convergence. It is shown in [2, pg. 335] that it is just this condition, $\rho(B_2) < 1$, that implies $\sigma(-B)$ is a subset of the interior of cardioid $\mathcal{C}$. In the special case where $\phi(\cdot)$ is a constant, however, bounding $\sigma(-B)$ by $\mathcal{C}$ is too crude. In fact, we note that for constant $\phi(Z) = s \neq -1$, we have

$$\rho(B_2) < 1 \Rightarrow \sigma(-B) \in \{Z: |Z-1| < 2\} , \qquad (2.4)$$

the proof of which is in [4, Prop. 3.3].

3. The case $\phi(Z)$ is constant.

Henceforth, we assume that analytic $\phi(Z) = s \neq -1$ for fixed complexes and all complex Z. In operator terms, $\phi(B) = sI$, for all operators B on H. One immediate simplication is that A_3 is easy to compute, given the representation $A = A_1 + A_2 + A_3$ with A_3 defined in (2.2) as Hypothesis 3 of Theorem 2.1. In fact, equation (1.2b) for the three-part splitting (using (2.2) with $\phi(Z) = s$) reduces to

$$\begin{aligned} x_{k+2} &= -A_1^{-1}A\left(\frac{1}{s+1}x_{k+1} + \frac{s}{s+1}x_k\right) \\ &\quad + ((1-s)x_{k+1} + sx_k) + A_1^{-1}y_o , \end{aligned} \tag{3.1}$$

where complex $s \neq -1$, $A = A_1 + A_2 + A_3 = A_1(I-B) = A_1 + A_2'$.

Remark. If complex s is chosen to be zero, then (3.1), the equivalent to the three-part expression (1.2b), reduces to the two-part splitting (1.2a).

We give an analysis of three-part convergence versus two-part convergence for $A = A_1(I-B)$ in the following two case of the iteration operator B, viz., when the spectrum of B, $\sigma(B)$, is real, and when it is pure imaginary.

3.1 Case (A): $A = A_1(I-B)$, $\sigma(B)$ is real ($\phi(Z) = s = \text{constant}$).

We have already noted that if $\rho(B_2) < 1$, for B_2 given in (1.3), then, necessarily, $\sigma(-B)$ is a subset of the interior of cardioid $\mathcal{C}$ of (2.1), and if analytic ϕ is constant, then $\sigma(-B)$ is a subset of the open disc in the complex plane centered at real 1, having radius 2 (cf. (2.4)). If, moreover, $\sigma(B)$ is real, then this reduces to the condition that there exists real $\alpha, \beta \in \sigma(-B)$ such that for all $\lambda \in \sigma(-B)$, $-1 < \alpha \leq \lambda \leq \beta < 3$. The next theorem responds to the following:

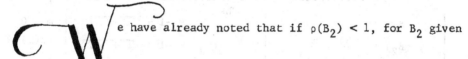

1) What is the optimal real s, call it s_o, we may assign to $\phi(-B) = sI$ so as to give optimal $\rho(B_2)$, the minimal spectral radius of B_2 of (1.3)? Note that B_2 depends on the choice of A_3 from (1.3), and A_3, in turn, depends on $\phi(-B) = sI$ from (2.2).

2) Given optimal s_o, what is the explicit value of the optimal, or minimal $\rho(B_2)$ as a function of the spectral radius $\rho(B)$?

Here is that theorem now.

Theorem 3.1. Given $A = A_1(I-B)$, $\sigma(B)$ real (where A_1^{-1} is easy to compute). Suppose $\alpha, \beta \in \sigma(-B)$ where fore all $\lambda \in \sigma(-B)$,

$$-1 < \alpha \leq \lambda \leq \beta < 3. \tag{3.2}$$

Let m denote the midpoint of $\sigma(-B)$, i.e.,

$$m = \frac{\alpha + \beta}{2} . \tag{3.3}$$

Then, according to whether $m > 0$ or $m < 0$, we have

<table>
<tr><td>If $m \leq -1 + \sqrt{1 + \rho(B)}$</td><td>If $m \geq -1 + \sqrt{1 + \rho(B)}$</td><td rowspan="2">$m > 0$
$(\rho(B)=\beta)$</td></tr>
<tr><td>Then
$s_o = m$
$\rho_o(B_2) = \frac{\rho(B) - m}{1 + m}$</td><td>Then
$s_o = -1 + \sqrt{1 + \rho(B)}$
$\rho_o(B_2) = -1 + \sqrt{1 + \rho(B)}$</td></tr>
</table>

(3.4)

<table>
<tr><td>If $m \geq -1 + \sqrt{1 - \rho(B)}$</td><td>If $m \leq -1 + \sqrt{1 - \rho(B)}$</td><td rowspan="2">$m < 0$
$\rho(B)=-\alpha$</td></tr>
<tr><td>Then
$s_o = m$
$\rho_o(B_2) = \frac{\rho(B) + m}{1 + m}$</td><td>Then
$s_o = -1 + \sqrt{1 - \rho(B)}$
$\rho_o(B_2) = 1 - \sqrt{1 - \rho(B)}$</td></tr>
</table>

By setting $\phi(-B) = s_o I$ in (2.2), we construct that B_2 in (1.3) whose spectral radius is <u>minimal</u> relative to all $\phi(-B) = sI$, as s varies over all real s.

Moreover, in all four cases in the table, above, three-part convergence is better than two-part convergence in the sense that

$$\begin{aligned} &\rho(B_2) < \rho(B), \text{ or} \\ &R > R' \text{ (cf. Def. 1.2).} \end{aligned} \tag{3.5}$$

<u>On the proof of Theorem 3.1</u>. Complete details appear in [4, section 5]. The proof depends on several technical lemmas whose end objective is to establish that $\rho(B_2) = f \vee g$, the maximum of a pair of functions f and g, each defined for real s (recall that $\phi(-B) = sI$). The optimal (smallest) value of $\rho(B_2)$, which we denote $\rho_o(B_2)$, is computed as the minimum of $f \vee g$. These are the values given for $\rho_o(B_2)$ in the table above. ■

<u>Remark</u>. Observe that in case the midpoint $m < 0$, then it is automatic that $\rho(B) < 1$, so that in the table (3.4), the expression $\sqrt{1 - \rho(B)}$ is always well defined. Also, when $m \neq 0$, we produce an s_o where $\rho_o(B_2)$ in all four categories is always <u>less</u> than $\rho(B)$. This is easy to confirm directly from the table (3.4).

<u>Example 3.2</u>. Consider the 3x3 matrix A and its decompositions

$$A = \begin{bmatrix} 1 & 0 & 1 \\ 1 & 1 & 0 \\ 0 & 1 & 1 \end{bmatrix}$$

$$= A_1(\underbrace{I - \mathcal{L}_\omega}_{B}) \qquad \text{(SOR decomposition)} \tag{3.6}$$

$$= I - \underbrace{\begin{bmatrix} 0 & 0 & -1 \\ -1 & 0 & 0 \\ 0 & -1 & 0 \end{bmatrix}}_{J} \qquad \text{(Jacobi decomposition)}$$

where for all ω, $0 < \omega < 2$, $\mathcal{L}_\omega$, the SOR iteration matrix (cf. (1.2c)) given by

$$B = -\mathcal{L}_\omega = \begin{bmatrix} \omega-1 & 0 & \omega \\ -\omega(\omega-1) & \omega-1 & -\omega^2 \\ \omega^2(\omega-1) & -\omega(\omega-1) & \omega^3+\omega+1 \end{bmatrix} . \qquad (3.7)$$

Note that (3.6) presents two two-part splittings for A, viz., the SOR decomposition with iteration matrix, $B = \mathcal{L}_\omega$, and the point-Jacobi decomposition with iteration matrix J. Note that J is consistently ordered, cyclic of index three with eigenvalues equal to the cube roots of -1. Under these conditions, we may use $\sigma(J)$ the spectrum of J, to find $\sigma(\mathcal{L}_\omega)$, the spectrum of $\mathcal{L}_\omega$ [8, Theorem 4.3]. In fact, let ω, $0 < \omega < 2$ define $\mathcal{L}_\omega$ in (3.7). Let J be given as in (3.6). Then

for all $\lambda \in \sigma(J)$, we have $\mu \in \sigma(\mathcal{L}_\omega)$ if and only if

$$(\lambda + \omega - 1)^3 + \lambda^2\omega^3 = 0 . \qquad (3.8)$$

Let us select

$$\omega = \omega_o = 0.897107 \qquad (3.9)$$

which, from (3.8) yields the spectrum of the SOR iteration matrix $\mathcal{L}_{\omega_o}$, i.e., for $\omega_o = 0.897107$,

$$\sigma(-\mathcal{L}_{\omega_o}) = \{-0.026473,\ 0.211778,\ 0.211791\},$$

so that the hypotheses (3.2) and (3.3) of Theorem 3.1 write themselves as follows:

$$\begin{aligned} \alpha &= -0.026473 \\ \beta &= \ \ 0.211791 = \rho(\mathcal{L}_{\omega_o} = B) \qquad (3.10) \\ m &= \ \ 0.092659 . \end{aligned}$$

From this, the SOR convergence rate for the two-part splitting of (3.6) is

$$R' = -\log_{10}(\rho(\mathcal{L}_{\omega_0})) \approx 0.674. \tag{3.11}$$

Note. By trial and error, it seems the ω_0 of (3.9) is optimal, i.e., produces the smallest $\rho(\mathcal{L}_\omega)$. In any case, the fact that the midpoint m of (3.10) is not zero, guarantees that improved convergence must result in passing to the three-part splitting. In particular, from (3.10) $m^2 < -\alpha$ and $m > 0$. From (3.4) in Theorem 3.1, we see that optimal s_0 is

$$s_0 = m \approx 0.092659,$$

while optimal (smallest) spectral radius $\rho(B_2)$ is

$$\rho(B_2) = \frac{\rho(B) - m}{1 + m} \approx 0.109,$$

which defines the three-part convergence rate

$$R = -\log_{10}(\rho(B_2)) \approx 0.962 \tag{3.12}$$

(compare with (3.11)).

Finally, note that the three-part convergence rate (3.12) predicts (asymptotically) $1/R \approx 1$ iteration for each decimal place of accuracy, while the two-part SOR rare R' of (3.11) predicts $1/R' \approx 1.5$ iterations per additional decimal place of accuracy. We found in actual numerical examples with various values of y_0 for $Ax = y_0$, and for various initial vectors x_0, that these asymptotic convergence rates manifested themselves after only seven or eight iterations.

3.2 Case (B): $A = A_1(I-B)$, $\sigma(B)$ is imaginary ($\phi(z) = s$ = constant

In the previous section ($\sigma(B)$ real), any two part splitting $A = A_1 + A_2 = A(I-B)$, whose iteration operator B had an unbalanced spectrum about the origin ($m \neq 0$) would always yield to improvement convergence rate by passing to the three-part splitting (Theorem 3.1). This section considers the case when $\sigma(B)$ is on the imaginary axis which proves to be far simpler than the real case. First, observe that since analytic $\phi(Z) \equiv s$ is constant, $\sigma(B)$, the spectrum of B lies inside the circle of radius two, centered at real 1 (cf. (2.4)). This means that all λ in for $\sigma(B)$ imaginary, we have $|\lambda| < \sqrt{3}$. The imaginary $\sigma(B)$ proves simpler than the real $\sigma(B)$ because whenever $\rho(B) < \sqrt{3}$ (so that the two part sequence $\{x_k'\}$ may diverge), the three-part splitting will always produce a convergent sequence $\{x_k\}$. Moreover, no restrictions on the imaginary midpoint apply. The following theorem states the details for passing to the optimal three-part splitting (1.2b) via construction of A_3 as per (2.2) for specifically defined constant $\phi(-B) = sI$.

Theorem 3.3. Given $A = A_1(I-B)$, $\sigma(B)$ imaginary (where A_1^{-1} is easy to compute). Suppose $\rho(B) < \sqrt{3}$. Then

case (a): If $\frac{\sqrt{5}-1}{2} \leq \rho(B)^2 < 3$ $(0.786 \leq \rho(B) < 1.732)$ then the optimal $s = s_o$ is the unique real solution in $(-\rho(B), \rho(B))$ to the polynomial

$$s_o^4 + 2s_o^3 - \rho(B)^2 = 0 . \qquad (0 \leq |s_o| < \rho(B))$$

Moreover, the optimal (smallest) $\rho(B_2)$ is then given by

$$\rho(B_2)_{\text{optimal}} = |s_o| .$$

<u>case (b)</u>: If $0 < \rho(B)^2 \leq \frac{\sqrt{5}-1}{2}$ $(0 < \rho(B) \leq 0.786)$, then the optimal $s = s_o$ is given by

$$s_o = \rho(B)^2,$$

in which case the optimal $\rho(B_2)$ is given by

$$\rho(B_2)_{\text{optimal}} = \frac{\rho(B)}{\sqrt{1 + \rho(B)^2}} .$$

For case (a) and case (b) above,

$$\rho(B_2) < 1, \text{ and } \rho(B_2) < \rho(B).$$

<u>On the proof of Theorem 3.3</u>. The complete proof requires the structure theorem of three-part splittings for constant $\phi(Z) = s$ [4, Theorem 4.2], which we do not present here. ■

<u>Remark</u>. Observe from the statement of Theorem 3.3, case (a), that even if $A = A_1(I - B)$, where $\rho(B)$ is near to unity (so that $\{x_k'\}$ converges slowly, if at all), then the three-part splitting guarantees a $\rho(B_2)$ very near to s_o where $s_o^4 + 2s_o^3 - 1 = 0$, i.e., $\rho(B_2) \approx 0.7167$ (so that $\{x_k\}$ converges). That is, although the two-part convergence rate $R' = \log_{10}(\rho(B)) \approx 0$, we have the three-part convergence rate $R = -\log_{10}(\rho(B_2)) \approx 0.14$. This fact holds regardless of the size of the system. To illustrate, consider the following example.

Example 3.4

The class $\mathcal{A}$ of n x n matrices, n varies, is defined by

$\mathcal{A} = \{A(n): n = 1,2,3,\ldots\}$ where each A(n) is the tridiagonal n x n matrix

$$A(n) = \begin{bmatrix} 2 & 1 & & & 0 \\ -1 & 2 & 1 & & \\ & -1 & 2 & \ddots & \\ & & \ddots & \ddots & 1 \\ 0 & & & -1 & 2 \end{bmatrix} = 2I_n(I_n - B(n)) .$$

Accordingly, the n x n skew-symmetric iteration matrix B(n) for the two-part splitting $A(n) = 2I_n + A_2' = 2I_n(I_n - B(n))$ is

$$B(n) = \tfrac{1}{2} \begin{bmatrix} 0 & -1 & & & 0 \\ 1 & 0 & -1 & & \\ & 1 & 0 & \ddots & \\ & & \ddots & \ddots & -1 \\ 0 & & & 1 & 0 \end{bmatrix} .$$

Now the n-element, imaginary spectrum $\sigma(B(n))$, which depends on n, is

$$\sigma(B(n)) = \left\{ i \cos\left(\frac{k\pi}{n+1}\right) : k = 1,2,\ldots n \right\}$$

so that

$$\rho(B(n)) = \cos\left(\frac{\pi}{n+1}\right) \uparrow 1 \text{ as } n \uparrow \infty .$$

(See [1, page 380, Ex. 8, 9] for a related example.)

Now for $n \geq 4$, case (a) of Theorem (3.3) applies. That is, suppose $A(n)x = y_o$. Then as $n \to \infty$, the three-part splitting will produce a convergent sequence $\{x_k\}$, which eventually has every seventh term ($1/R \approx 7$) yielding one more decimal place of accuracy, while the two-part sequence $\{x_k'\}$ (given skew-symmetric iteration matrix $B(n)$) becomes slower and slower in its convergence.

Actual computations reveal that even when n is large, the asymptotic convergence rate prevails after only about a dozen iterations.

4. Some open questions

Recall that our investigation of three-part splittings (1.1b) for invertible operator A on H relied on the imbedding (1.3) as a two-part splitting on the larger space $H \oplus H$; for this reason, convergence rate analysis of $\{x_k\}$ versus that of $\{x_k'\}$ (cf. Def. 1.1) is effected by comparing the spectral radius of B_2 of (1.3) with that of B in (1.5).

Given fixed iteration operator B for $A = A_1(I-B)$ (equivalently, invertible A_1 is fixed) therefore, we constructed B_2 by first analytically defining $\phi(B)$ which, in turn, fixes A_3 (cf. (2.2) of Theorem 2.1), and hence, also fixes A_2 in the expression $A = A_1 + A_2 + A_3$.

This paper proceeds under two restrictions, namely that the analytic function ϕ is a constant s (in operator terms, $\phi B \equiv sI$), and that the operators $A = A_1 + A_2'$ and A_1 are both invertible; A_1^{-1} is supposedly easy to compute. Some work has been done in each of these areas where these two restrictions are relaxed. We briefly mention these results, while indicating paths of further exploration.

When analytic $\phi(\cdot)$ is not constant. In [2, p. 336], linear analytic $\phi(Z) = p_1 z + p_2$ is explored under the assumption that $\sigma(-B)$ lies in the circle centered at 1, having radius two. This constraint on $\sigma(-B)$ is necessary if ϕ is given in advance as a constant function (see Remarks at the end of section 2).

In [3], non-constant ϕ is also studied. The main difficulty is in the computation of A_3 in (2.2) which involves the inverse of $(\phi(-B) + I)$; in [2], this entails the computation of $(-p_1B + (p_2 + 1)I)^{-1}$ which may be as difficult as the computation for A^{-1}, the inverse of the original operator A. (Of course, in the present paper, where $\phi(B) \equiv sI$, then $(\phi(-B) + I)^{-1}$ is just the scaler operator $(s + 1)^{-1}I$. Hence the questions:

4.1 Given the operator A, and the invertible component A_1 (so that $A = A_1 + A_2'$), and given the explicit construction of A_3 for analytic ϕ in (2.2), what is the convergence rate of the three-part sequence $\{x_k\}$ defined by $A = A_1 + A_2 + A_3$, compared to that of the two-part sequence $\{x_k'\}$ defined by $A = A_1 + A_2'$ (necessarily, $A_2' = A_2 + A_3$), when analytic ϕ is not constant?

4.2 Moreover, under what circumstances, or for which ϕ, are the operator terms A_2 and A_3 easy to compute? (More exactly, since from (1.2b), $x_{k+2} = -A_1^{-1}A_2x_{k+1} - A_1^{-1}A_3x_k + A_1^{-1}y_o$, we require easy computation only of the operator products $A_1^{-1}A_2$, $A_1^{-1}A_3$ and, of course, A_1^{-1}.

When A is not invertible. In a paper of Michael Neumann [6], three-part splittings are studied for non-invertible operators on finite-dimensional Hilbert space H, i.e., for rectangular matrices A. Given that A is m x n matrix (an operator sending n-space to m-space by left-hand multiplication on the columns), and given y_o in m-space, we are to find x in n-space such that $Ax = y_o$. Either a solution vector x exists, or else the system is not solvable, i.e., an approximate or best solution may be obtained. In the second instance, Neumann writes

$$A = A_1 + A_2',$$

where the splitting is <u>subproper</u>, i.e.,

$$rg(A) \subseteq rg(A_1) \text{ and } n(A) \supseteq n(A_1),$$

where rg(•) and n(•) indicate range and nullspace, respectively. If equality obtains, the splitting is <u>proper</u>. The idea then is to compose the iteration matrix

$$B = -A_1^+ A_2'$$

where A_1^+ is the Moore-Penrose generalized inverse of A_1. Thus, a counterpart to matrix B_2 of (1.3) is possible. Under these more general conditions, Neumann considers two cases, viz., when $A_1^+ A_2'$ is or is not weakly convergent ($A_1^+ A_2'$ is <u>weakly convergent</u> means the splitting $A = A_1 + A_2'$ is subproper, $\lambda \in (A_1^+ A_2') \Rightarrow |\lambda| \leq 1$, and if $|\lambda| = 1$, then $\lambda = 1$, and the Jordan blocks corresponding to $\lambda = 1$ are of order one).

<u>Case (a)</u> [6]. $-B = A_1^+ A_2'$ is not weakly convergent: the constraint here is that $\sigma(-B) \subset \mathcal{C}^o \cup \{-1\}$, where $\mathcal{C}^o$ is the interior of the cardioid $\mathcal{C}$ of (2.1).

<u>Case (b)</u> $-B = A_1^+ A_2'$ is weakly convergent. Then it is further assumed that $\sigma(-B) \subset \{Z:Re(Z) \geq 0\} \cup \{-1\}$.

Among many other things, Newmann exhibits certain analytic ϕ which effect A_3 in the construction of the generalized three-part splitting

$$A = A_1 + A_2 + A_3.$$

The question is raised, therefore

4.3 <u>Can specificially computed convergence rates be given for non-invertible matrices A under a three-part splitting, relative to rates under a two-part splitting?</u>

4.4 Can the results of Neumann be extended to the operator case for infinite-dimensional Hilbert space H?

References

1. Dahlquist, G. and Björck, Å., Numerical methods, Prentice-Hall, Englewood Cliffs, N. J. (1974).

2. de Pillis, J., k-part splittings and operator-parameter over-relaxation, J. Math. Anal. Appl., 53 (1976), pp. 313-342.

3. _____, Graphical techniques and 3-part splittings for linear systems, J. Approx. Theory, 17 (1976), pp. 44-56.

4. _____, Faster convergence for iterative solutions to systems via three-part splittings (to appear).

5. Donnelly, J. D. P., Periodic chaotic relaxation, Linear Algebra and Appl. 4 (1971), pp. 117-128.

6. Neumann, M. 3-part splittings for singular and rectangular linear systems (to appear).

7. Ortega, J. M. and Rheinboldt, , Iterative solution of non-linear equations in several variables. Academic Press, New York (1970).

8. Varga, R. S., Matrix iterative analysis. Prentice-Hall, Englewood Cliffs, N. J. (1962).

9. Young, D. M., Iterative solution of large linear systems, Academic Press, New York (1971).

NONSELFADJOINT SPECTRAL APPROXIMATION AND THE FINITE ELEMENT METHOD

William Kolata
American University, Washington D.C. 20016

John Osborn*
University of Maryland, College Park, Md. 20742

1. Introduction.

The approximation of eigenvalues and principal vectors for nonselfadjoint differential and integral operators has been an active field of research in the last several years. A satisfactory spectral approximation theory has been developed for nonselfadjoint operators and error bounds have been obtained for various approximation methods.

In this paper we will outline some of the results obtained on spectral approximation with emphasis upon the approximation of the eigenvalues and principal vectors of nonselfadjoint elliptic differential operators by the finite element method. We will concentrate our attention, for the most part, upon the results obtained in [17] and [12]. But we will also attempt to indicate the connections between this work and other related work and to point out some generalizations.

---

*The work of this author was partially supported by NSF Grant MCS76-06063.

2. Spectral Approximation for Compact Operators.

Suppose that A is a compact operator on a complex Hilbert space H. As is customary, we denote by $\sigma(A)$ and $\rho(A)$ the spectrum and resolvent set of A. For any $z \in \rho(A)$, $R_z(A) \equiv (z-A)^{-1}$ is the resolvent operator.

According to the Riesz-Schauder theory, $\sigma(A)$ is a countable or finite set of complex numbers with a cluster point possible only at zero. Each nonzero element of $\sigma(A)$ is an eigenvalue and zero may or may not be an eigenvalue. For any nonzero μ in $\sigma(A)$ there is a smallest positive integer $\alpha \equiv \alpha(\mu)$, called the ascent of $\mu - A$, such that $N((\mu-A)^{\alpha}) = N((\mu-A)^{\alpha+1})$, where N denotes the null space. The subspace $N((\mu-A)^{\alpha})$ is finite dimensional, its elements are called principal vectors (or generalized eigenvectors), and its dimension, denoted by $m = m(\mu)$, is called the algebraic multiplicity of μ. The order of a principal vector φ is the smallest positive integer j such that φ is in $N((\mu-A)^j)$. A principal vector of order 1 is an eigenvector, and the dimension of the subspace of eigenvectors, namely $N(\mu-A)$, is called the geometric multiplicity of μ. In general, the algebraic multiplicity is greater than or equal to the geometric multiplicity and the two multiplicities are equal if A is self-adjoint. Let A^* denote the adjoint of A. A^* is compact and $\sigma(A^*)$ consists of the complex conjugates of members of $\sigma(A)$. The dimensions of $N((\bar{\mu} - A^*)^j)$ and $N((\mu-A)^j)$ are equal for all j and thus the ascent of $\bar{\mu} - A^*$ equals the ascent of $\mu - A$ and the algebraic multiplicity of μ as an eigenvalue of A equals that of $\overline{\mu}$ as an eigenvalue of A^*.

Throughout this section we consider a compact operator $T: H \to H$ and a family of compact operators $T_h: H \to H$, $0 < h \le 1$, such that $\|T_h - T\| \to 0$ as $h \to 0$. Let μ be a fixed nonzero eigenvalue of T, and suppose that Γ is a circle centered at μ that lies in $\rho(T)$ and contains no other elements of $\sigma(T)$. Then the operator

$E = \frac{1}{2\pi i}\int_\Gamma R_z(T)\,dz$ is a projection onto $R(E) \equiv N((\mu-T)^\alpha)$,

the space of principal vectors associated with μ. Since $T_h \to T$ in norm, $\Gamma \subseteq \rho(T_h)$ for h sufficiently small and

$E_h = \frac{1}{2\pi i}\int_\Gamma R_z(T_h)\,dz$ is a projection onto $R(E_h)$, the direct sum of the principal vectors of T_h associated with eigenvalues lying inside Γ. E^* is the projection onto $R(E^*)$, the space of principal vectors of T^* associated with $\overline{\mu}$, and E_h^* is the projection onto $R(E_h^*)$, the direct sum of the spaces of principal vectors of T_h^* associated with the complex conjugates of the eigenvalues of T_h in Γ.

The projections $E_h \to E$ in norm, and so for h sufficiently small $\dim R(E_h) = \dim R(E) = m$. Thus, counting according to algebraic multiplicity, there are m eigenvalues of T_h in Γ which we denote $\mu_1(h),\ldots,\mu_m(h)$. For each j, $\mu_j(h) \to \mu$ as $h \to 0$.

For any two subspaces M and N of H let $\delta(M,N) = \sup_{\substack{x\in M\\ \|x\|=1}} \operatorname{dist}(x,N)$. Let $\hat{\delta}(M,N) = \max[\delta(M,N),\delta(N,M)]$; this is called the gap between M and N. Our first theorem provides an estimate for the gap between $R(E)$ and $R(E_h)$. In what follows the symbol C will denote a positive constant whose value may be different in different contexts but will always be independent of h.

Theorem 1. For all sufficiently small h,

$$\hat{\delta}(R(E),R(E_h)) \leq C\|(T-T_h)|_{R(E)}\|, \text{ where}$$

$(T-T_h)|_{R(E)}$ denotes the restriction of $T-T_h$ to $R(E)$.

As pointed out previously each of the eigenvalues $\mu_1(h),\ldots,\mu_m(h)$ are close to μ for small h. It turns out, however, that the arithmetic mean of $\{\mu_j(h)\}$ is generally a better approximation to μ. Thus we define

$$\hat{\mu}(h) = \frac{1}{m}\sum_{j=1}^{m} \mu_j(h).$$

Our next theorem gives an estimate for $|\mu-\hat{\mu}(h)|$.

Theorem 2. Let $\varphi_1,\ldots,\varphi_m$ be any basis for $R(E)$. Then there is a basis $\varphi_1^*,\ldots,\varphi_m^*$ for $R(E^*)$ such that

$$|\mu-\hat{\mu}(h)| \leq \frac{1}{m}\sum_{j=1}^{m}|((T-T_h)\varphi_j,\varphi_j^*)| + C\|(T-T_h)|_{R(E)}\|\cdot\|(T^*-T_h^*)|_{R(E)}\|.$$

In addition to estimating $|\mu-\hat{\mu}(h)|$ we can estimate $|\mu-\mu_j(h)|$ for each j as follows.

Theorem 3. Suppose α is the ascent of $\mu - T$. Let $\varphi_1,\ldots,\varphi_m$ be any basis for $R(E)$. Then there is a basis $\varphi_1^*,\ldots\varphi_m^*$ for $R(E^*)$ such that

$$|\mu-\mu_j(h)|^\alpha \le C\{\sum_{i,j=1}^{m} |((T-T_n)\varphi_i,\varphi_j^*)| + \|(T-T_h)|_{R(E)}\|\cdot\|(T^*-T_n^*)|_{R(E^*)}\|\}.$$

Our final theorem in this section concerns the approximation of principal vectors in $R(E)$ by principal vectors in $R(E_h)$.

Theorem 4. For each h let $\mu(h)$ be an eigenvalue of T_h and w_h a unit vector in $R(E_h)$ satisfying $(\mu(h)-T_h)^k w_h = 0$ for some positive integer $k \le \alpha$. Then for any integer ℓ with $k \le \ell \le \alpha$, there is a vector u_h in $R(E)$ such that $(\mu-T)^\ell u_h = 0$ and

$$\|u_h-w_h\| \le C\|(T-T_h)|_{R(E)}\|^{(\ell-k+1)/\alpha}.$$

Thus we see that a principal vector of order $\le k$ from $R(E_h)$ can draw close to a principal vector of order $\le \ell$ from $R(E)$ where $k \le \ell \le \alpha$ and the rate of approach is governed by the exponent $(\ell-k+1)/\alpha$.

The proof of the results in this section are presented in [17]. We note that in [17] the more general situation of collectively compact convergence of a family of compact operators on a Banach space is treated.

3. Variationally Posed Eigenvalue Problems.

We present in this section a simplified version of the results obtained in [12]. We will be concerned with eigenvalue problems posed in terms of sesquilinear forms $a(\cdot,\cdot)$, $b(\cdot,\cdot)$ defined on a Hilbert space H. A complex number λ will be called an eigenvalue of $a(\cdot,\cdot)$ relative to $b(\cdot,\cdot)$ if there is a nonzero u in H such that

$$a(u,v) = \lambda b(u,v) \quad \text{for all } v \text{ in } H.$$

We refer to such problems as variationally posed eigenvalue problems.

Suppose H is a complex Hilbert space and $a(\cdot,\cdot)$, $b(\cdot,\cdot)$ are bounded sesquilinear forms defined on $H \times H$. Assume that $a(\cdot,\cdot)$ is strongly coercive, i.e., $\operatorname{Re} a(u,u) \geq \alpha\|u\|^2$ for all u in H where α is a positive constant.

By the Riesz representation theorem, there are bounded operators A, B on H such that $a(u,v) = (Au,v)$ and $b(u,v) = (Bu,v)$ for all u,v in H. By coercivity, it follows that A^{-1} is a bounded operator defined on H. Let $T = A^{-1}B$ and $T_* = A^{*-1}B^*$. It is easy to see that $a(Tu,v) = b(u,v) = a(u,T_*v)$ for all u,v in H. We assume that T (and thus T_*) is a compact operator. It is useful to view T_* as the "adjoint" of T with respect to the form $a(\cdot,\cdot)$ ($a(Tu,v) = a(u,T_*v)$ for all u,v in H). In deriving our error estimates we make use of this nonstandard adjoint.

Suppose λ is an eigenvalue of $a(\cdot,\cdot)$ with respect to $b(\cdot,\cdot)$ with eigenvector φ^1. For an integer $j > 1$, φ^j is a principal vector of order j associated with λ if $a(\varphi^j,v) = \lambda b(\varphi^j,v) - \lambda a(\varphi^{j-1},v)$ for all v in H, where φ^{j-1} is a principal vector of order $j-1$. A nonzero vector ψ^1 is an adjoint eigenvector associated with λ if $a(u,\psi^1) = \lambda b(u,\psi^1)$ for all u in H. Similarly ψ^j is an adjoint principal vector of order $j > 1$ if $a(u,\psi^j) = \lambda b(u,\psi^j) - \lambda a(u,\psi^{j-1})$ for all u in H, where ψ^{j-1} is a principal vector of order $j-1$. It is easily seen that λ is an eigenvalue of $a(\cdot,\cdot)$ with respect to $b(\cdot,\cdot)$ if and only if $\mu = 1/\lambda$ is an eigenvalue of T. Moreover, the principal vectors associated with λ are the same as the principal vectors of T associated with μ. The adjoint principal vectors associated with λ are the same as the principal vectors of T_* associated with $\bar{\mu}$.

Let μ be an eigenvalue of T with ascent α and algebraic multiplicity m. Let $E = \frac{1}{2\pi i}\int_\Gamma R_z(T)dz$ be the projection onto $R(E) = N((\mu-T)^\alpha)$, the space of principal vectors associated with μ. Then E_*, the adjoint of E with respect to $a(\cdot,\cdot)$, is a projection onto $R(E_*) = N((\bar{\mu}-T_*)^\alpha)$, the space of adjoint principal vectors associated with $\bar{\mu}$.

In order to construct approximations for $\lambda = 1/\mu$ and for the associated principal vectors in $R(E)$, assume that we have a family of finite dimensional subspaces $\{S_h\}$, $0 < h \le 1$, of H such that

$$\lim_{h\to 0}\ \inf_{\chi\in S_h} \|u-\chi\| \longrightarrow 0 \quad \text{for all } u \text{ in } H.$$

and restrict the original variationally posed problem to $S_h \times S_h$.

Thus, we seek complex numbers λ_h such that for some nonzero u_h in S_h,

$$a(u_h,\chi) = \lambda_h b(u_h,\chi) \quad \text{for all} \quad \chi \in S_h .$$

We now view the λ_h's as approximations to the eigenvalues of the original problem. The principal vectors associated with these λ_h will supply approximations for the principal vectors of the original problem. If the forms $a(\cdot,\cdot)$ and $b(\cdot,\cdot)$ are associated with elliptic differential operators and the family $\{S_h\}$ is formed by piecewise polynomials defined on a grid of diameter h, we have an example of a finite element approximation of the corresponding differential eigenvalue problem.

To put us in the context of Section 2, we define first a projection P_h of H onto S_h by

$$a(P_h u,\chi) = a(u,\chi) \quad \text{for all} \quad \chi \quad \text{in} \quad S_h$$

and then set $T_h = P_h T$. The operator T_h has finite rank and $a(T_h u,\chi) = b(u,\chi)$ for all χ in S_h. Moreover, $T_h \to T$ in norm as $h \to 0$. We note that the relationship of the eigenvalues, principal vectors, and adjoint principal vectors of the restricted variationally posed problem to the eigenvalues and principal vectors of T_h and $(T_h)_*$ is the same as that for the original variational problem, T and T_*. Thus, for example, λ_h is an eigenvalue of the restricted variationally posed problem if and only if $\mu_h = 1/\lambda_h$ is an eigenvalue of T_h.

Suppose now that $\mu = 1/\lambda$ is an eigenvalue of T with ascent α and algebraic multiplicity m. Then, as in Section 2, there are m eigenvalues $\mu_1(h) = 1/\lambda_1(h),\ldots,\mu_m(h) = 1/\lambda_m(h)$ of T_h, counted according to algebraic multiplicity, converging to μ. If E_h is the spectral projection onto $R(E)$, the direct sum of the spaces of principal vectors of T_h associated $\mu_1(h),\ldots,\mu_m(h)$, then E_{h*} is the spectral projection onto $R(E_{h*})$, the direct sum of spaces of principal vectors of T_{h*} associated with $\bar{\mu}_1(h)_1\ldots,\bar{\mu}_m(h)$. For h sufficiently small, $\dim R(E) = \dim R(E_h) = m = \dim R(E_*) = \dim R(E_{h*})$.

Let

$$\varepsilon_h = \sup_{\substack{\varphi\in R(E)\\ \|\varphi\|=1}} \inf_{\chi\in S_h} \|\varphi-\chi\|$$

and

$$\varepsilon_h^* = \sup_{\substack{\psi\in R(E_*)\\ \|\psi\|=1}} \inf_{\chi\in S_h} \|\psi-\chi\|.$$

The estimates in this section will be expressed in terms of these numbers which measure how well the principal vectors and adjoint principal vectors can be approximated by elements in S_h.

<u>Theorem 1</u>. For h sufficiently small,

$$\hat{\delta}(R(E),R(E_h)) \le C\varepsilon_h, \text{ and}$$

$$\hat{\delta}(R(E_*),R(E_{h*})) \le C\|T-T_h\|\varepsilon_h^*.$$

Theorem 2. For h sufficiently small,

$$|\mu - \hat{\mu}(h)| \leq C\varepsilon_h \varepsilon_h^*.$$

This estimate can be expressed in terms of λ and $\lambda_j(h)$. In fact, $\mu - \hat{\mu}(h) = \frac{1}{m}\sum_{j=1}^{m}\left(\frac{\lambda_j(h)-\lambda}{\lambda_j(h)\lambda}\right)$.

Theorem 3. For h sufficiently small, and each $j = 1,\dots,m$

$$|\mu - \mu_j(h)|^{\alpha} = \left|\frac{\lambda_j(h)-\lambda}{\lambda_j(h)\lambda}\right|^{\alpha} \leq C\varepsilon_h\varepsilon_h^*.$$

Theorem 4. For each h, let $\mu(h)$ be an eigenvalue of T_h and w_h a unit vector in $R(E_h)$ satisfying $(\mu(h)-T_h)^k w_h = 0$ for some positive integer $k \leq \alpha$. Then for any integer ℓ with $k \leq \ell \leq \alpha$, there is a vector φ_h in $R(E)$ such that $(\mu-T)^{\ell}\varphi_h = 0$ and

$$\|\varphi_h - w_h\| \leq C\varepsilon_h^{(\ell-k+1)/\alpha}.$$

The results in this section are proved in [12] in the more general context in which the form $a(\cdot,\cdot)$ is required only to satisfy the "inf - sup" condition considered in [12]; see also [1, p. 112-186]. Although many applications can be treated with the results as presented in this section, there are applications that require the additional generality considered in [12]. For example Babuška and Osborn [3] analyze a method for the calculation of eigenvalues of differential equations with discontinuous coefficients which arise in the study of composite materials.

The coercivity condition does not hold for the appropriate form but the "inf - sup" condition does hold. They show that the method has a higher rate of convergence than the standard Ritz method.

When $a(\cdot,\cdot)$ and $b(\cdot,\cdot)$ are hermitian and so T is selfadjoint lower bounds for the error in the eigenvalue approximation are also derived in [12]. These lower bounds show that, at least in the selfadjoint case, the rate of convergence estimates for the eigenvalue error are optimal. It would be interesting to see under what conditions lower bounds for the error are possible for nonselfadjoint problems.

It is very convenient to have the error estimates written in terms of ε_h and ε_h^* because the rate of convergence is immediate upon investigation of how well the principal vectors (and adjoint principal vectors) can be approximated by elements of S_h. When using the finite element method this boils down to knowing the smoothness of the principal vectors (and adjoint principal vectors) and the approximability properties of the finite element subspaces.

We give two examples in Section 4 to illustrate this point.

4. Applications

Let Ω be a bounded domain in $\mathbb{R}^N$ with boundary $\partial\Omega$ which we assume to be of class C^∞. For any $s \geq 0$, $H^s(\Omega)$ and $H^s(\partial\Omega)$ will denote the usual Sobolev spaces. The norms on these spaces will be denoted $\|\cdot\|_s$ and $<\cdot>_s$, respectively. The inner product on $H^0(\Omega) \equiv L_2(\Omega)$ is denoted by $(\cdot,\cdot)$ and the inner product on $H^0(\partial\Omega) \equiv L_2(\partial\Omega)$ by $<\cdot,\cdot>$. The space $H_0^1(\Omega)$ is the subspace of $H^1(\Omega)$ consisting of those functions that vanish on $\partial\Omega$.

Suppose that Lu is a uniformly strongly elliptic operator defined by $Lu = -(a^{ij}u,_j),_i + b^i u,_i + cu$ where a^{ij}, b^i and c are assumed to be in $C^\infty(\bar{\Omega})$ and $a^{ij} = a^{ji}$. Thus, there is a constant $a_0 > 0$ such that $\operatorname{Re} a^{ij}(x)\xi_i\xi_j \geq a_0\xi_i\xi_i$ for all real $\xi_1 \ldots \xi_N$. Let M be the first order differential operator given by $Mu = d^i u,_i + eu$ where d^i and e are in $C^\infty(\bar{\Omega})$. Finally let $\frac{\partial u}{\partial\nu}$ denote the conormal derivative, that is $\frac{\partial u}{\partial\nu} = a^{ij}n_j u,_i$ where $n = (n_1 \ldots n_N)$ is the unit outer normal to the boundary.

We consider two elliptic eigenvalue problems. Namely, a Dirichlet problem, $Lu = \lambda Mu$, $u = 0$ on $\partial\Omega$, and a Steklov problem, $Lu = 0$, $\frac{\partial u}{\partial\nu} + \beta u = \lambda u$ on $\partial\Omega$, where $\beta \in C^\infty(\partial\Omega)$. In both of these problems we make the further assumption that $\operatorname{Re} c > \frac{a_0}{2} + \frac{b^2}{2a_0}$, where $b = \max_i |b^i(x)|$. This assumption is not necessary in the Dirichlet problem if $d^i = 0$, $1 \leq i \leq N$, and e is a nonzero constant since we can always add a large enough positive constant γ to c and in doing so only shift the eigenvalues by γ/e. For similar reasons we can assume in the Steklov problem that $\operatorname{Re}\beta > 0$.

We get a weak or variational form for these problems by appropriately defining in each case the forms $a(\cdot,\cdot)$, $b(\cdot,\cdot)$ and the space H. For the Dirichlet problem, let $a(u,v) = \int_\Omega (a^{ij}u,_i\bar{v},_j + b^i u,_i\bar{v} + cu\bar{v})dx$, $b(u,v) = \int_\Omega (d^i u,_i\bar{v} + eu\bar{v})dx$ and $H = H_0^1(\Omega)$. For the Steklov problem, let $a(u,v) = \int_\Omega (a^{ij}u,_i\bar{v},_j + b^i u,_i\bar{v} + cu\bar{v})dx + \int_{\partial\Omega} \beta u\bar{v}dx$, $b(u,v) = \int_{\partial\Omega} u\bar{v}ds = \langle u,v\rangle$ and $H = H^1(\Omega)$. In both cases it is easy to see that $a(\cdot,\cdot)$ and $b(\cdot,\cdot)$ are bounded forms on H (in the Steklov problem one uses the trace inequality $\langle u\rangle_{1/2} \le C\|u\|_1$, cf. [14]). Our assumptions immediately imply in both cases that $\operatorname{Re} a(u,u) \ge \frac{a_0}{2}\|u\|_1^2$ for all $u \in H$.

When we replace the differential eigenvalue problems by their associated variational form, $a(u,v) = \lambda b(u,v)$, we get an equivalent problem. The proof of this equivalence in one direction follows by integration by parts, and in the other direction by regularity theorems [14]. In fact, in both cases, if T is the solution operator given by $a(Tu,v) = b(u,v)$ for all u,v in H, then T and T_* are bounded maps of $H = H_0^1(\Omega)$ into $H_0^1(\Omega) \cap H^2(\Omega)$ for the Dirichlet problem, and $H = H^1(\Omega)$ into $H^2(\Omega)$ for the Steklov problem. For any $s > 1$, T and T_* are bounded maps of $H_0^1(\Omega) \cap H^s(\Omega)$ into $H_0^1(\Omega) \cap H^{s+1}(\Omega)$ in the first case, and $H^s(\Omega)$ into $H^{s+1}(\Omega)$ in the second. From this it follows, by the Rellich compactness theorem, that T is compact, and that the principal vectors of T and T_* are in $H^s(\Omega)$ for any $s \ge 1$.

A family of finite dimensional subspaces $\{S_h\}$, $0 < h \le 1$, is said to be of class $S_{k,r}$, for integers k and r such that $0 \le k \le r$, if $S_h \subseteq H^k(\Omega)$ and there is a constant C independent of h and v such that $\inf_{\chi \in S_h} \sum_{j=0}^{k} h^j \|v-\chi\|_j \le Ch^t \|v\|_t$ for all $0 < h \le 1$ and v in $H^t(\Omega)$ with $k \le t \le r$. The family is said to be of class $\overset{\circ}{S}_{k,r}$ if the estimate above is required to hold only for $v \in H_0^1(\Omega) \cap H^t(\Omega)$. Many examples of such families have been studied. Each of the usual families used in the finite element method is of class $S_{k,r}$ or $\overset{\circ}{S}_{k,r}$ for some k and r [1]. For example, when $N = 2$, the restrictions to Ω of piecewise linear functions defined on a uniform triangulation of the plane with h the diameter of each triangle form a family of class $S_{1,2}$.

For the Dirichlet problem let $\{S_h\}$ be a family of subspaces of class $\overset{\circ}{S}_{1,r}$ and assume in addition that $S_h \subset H_0^1(\Omega)$. Such families of functions are difficult to construct because the functions are required to vanish on $\partial\Omega$, but several methods have been developed for bypassing this difficulty, such as the least squares method of Bramble and Schatz [6], the methods of Nitsche [15, 16], and the Lagrange multiplier method of Babuška [2]. All of these methods are amenable to the approach in [17] and at least the last is amenable to the approach in [12].

For the Steklov problem, let $\{S_h\}$ be a family of subspaces of class $S_{1,r}$. There is no difficulty in constructing such S_h caused by the boundary conditions in this problem.

It follows from regularity results (cf. [14]) that the principal vectors as well as the adjoint principal vectors of both problems are in $H^k(\Omega)$ for any $k \ge 0$ and so in particular in $H^r(\Omega)$. Therefore, if $\mu = 1/\lambda$ is an eigenvalue with associated spaces of principal vectors, $R(E)$, and adjoint principal vectors, $R(E_*)$, then, because $\{S_h\}$ is of class $S_{1,r}$ (or $\mathring{S}_{1,r}$),

$$\varepsilon_h = \sup_{\substack{\varphi\in R(E)\\ \|\varphi\|_1=1}} \inf_{X\in S_h} \|\varphi - X\|_1 \le C \sup_{\substack{\varphi\in R(E)\\ \|\varphi\|_1=1}} \|\varphi\|_r\, h^{r-1}$$

and

$$\varepsilon_h^* = \sup_{\substack{\psi\in R(E_*)\\ \|\psi\|_1=1}} \inf_{X\in S_h} \|\psi - X\|_1 \le C \sup_{\substack{\psi\in R(E_*)\\ \|\psi\|_1=1}} \|\psi\|_r\, h^{r-1}.$$

Consequently, $|\mu - \hat{\mu}(h)| \le Ch^{2r-2}$, $|\mu - \mu_j(h)| \le Ch^{(2r-2)/\alpha}$, $\hat{\delta}(R(E),R(E_h)) \le Ch^{r-1}$, and $\hat{\delta}(R(E_*), R(E_{h*}) \le C\|T-T_h\|h^{r-1}$.

5. Survey of other results

In this section we discuss the work of several authors on spectral approximation, with particular emphasis on results that are applicable to the finite element method.

In a series of papers Vainikko [20, 21, 22] proves a general theorem on the convergence of eigenvalues and principal vectors. He considers the approximation of an operator T by an operator P_hT, where P_h is a projection. Theorems 3 and 4 of Section 2 are analogous to his. See also Petryshyn [18].

In a paper, concerned with selfadjoint problems, Birkhoff, de Boor, Swartz, and Wendroff [4] prove an inequality from which lower bounds for the eigenvalues can be derived. These bounds are expressed in terms of the Rayleigh-Ritz approximate eigenvalues corresponding to a given finite dimensional subspace and the approximability properties of this subspace.

Spectral approximation results that fit naturally into the context of the finite element method were developed by Bramble and Osborn [5]. They consider a particular type of compact operator defined on a Sobolev space and obtain rate of convergence estimates for the approximation of eigenvalues and principal vectors by Ritz-Galerkin methods. They applied these results to several finite element techniques for general (not necessarily selfadjoint) elliptic eigenvalue problems, including the least square method of Bramble and Schatz [6], methods of Nitsche [15, 16], and the Lagrange multiplier method of Babuška [2]. The rate of convergence estimates are expressed in terms of L_2 norms and Sobolev norms of negative order.

The results presented in Section 2 are developed in [17] in the more general context of collectively compact convergence of a family of compact operators on a Banach space. These results are easily applicable to a variety of finite element methods as well as to several other types of problems, such as the approximation of eigenvalues of integral operators by quadrature. These results have been extended to different kinds of convergence by Chatelin and Lemordant [8].

Results of the type presented in Section 3 (specifically Theorems 1 and 3) were proved by Babuška [1] and Fix [10] in the case of a simple eigenvalue. The proof of the estimates in Section 3 in the general case of a multiple eigenvalue was given by Kolata [12].

The approach used by Osborn [17] has been generalized to noncompact operators by Descloux, Nassif and Rappaz [9]. Generalizations to noncompact operators have also been developed by Mills [13] and Chatelin [7].

In a sequence of papers, Grigorieff [11] has presented a spectral approximation theory in the context of the notion of discrete convergence in the sense of Stümmel [19]. Rate of convergence estimates are given as well as results on the asymptotic development of eigenvalues and principal vectors. Besides being applicable to Galerkin methods, the theory is applicable in various other situations, such as simultaneous domain and coefficient perturbation.

Bibliography

1. Babuška, I.; Aziz, A.K.: Survey lectures on the mathematical foundations of the finite element method. The mathematical foundations of the finite element method with applications to partial differential equations, Academic Press, New York, 5 - 359 (1973).

2. Babuška, I.: The finite element method with Lagrangian multipliers. Numer. Math. 20, 179-192 (1973).

3. Babuška, I.; Osborn, J.E.: Numerical treatment of eigenvalue problems for equations with discontinuous coefficients Tech. Note. BN-853, Inst. for Physical Science and Technology, Univ. of MD., April, 1977.

4. Birkhoff, G.; de Boor, C.; Swartz, B.; Wendroff, B.: Rayleigh-Ritz approximation by piecewise cubic polynomials, SIAM J. Numer. Anal. 3, 188-203 (1966).

5. Bramble, J.H.; Osborn, J.E.: Rate of convergence estimates for nonselfadjoint eigenvalue approximation, Math. Comp. 27, 525-549 (1973).

6. Bramble, J.H.; Schatz, A.H.: Rayleigh-Ritz-Galerkin methods for Dirichlet's problem using subspaces without boundary conditions. Comm. Pure Appl. Math. 23, 653-675 (1970).

7. Chatelin, F.: Approximation spectrale d'un operateur borne par la methode de Galerkin et la variante de Sloan (preprint).

8. Chatelin, F.; Lemordant, J.: Error bounds in the approximation of eigenvalues of differential and integral operators, J. Math. Anal. App. (to appear).

9. Descloux, J.; Nassif, N.; Rappaz, J.: Spectral approximations with error bounds for non compact operators, report. Départment de Mathématique, Ecole Polytechnique Fédérale de Lausanne, March 1977.

10. Fix, G.M.: Eigenvalue approximation by the finite element method, Advances in Math. 10, 300-316 (1973).

11. Grigorieff, R.D.: Diskrete approximation von eigenvertproblemen I. Qualitative Konvergenz, Numer. Math. 24, 355-374 (1975), II. Konvergenzordung, ibid., 415-433, III. Asymptotische entwicklungen, ibid. 25, 79-97 (1976).

12. Kolata, W.G.: Approximation in variationally posed eigenvalue problems (preprint).

13. Mills, W.H.: Non-compact operator eigenvalue convergence (preprint).

14. Necas, J.: Les Méthodes Directes en Théorie de Équations Elliptiques, Masson et Cie, Paris (1967).

15. Nitsche, J.: Über ein Variationsprinzip zur Lösung von Dirichlet-Problemen bei Verwendung von Teilräumen, die keinen Randbedingungen unterworfen sind. Abh. Math. Sem. Univ. Hamburg 37 (1970/71).

16. Nitsche, J.: A projection method for Dirichlet problems using subspaces with nearly zero boundary conditions (preprint).

17. Osborn, J.E.: Spectral approximation for compact operators, Math. Comp. 29, 712-725 (1975).

18. Petryshyn, W.V.: On the eigenvalue problem $Tu - \lambda Su = 0$ with unbounded and nonsymmetric operators T and S. Philos. Trans. Roy. Soc. London Ser. A., No. 1130, v. 262, 413-458.

19. Stümmel F.: Diskrete Konvergenz linearer Operatoren I, Math. Ann. 190, 45-92 (1970); II. Math. Z. 120, 231-264 (1971).

20. Vainikko, G.M.: Asymptotic error bounds for projection methods in the eigenvalue problem, U.S.S.R. Comput. Math. and Math. Phys., v. 4, 9-36 (1964).

21. Vainikko, G.M.: On the rate of convergence of certain approximation methods of Galerkin type in an eigenvalue problem, Amer. Math. Soc. Transl. (2), 86, 249-259 (1970).

22. Vainikko, G.M.: Rapidity of convergence of approximation methods in the eigenvalue problem, U.S.S.R. Comput. Math. and Math. Phys., v. 7, 18-32 (1967).

HERMITE METHODS FOR THE NUMERICAL SOLUTION OF ORDINARY INITIAL VALUE PROBLEMS

Luis Kramarz, Emory University

1. Introduction. In [4] we consider methods using piecewise-continuous functions to approximate the solution of the initial value problem

(1.1) $$y^{(s)}(x) = f(x, y(x), \ldots, y^{(m)}(x)), \quad 0 \leq x \leq b,$$

(1.2) $$y^{(i)}(0) = g_i, \quad 0 \leq i \leq s-1,$$

where f is a real-valued function continuous in $D = [0, b] \times R^{m+1}$, with $R = (-\infty, \infty)$. Rates of convergence were obtained for methods based on Hermite interpolation for which the interpolation formula used up to q consecutive derivatives at any point, where

(1.3) $$0 \leq q \leq s-m-1.$$

By restricting the choice of points, we can get around condition (1.3). Rates of convergence are obtained for a general case and improved rates are given when $m = 0$. In particular, we extend the methods of Loscalzo [6] for first-order problems and strengthen the error bounds in [6] and [9] for the higher derivatives. In addition, we also extend the methods of Hung [2] for second-order problems and some of the methods of Wittenbrink [10] as applied to (1.1), (1.2). Global results for the first family of Ehle's [1] first class of L-acceptable methods for first-order problems are also given.

We point out that for first-order problems, the methods are equivalent to a subclass of the implicit Runge-Kutta methods with multiple nodes of Kastlunger and Wanner [3]. Finally, a numerical example is given to illustrate the rates of convergence.

2. Preliminary theory. Consider a partition D_1 of $[0, 1]$ given by

(2.1) $$D_1 : 0 = \gamma_1 < \gamma_2 < \ldots < \gamma_p = 1 , \quad p \geq 2 ,$$

and let q be an integer, $q \geq 1$. Let $\{r_i\}_{i=1}^{p}$ be a set of nonnegative integers satisfying

$$0 \leq r_p - r_1 \leq 1 ,$$

and $$r_i \leq r_p = q , \qquad 2 \leq i \leq p-1 .$$

For $G \in C^q[0, 1]$ define QG to be the polynomial of degree $\leq \tilde{p}-1$ such that

$$(QG)^{(i)}(\gamma_k) = G^{(i)}(\gamma_k) , \quad 1 \leq k \leq p , \quad 0 \leq i \leq r_k ,$$

where $\tilde{p} = p + \sum_{i=1}^{p} r_i$. We can write

(2.2) $$(QG)(t) = \sum_{k=1}^{p} \sum_{i=0}^{r_k} G^{(i)}(\gamma_k)\, \ell_{k,i}(t) , \quad 0 \leq t \leq 1 ,$$

where $\ell_{k,i}$ is a polynomial of degree $\leq \tilde{p}-1$ and $\ell_{k,i}^{(v)}(\gamma_u) = \delta_{iv}\delta_{ku}$, $1 \leq u \leq p$, $0 \leq v \leq r_u$. Here δ_{iv} is the Kronecker delta. An explicit expression for $\ell_{k,i}$ is given in [7].

Now define the operator $\tilde{Q}$ on $C[0, 1]$ as follows: if $z \in C[0, 1]$, let

$$S(t) = \int_0^t \frac{(t-u)^{q-1}}{(q-1)!} z(u)\, du , \quad 0 \leq t \leq 1 .$$

Then define $(\tilde{Q}z)(t) = (QS)^{(q)}(t)$, a polynomial of degree $\leq \tilde{p} - q - 1$.

Let $\{\Delta_n\}$ be a sequence of partitions of $[0, b]$ given by

$$\Delta_n : 0 = x_0 < x_1 < \ldots < x_n = b ,$$

and let $|\Delta_n| = \max_j \Delta x_j \to 0$ as $n \to \infty$, where $\Delta x_j = x_j - x_{j-1}$. We say that $\{\Delta_n\}$ is quasi-uniform if $\max_j |\Delta_n| / \Delta x_j \leq A$ for some constant A.

We then define sequences of operators $\{P_n\}$ and $\{\tilde{P}_n\}$ on $C^q[0, b]$ and $C[0, b]$ respectively by

$$(2.3) \qquad (P_n g)(x) = (QG_j)\left(\frac{x - x_{j-1}}{\Delta x_j}\right),$$

and

$$(2.4) \qquad (\tilde{P}_n g)(x) = (\tilde{Q}G_j)\left(\frac{x - x_{j-1}}{\Delta x_j}\right), \quad x_{j-1} \le x \le x_j, \quad 1 \le j \le n,$$

where $\quad G_j(t) = g(x_{j-1} + t\Delta x_j), \quad 0 \le t \le 1.$

The following lemma will be useful in obtaining rates of convergence.

Lemma 2.1. ([4, Lemma 5.4]) There is a constant C independent of n such that

(i) for all $g \in C^q[0, b]$,

$$\sup_{x_{j-1}<x<x_j} |(P_n g - g)(x)| \le C(\Delta x_j)^q \, \omega(g^{(q)}, \Delta x_j / (\tilde{p} - q - 1));$$

(ii) for all $g \in C^{q+u}[0, b]$, where $1 \le u \le \tilde{p} - q - 1$,

$$\sup_{x_{j-1}<x<x_j} |(P_n g - g)^{(i)}(x)| \le C(\Delta x_j)^{q+u-i} \sup_{x_{j-1}<x<x_j} |g^{(q+u)}(x)|,$$

$$0 \le i \le q;$$

(iii) for all $g \in C^{\tilde{p}}[0, b]$,

$$\sup_{x_{j-1}<x<x_j} |(P_n g - g)^{(i)}(x)| \le C(\Delta x_j)^{\tilde{p}-i} \sup_{x_{j-1}<x<x_j} |g^{(\tilde{p})}(x)|, \quad 0 \le i \le \tilde{p} - 1.$$

Even though the operators P_n are used in practice, we need to consider the operators $\tilde{P}_n$ in the analysis. Next we establish some relationships between P_n and $\tilde{P}_n$.

Lemma 2.2 For any $G \in C^q[0, 1]$,

$$\int_0^t \frac{(t-u)^{s-1}}{(s-1)!} (QG)(u)\, du = \sum_{i=0}^{q-1} G^{(i)}(0) \frac{t^{i+s}}{(i+s)!} + \int_0^t \frac{(t-u)^{s+q-1}}{(s+q-1)!} \tilde{Q}(G^{(q)})(u)du.$$

<u>Proof</u>. The definition of $\tilde{Q}$ implies that $(QG)^{(q)}(t) = \tilde{Q}(G^{(q)})(t), \quad 0 \le t \le 1.$

Hence

$$(QG)(t) = \sum_{i=0}^{q-1} G^{(i)}(0) \frac{t^i}{i!} + \int_0^t \frac{(t-u)^{q-1}}{(q-1)!} \tilde{Q}(G^{(q)})(u)du, \quad 0 \le t \le 1,$$

from which the conclusion follows.

Lemma 2.3. For any $g \in C^q[0, b]$,

$$\int_0^x \frac{(x-t)^{s-1}}{(s-1)!} (P_n g)(t)dt = \sum_{i=0}^{q-1} g^{(i)}(0) \frac{x^{i+s}}{(i+s)!} + \int_0^x \frac{(x-t)^{s+q-1}}{(s+q-1)!} \widetilde{P}_n(g^{(q)})(t)dt,$$

$$0 \leq x \leq b.$$

Proof. Let $G_j(t) = g(x_{j-1} + t\Delta x_j)$, $0 \leq t \leq 1$. The definition of $\widetilde{P}_n$ implies

$$(2.5)\quad (\Delta x_j)^q \widetilde{P}_n(g^{(q)})(x) = \widetilde{Q}(G_j^{(q)})\left(\frac{x - x_{j-1}}{\Delta x_j}\right), \quad x_{j-1} < x < x_j.$$

Also, from the definition of P_n and since $(QG_j)^{(q)} = \widetilde{Q}(G_j^{(q)})$, we have

$$(2.6)\quad (P_n g)^{(q)}(x) = \frac{1}{(\Delta x_j)^q} (QG_j)^{(q)}\left(\frac{x - x_{j-1}}{\Delta x_j}\right) = \widetilde{P}_n(g^{(q)}(x)), \quad x_{j-1} < x < x_j.$$

Using Lemma 2.2 and (2.5) and letting $u = (t - x_{j-1})/\Delta x_j$, $z = (x - x_{j-1})/\Delta x_j$ we have

$$(2.7)\quad \int_{x_{j-1}}^x \frac{(x-t)^{s-1}}{(s-1)!} (P_n g)(t)dt = (\Delta x_j)^s \int_0^z \frac{(z-u)^{s-1}}{(s-1)!} (QG_j)(u)du$$

$$= \sum_{i=0}^{q-1} g^{(i)}(x_{j-1}) \frac{(x-x_{j-1})^{i+s}}{(i+s)!} + \int_0^z \frac{(z-u)^{s+q-1}}{(s+q-1)!} (\Delta x_j)^{s+q} \widetilde{P}_n(g^{(q)})(x_{j-1} + u\Delta x_j)du$$

$$= \sum_{i=0}^{q-1} g^{(i)}(x_{j-1}) \frac{(x-x_{j-1})^{i+s}}{(i+s)!} + \int_{x_{j-1}}^x \frac{(x-t)^{s+q-1}}{(s+q-1)!} \widetilde{P}_n(g^{(q)})(t)dt, \quad x_{j-1} \leq x \leq x_j.$$

Suppose $x \in [x_{\ell-1}, x_\ell]$. Then by (2.6) and (2.7),

$$\int_0^x \frac{(x-t)^{s+q-1}}{(s+q-1)!} \widetilde{P}_n(g^{(q)})(t)dt = \sum_{i=1}^{\ell-1} \left[\int_{x_{i-1}}^{x_i} \frac{(x-t)^{s+q-1}}{(s+q-1)!} (P_n g)^{(q)}(t)dt\right]$$

$$+ \int_{x_{\ell-1}}^x \frac{(x-t)^{s-1}}{(s-1)!} (P_n g)(t)dt - \sum_{i=0}^{q-1} g^{(i)}(x_{\ell-1}) \frac{(x-x_{\ell-1})^{i+s}}{(i+s)!}.$$

The result follows integrating by parts the integrals involving $(P_n g)^{(q)}$, and using $(P_n g)^{(v)}(x_u) = g^{(v)}(x_u)$, $0 \leq v \leq q-1$, $u = i-1, i$.

3. Existence of approximate solutions and error bounds. We assume that the problem (1.1), (1.2) has a unique solution $y \in C^s[0, b]$, which is the case if, for example, f satisfies a uniform Lipschitz condition in D.

To simplify the notation, let $f(x; y) \equiv f(x, y(x), \ldots, y^{(m)}(x))$. Problem (1.1), (1.2) is equivalent to the operator equation

$$(3.1) \qquad (I-K)y = F,$$

where

$$(3.2) \qquad F(x) = \sum_{i=0}^{s-1} g_i \frac{x^i}{i!}, \quad 0 \leq x \leq b,$$

and

$$(3.3) \qquad (Ku)(x) = \int_0^x \frac{(x-t)^{s-1}}{(s-1)!} (Tu)(t)\,dt, \quad 0 \leq x \leq b,$$

with
$$(Tu)(x) = f(x; u).$$

The approximate solution y_n is defined by

$$(3.4) \qquad (I-K_n)y_n = F,$$

where

$$(3.5) \qquad (K_n u)(x) = \int_0^x \frac{(x-t)^{s-1}}{(s-1)!} (P_n Tu)(t)dt.$$

To carry out the analysis for the existence and uniqueness of a solution of equation (3.4) define

$$(3.6) \qquad g_i = (f(\cdot\,; y))^{(i-s)}(0), \quad s \leq i \leq s+q-1,$$

and

$$(3.7) \qquad (\tilde{T}u)(x) = \tilde{f}(x; u) \equiv (f(\cdot\,; u))^{(q)}(x), \quad 0 \leq x \leq b.$$

We assume that $f \in C^{q+2}(\mathcal{N})$, where

$$(3.8) \qquad \mathcal{N} = \{(x, z_0, z_1, \ldots, z_m) : 0 \leq x \leq b, \ |z_k - y^{(k)}(x)| \leq \delta, \ 0 \leq k \leq m, \ \delta > 0\}$$

is a neighborhood of the exact solution y of (1.1), (1.2).

The solution y of (1.1), (1.2) satisfies

$$y^{(s+q)}(x) = \tilde{f}(x;y), \quad 0 \leq x \leq b, \tag{3.9}$$

$$y^{(i)}(0) = g_i, \quad 0 \leq i \leq s+q-1. \tag{3.10}$$

Hence y also satisfies

$$(I-\tilde{K})y = \tilde{F}, \tag{3.11}$$

where

$$\tilde{F}(x) = \sum_{i=0}^{s+q-1} g_i \frac{x^i}{i!}, \quad 0 \leq x \leq b, \tag{3.12}$$

and

$$(\tilde{K}u)(x) = \int_0^x \frac{(x-t)^{s+q-1}}{(s+q-1)!} (\tilde{T}u)(t)dt, \quad 0 \leq x \leq b. \tag{3.13}$$

Equation (3.11) is set in the Banach space $C^{s+q-1}[0, b]$ with norm

$$\|g\| = \sup_{0 \leq x \leq b} \sum_{i=0}^{s+q-1} |g^{(i)}(x)|.$$

Now introduce the equation

$$(I-\tilde{K}_n)\tilde{y}_n = \tilde{F}, \tag{3.14}$$

also set in $C^{s+q-1}[0, b]$, with

$$(\tilde{K}_n u)(x) = \int_0^x \frac{(x-t)^{s+q-1}}{(s+q-1)!} (\tilde{P}_n \tilde{T}u)(t)dt, \quad 0 \leq x \leq b.$$

The following theorem is a special case of [4, Thm. 4.2].

Theorem 3.1. Consider equations (3.11) and (3.14) and suppose $\tilde{f} \in C^2(\mathcal{N})$. Let y be a solution of (3.9), (3.10). Then there is some $N > 0$ such that for all $n \geq N$ there exists r_n such that equation (3.14) has a unique solution $\tilde{y}_n \in B(y, r_n) \equiv \{z \in C^{s+q-1}[0, b] : \|z-y\| \leq r_n\}$. The Newton iterates $\tilde{y}_{n,i}$ are defined in $B(y, r_n)$ for $n \geq N$ and converge to $\tilde{y}_n$. In addition, there is a constant C independent of n such that

$$\|y-\tilde{y}_n\| \leq C\|\tilde{K}y-\tilde{K}_n y\|.$$

Theorem 3.2. Consider equations (3.1) and (3.4) and suppose $f \in C^{q+2}(\mathcal{N})$. Let y be a solution of (1.1), (1.2). Then there is some $N > 0$ such that for all $n \geq N$ there exists r_n such that equation (3.4) has a unique solution $y_n \in B(y, r_n) \equiv \{z \in C^{s+q-1}[0, b] : \|z - y\| \leq r_n\}$, which is a polynomial of degree $\leq \tilde{p} + s - 1$ in each subinterval $[x_{j-1}, x_j]$ and which satisfies (1.2) and

$$(3.15) \quad y_n^{(s+v)}(x_{j-1} + \gamma_k \Delta x_j) = (f(\cdot\, ; y_n))^{(v)}(x_{j-1} + \gamma_k \Delta x_j), \quad 1 \leq k \leq p,$$
$$0 \leq v \leq r_k, \quad 1 \leq j \leq n.$$

The Newton iterates $y_{n,i}$ obtained from equation (3.4) are defined in $B(y, r_n)$ for $n \geq N$ and converge to y_n if the initial function $y_{n,0}$ satisfies (3.10). Moreover, there is a constant C independent of n such that if $y \in C^{s+q+u}[0,b]$, where $2 \leq u \leq \tilde{p} - q$ then

$$\sup_{0 \leq x \leq b} |(y - y_n)^{(i)}(x)| \leq C|\Delta_n|^{u+1}, \quad 0 \leq i \leq s + q - 1 .$$

In addition, if $y \in C^{s+\tilde{p}}[0, b]$ and $\{\Delta_n\}$ is quasi-uniform, then

$$\max_{1 \leq j \leq n} \sup_{x_{j-1} < x < x_j} |(y - y_n)^{(s+i)}(x)| \leq C|\Delta_n|^{\tilde{p}-i}, \quad q \leq i \leq \tilde{p} - 1 .$$

<u>Proof</u>. By Theorem 3.1, equation (3.14) has a unique solution $\tilde{y}_n \in B(y, r_n)$ for $n \geq N$. By Lemma 2.3 and since $\tilde{y}_n$ satisfies (3.10) we can show that

$$\tilde{y}_n = \tilde{F} + \tilde{K}_n \tilde{y}_n = F + K_n \tilde{y}_n ,$$

that is, $\tilde{y}_n$ also satisfies equation (3.4). Conversely, each solution of (3.4) is also a solution of (3.14). Hence, equation (3.4) has a unique solution $y_n \in B(y, r_n)$ for $n \geq N$. It is also clear that y_n satisfies (1.2) and also (3.15), since

$$y_n^{(s)}(x) = (P_n f(\cdot\, ; y_n))(x), \quad x_{j-1} \leq x \leq x_j, \quad 1 \leq j \leq n .$$

To prove the statement about Newton's method, let $y_{n,0} \in B(y, r_n)$ and such that $y_{n,0}$ satisfies (3.10). By Theorem 3.1, the iterates

$$(3.16) \quad \tilde{y}_{n,r+1} = \tilde{F} + \tilde{K}_n \tilde{y}_{n,r} - \tilde{K}_n'(\tilde{y}_{n,r})(\tilde{y}_{n,r} - \tilde{y}_{n,r+1}), \quad r = 0, 1, 2, \ldots,$$
$$\tilde{y}_{n,0} = y_{n,0},$$

are defined in $B(y, r_n)$ and converge to $\tilde{y}_n (\equiv y_n)$.

However, by Lemma (2.3) and since $\tilde{y}_{n,r}$ and $\tilde{y}_{n,r+1}$ satisfy (3.10), it follows that

$$(3.17) \quad \tilde{y}_{n,r+1} = F(x) + K_n \tilde{y}_{n,r} - K_n'(\tilde{y}_{n,r})(\tilde{y}_{n,r} - \tilde{y}_{n,r+1}), \quad r = 0, 1, 2, \ldots .$$

Here we have also used the fact [5, p. 77] that if $u, v \in C^{s+q-1}[0, b]$, then

$$(\tilde{T}'(u)v)(x) = (T'(u)v)^{(q)}(x) .$$

Moreover, each solution of (3.17) is also a solution of (3.16). Hence (3.17) has a unique solution $\tilde{y}_{n,r+1}$, and $\{\tilde{y}_{n,r}\}$ converges to y_n.

By Theorem 3.1 and Lemma 2.3 we have

$$\|y - y_n\| \leq C \left\| \int_0^{(\cdot)} \frac{(\cdot - t)^{s+q-1}}{(s+q-1)!} (\tilde{P}_n(y^{(s+q)}) - y^{(s+q)})(t)dt \right\|$$
$$= C \left\| \int_0^{(\cdot)} \frac{(\cdot - t)^{s-1}}{(s-1)!} (P_n(y^{(s)}) - y^{(s)})(t)dt \right\|$$
$$\leq C \max_{0 \leq i \leq q-1} \sup_{0 \leq x \leq b} |(P_n(y^{(s)}) - y^{(s)})^{(i)}(x)| .$$

An application of Lemma 2.1 gives the first error bounds of the theorem. To obtain the second error estimates of the theorem, we start from

$$(3.18) \quad (y - y_n)^{(s+i)}(x) = (y^{(s)} - P_n(y^{(s)}))^{(i)}(x) + (P_n(f(\cdot\,; y) - f(\cdot\,; y_n)))^{(i)}(x),$$
$$x_{j-1} < x < x_j, \quad q - 1 \leq i \leq \tilde{p} - 1 .$$

If $i = q - 1$, then by the previous estimates and Lemma 2.1, equation (3.18) implies

$$\sup_{x_{j-1}<x<x_j} |(P_n(f(\cdot\,;y) - f(\cdot\,;y_n)))^{(q-1)}(x)| = O(|\Delta_n|^{\tilde{p}-q+1}).$$

Applying Markov's inequality [8] in $[x_{j-1}, x_j]$ we have

(3.19) $$\sup_{x_{j-1}<x<x_j} |(P_n(f(\cdot\,;y) - f(\cdot\,;y_n)))^{(q-1+u)}(x)| = O(|\Delta_n|^{\tilde{p}-q+1-u}),$$

$$0 \le u \le \tilde{p} - q.$$

The estimates follow from (3.18), using (3.19) and Lemma 2.1.

The rates of convergence given in Theorem 3.2 in general are not the best possible for the lower derivatives. We now derive better estimates for the important case $m = 0$.

Theorem 3.3. Assume all the hypothesis of Theorem 3.2 but let $m = 0$. In addition, let $\{\Delta_n\}$ be quasi-uniform. Then if $f \in C^{\tilde{p}}(\mathscr{N})$ there is a constant C independent of n such that

$$\sup_{0\le x\le b} |(y - y_n)^{(i)}(x)| \le C|\Delta_n|^{\tilde{p}}, \quad 0 \le i \le s,$$

and

$$\max_{1\le j\le n} \sup_{x_{j-1}<x<x_j} |(y - y_n)^{(s+i)}(x)| \le C|\Delta_n|^{\tilde{p}-i}, \quad 1 \le i \le \tilde{p} - 1.$$

However,

$$\max_{0\le j\le n} \{|(y - y_n)^{(s+i)}(x_j)|\} \le C|\Delta_n|^{\tilde{p}}, \quad 1 \le i \le r_1.$$

Proof. Since

(3.20) $$y_n(x) - y(x) = \int_0^x \frac{(x-t)^{s-1}}{(s-1)!} [P_n f(\cdot, y_n) - f(\cdot, y_n) + f(\cdot, y_n) - f(\cdot, y)](t)dt,$$

then

$$|y_n(x) - y(x)| \le C_1 \|P_n f(\cdot, y_n) - f(\cdot, y_n)\|_n + C_2 \int_0^x |y_n(t) - y(t)|dt,$$

for some appropriate positive constants C_1 and C_2. Here we use the notation

$$\|g\|_n = \max_{1\le j\le n} \sup_{x_{j-1}<x<x_j} |g(x)|.$$

By Gronwall's inequality, it follows that

$$(3.21)\quad |y_n(x) - y(x)| \le C_1 \| P_n f(\cdot, y_n) - f(\cdot, y_n) \|_n e^{C_2 b}, \quad 0 \le x \le b .$$

By Lemma 2.1 there is a constant C independent of n such that

$$(3.22)\quad \| (P_n f(\cdot, y_n) - f(\cdot, y_n))^{(i)} \|_n \le C |\Delta_n|^{\tilde{p}-i} \| (f(\cdot, y_n))^{(\tilde{p})} \|_n ,$$

$$0 \le i \le \tilde{p} - 1 .$$

But by Theorem (3.2) $\| y_n^{(i)} \|_n$ is uniformly bounded for each $i = 0, 1, \ldots, \tilde{p}$. Hence equations (3.21) and (3.22) imply that for some C independent of n,

$$(3.23)\quad \sup_{0 \le x \le b} |(y - y_n)(x)| \le C |\Delta_n|^{\tilde{p}} .$$

Differentiating both sides of (3.20) up to s times and using (3.22) and (3.23) we have

$$(3.24)\quad \sup_{0 \le x \le b} |(y - y_n)^{(i)}(x)| \le C |\Delta_n|^{\tilde{p}}, \quad 0 \le i \le s ,$$

where C is independent of n.

To obtain the second estimates of the theorem, we start from

$$(3.25)\quad (y_n - y)^{(s+i)}(x) = (P_n f(\cdot, y_n) - f(\cdot, y_n))^{(i)}(x) + (f(\cdot, y_n) - f(\cdot, y))^{(i)}(x) ,$$

$$x_{j-1} < x < x_j , \quad 0 \le i \le \tilde{p} - 1 ,$$

apply (3.22) and (3.24), and use a uniform bound on all the partial derivatives up to order $\tilde{p} - 1$ of f in $\mathcal{N}$. The final result is a consequence of the previous estimates and of

$$(y_n - y)^{(s+i)}(x_j) = (P_n f(\cdot, y_n))^{(i)}(x_j) - (f(\cdot, y))^{(i)}(x_j) = (f(\cdot, y_n) - f(\cdot, y))^{(i)}(x_j) ,$$

$0 \le i \le r_1$.

4. Extensions and improvements. Wright [11] pointed out that if $s = 1$ and $q = 0$ (collocation), the methods are equivalent to a subclass of the implicit

Runge-Kutta methods. Here we observe that if $s = 1$ and $q \geq 1$, the methods are equivalent to a subclass of the implicit Runge-Kutta methods with multiple nodes of Kastlunger and Wanner [3]. This is seen from

$$(4.1) \qquad y_n(x) = y_n(x_{j-1}) + \sum_{k=1}^{p} \sum_{i=0}^{r_k} C_{k,i}(x)(\Delta x_j)^{i+1}(f(\cdot, y_n))^{(i)}(x_{j-1} + \gamma_k \Delta x_j),$$

$$x_{j-1} \leq x \leq x_j,$$

where

$$C_{k,i}(x) = \int_0^z \ell_{k,i}(u)du$$

and

$$z = (x - x_{j-1})/\Delta x_j .$$

The Hermite methods of Loscalzo [6] for first-order problems correspond to $p = 2$, $\gamma_1 = 0$, $\gamma_2 = 1$, $r_1 = r_2 = q$. Loscalzo gave error bounds for the ith derivative, $0 \leq i \leq 2q+2$. Later, Varga [9] gave improved bounds for $1 \leq i \leq q+1$. The estimates of Theorem 3.3 improve the bounds of Varga for $1 \leq i \leq q+1$, and those of Loscalzo for $q+2 \leq i \leq 2q+2$.

Hung [2] uses a Hermite method with $p = 2$, $\gamma_1 = 0$, $\gamma_2 = 1$, $r_1 = r_2 = q = 1$, for second-order problems. His error bounds are contained in Theorems 3.2 and 3.3.

The first class of L-acceptable methods of Ehle [1] corresponds to $p = 2$, $\gamma_1 = 0$, $\gamma_2 = 1$, $r_1 = q-1$, $r_2 = q$ and to $p = 2$, $\gamma_1 = 0$, $\gamma_2 = 1$, $r_1 = q-2$, $r_2 = q$, but only the first case is covered by our theory. Global approximations were not considered in [1].

Wittenbrink [10] considered $\gamma_1 = 0$, $\gamma_p = 1$, $r_1 = r_p = q$, $r_i = 0$ for $2 \leq i \leq p-1$, using the theory of projection methods, and obtained higher rates of convergence for special choice of the points γ_i, $2 \leq i \leq p-1$.

The methods considered here allow $\gamma_1 = 0$, $\gamma_p = 1$, $0 \leq r_p - r_1 \leq 1$ and $r_i \leq r_p = q$ for $2 \leq i \leq p-1$, but do not take advantage of special choices of points γ_i.

5. Construction of the approximate solution. It is possible to construct y_n in $[x_{j-1}, x_j]$ by writing

$$y_n(x) = \sum_{i=0}^{\tilde{p}+s-1} C_i \frac{(x-x_{j-1})^i}{i!}, \quad x_{j-1} < x < x_j,$$

and obtaining a nonlinear system for the C_i's, $s + r_1 + 1 \leq i \leq \tilde{p} + s - 1$, from (3.15). However if $r_i \geq s - m$, $1 \leq i \leq p$, a much smaller nonlinear system can often be obtained by writing

$$(5.1) \qquad y_n(x) = \sum_{i=0}^{s-1} y_n^{(i)}(x_{j-1}) \frac{(x-x_{j-1})^i}{i!}$$
$$+ \sum_{k=1}^{p} \sum_{i=0}^{r_k} (\Delta x_j)^{i+s} (f(\cdot;y_n))^{(i)}(x_{j-1} + \gamma_k \Delta x_j) d_{kij}(x),$$
$$x_{j-1} \leq x \leq x_j,$$

with

$$d_{kij}(x) = \int_0^z \frac{(z-u)^{s-1}}{(s-1)!} \ell_{k,i}(u)\,du$$

and

$$z = (x - x_{j-1})/\Delta x_j, \quad x_{j-1} \leq x \leq x_j.$$

Then whenever $y_n^{(i)}(x_{j-1} + \gamma_k \Delta x_j)$ appears on the right-hand side of (5.1), we differentiate both sides of (5.1) i times and evaluate at $x_{j-1} + \gamma_k \Delta x_j$, obtaining a system of equations for these values. Since we can express $y_n^{(s+u)}(x_{j-1} + \gamma_k \Delta x_j)$, $0 \leq u \leq m - s + r_k$ in terms of $y_n^{(u)}(x_{j-1} + \gamma_k \Delta x_j)$, $0 \leq u \leq s - 1$ using equation (3.15), the system has only $(p-1)s$ unknowns. This has been the approach used in [1], [2] and [6]. In this formulation, the methods can be used as predictor-corrector methods by extrapolating y_n from $[x_{j-2}, x_{j-1}]$ to $[x_{j-1}, x_j]$ to predict the necessary values of y_n, then using (5.1) to correct them.

If Newton's method is applied to equation (3.4), we iterate to find y_n in each subinterval $[x_{j-1}, x_j]$ before proceeding to the next. The comments above apply when finding each iterate in $[x_{j-1}, x_j]$, and the extrapolation of y_n to the next subinterval can be used as initial estimate.

6. A numerical example. To illustrate the rates of convergence given by Theorem 3.3, consider

$$y'(x) = y^2(x), \quad 0 \leq x \leq 1, \quad y(0) = -1.$$

The approximate solution is obtained using $p = 2$, $\gamma_1 = 0$, $\gamma_2 = 1$, $r_1 = r_2 = q = 1$, and $\Delta x_j = h$. The results are given in Table 6.1, where

$$\ell_n^{(i)} \equiv \sup_{0 \leq x \leq 1} |(y - y_n)^{(i)}(x)|, \quad i = 0, 1, 2,$$

$$\ell_n^{(i)} \equiv \max_{1 \leq j \leq n} \sup_{x_{j-1} < x < x_j} |(y - y_n)^{(i)}(x)|, \quad i = 3, 4,$$

and

$$\ell_{n,\Delta}^{(2)} = \max_{1 \leq j \leq n} |(y - y_n)^{(2)}(x_j)|.$$

The column next to each column of errors represents the computed orders of convergence

$$\frac{\log(\ell_n(h_1) / \ell_n(h_2))}{\log(h_1 / h_2)}.$$

For convenience, we denote 1.52×10^{-3} by $1.52(-3)$. All the computed rates of convergence agree with the rates given in Theorem 3.3. Earlier, Loscalzo [6] had given rates of $O(h^4)$, $O(h^2)$, $O(h)$, $O(1)$, $O(1)$ for $\ell_n^{(i)}$, $0 \leq i \leq 4$, respectively. Later, Varga [9] gave improved rates of $O(h^3)$, $O(h^2)$ for $\ell_n^{(1)}$ and $\ell_n^{(2)}$, respectively.

TABLE 6.1 Illustration of error rates of Theorem 3.3

h	ℓ_n		$\ell_n^{(1)}$		$\ell_n^{(2)}$		$\ell_n^{(3)}$		$\ell_n^{(4)}$		$\ell_{n,\Delta}^{(2)}$	
1/4	7.18(-5)		5.60(-4)		7.66(-3)		3.55(-1)		1.01(1)		2.62(-4)	
1/8	4.53(-6)	3.99	5.05(-5)	3.47	1.32(-3)	2.54	1.16(-1)	1.61	6.05(0)	0.74	1.63(-5)	4.01
1/16	2.80(-7)	4.02	3.85(-6)	3.71	1.96(-4)	2.75	3.34(-2)	1.80	3.35(0)	0.85	1.03(-6)	3.98
1/32	1.74(-8)	4.01	2.67(-7)	3.85	2.68(-5)	2.87	9.02(-3)	1.89	1.77(0)	0.92	6.45(-8)	4.00
1/64	1.08(-9)	4.01	1.76(-8)	3.92	3.50(-6)	2.94	2.35(-3)	1.94	9.11(-1)	0.96	4.04(-9)	4.00
1/128	6.74(-11)	4.00	1.13(-9)	3.96	4.48(-7)	2.97	5.98(-4)	1.97	4.62(-1)	0.98	2.52(-10)	4.00
1/256	4.09(-12)	4.04	7.18(-11)	3.98	5.66(-8)	2.98	1.51(-4)	1.99	2.33(-1)	0.99	1.55(-11)	4.02

BIBLIOGRAPHY

1. B. L. Ehle, A-stable methods and Padé approximations to the exponential, SIAM J. Math. Anal., 4(1973), pp. 671-680.

2. H. Hung, The numerical solution of differential and integral equations by spline functions, Tech. Sum. Rep. 1053, Math. Res. Ctr., U.S. Army, Univ. of Wisconsin, Madison, 1970.

3. K. H. Kastlunger and G. Wanner, Runge-Kutta processes with multiple nodes, Computing, 9(1972), pp. 9-24.

4. L. Kramarz, Global approximations to solutions of initial value problems, submitted for publication.

5. L. Kramarz, Global approximations to solutions of initial value problems, Ph.D. Thesis, Georgia Inst. of Tech., Atlanta, 1977.

6. F. R. Loscalzo, On the use of spline functions for the numerical solution of ordinary differential equations, Tech. Sum. Rep. 869, Math. Res. Ctr, U.S. Army, Univ. of Wisconsin, Madison, 1968.

7. A. Spitzbart, A generalization of Hermite's interpolation formula, Amer. Math. Monthly, 67(1960), pp. 42-46.

8. J. Todd, A survey of numerical analysis, McGraw-Hill, New York, 1962.

9. R. S. Varga, Error bounds for spline interpolation, Approximation with special emphasis on spline functions, I. J. Schoenberg, ed., Academic Press, New York, 1969, pp. 367-388.

10. K. A. Wittenbrink, High order projection methods of moment and collocation type for nonlinear boundary value problems, Computing, 11(1973), pp. 255-274.

11. K. Wright, Some relationships between implicit Runge-Kutta, collocation, and Lanczos τ methods, and their stability properties, BIT, 10(1970), pp. 217-227.

ON LEAST SQUARES METHODS FOR LINEAR TWO-POINT BOUNDARY VALUE PROBLEMS

by

John Locker and P. M. Prenter
Colorado State University

1. Introduction

In this paper we present several variations on the theme of least squares methods for two-point boundary value problems. In particular, we consider the method of continuous least squares (CLS), the method of discrete least squares (DLS) and its connection with collocation (COLL), and the factorization of differential operators and its relationship to Galerkin schemes (GAL). Of special interest is the superconvergence phenomena for both CLS and DLS when using spline approximating subspaces and the corresponding optimal L^2 and L^∞ error estimates. A number of the theorems are new while part of the material amounts to new ways of looking at old themes.

We consider the problem of approximating solutions of the equation

$$Lu = f, \tag{1.1}$$

where L is an nth order differential operator in $L^2[a, b]$ and f is a given function belonging to $L^2[a, b]$. Let $H^n[a, b]$ denote the space of all functions $u \in C^{n-1}[a, b]$ such that $u^{(n-1)}$ is absolutely continuous on $[a, b]$ and $u^{(n)} \in L^2[a, b]$. Given a formal differential operator

$$\tau = \sum_{i=0}^{n} a_i(t)\left(\frac{d}{dt}\right)^i$$

and given n linearly independent boundary values $B_1, \ldots, B_n$ on $H^n[a,b]$, the differential operator L is defined by

$$\mathcal{D}(L) = \{u \in H^n[a, b]: B_i(u) = 0,\ i = 1, \ldots, n\},\quad Lu = \tau u.$$

To simplify the discussion we assume that the coefficients $a_i(t)$

belong to $C^\infty[a, b]$ for $0 \leq i \leq n$, although much weaker conditions suffice, and that $a_n(t) \neq 0$ on $[a, b]$.

We assume that the differential operator L is invertible, so for each $f \in L^2[a, b]$ there exists a unique $u \in \mathcal{D}(L)$ which is a solution of (1.1). Given a norm or semi-norm $\| \ \|$ and a finite-dimensional subspace $S \subset \mathcal{D}(L)$, we refer to an element $\tilde{u} \in S$ as a variational approximate to the unique solution u of (1.1) provided

$$(1.2) \qquad \|Lu - L\tilde{u}\| = \min_{v \in S} \|Lu - Lv\| .$$

Since this viewpoint includes continuous and discrete least squares schemes, collocation schemes, and certain Galerkin schemes, it is natural to think of most popular variational methods as simply least squares methods or methods of weighted residuals.

Continuous and discrete least squares methods are in a sense the third act of a trilogy: Galerkin-Collocation-Least Squares, whose first two acts have been quite extensively researched in the context of two-point boundary value problems. Particular attention is called to the work of Douglas and Dupont [6], Douglas, Dupont, and Wahlbin [7], and Wheeler [22] on Galerkin schemes, and to the work of Vainikko [21], Russell and Shampine [16], and DeBoor and Swartz [5] on collocation. Only recently has an in-depth study into least square methods, including rigorous L^2 and L^∞ error estimates, been undertaken for two-point boundary value problems. In particular, Sammon [18] has examined both continuous and discrete least squares applied to (1.1). He gives a quite general existence analysis and establishes L^2 error estimates which are order of best. More recently, the authors [12] have established superconvergence results and optimal L^2 and L^∞ error estimates for both continuous and discrete least squares which include Sammon's L^2 estimates and involve a much simpler analysis. Similar results including more general approximating subspaces have also been obtained by Ascher [1] using a much more complicated analysis.

An outline of our error analysis comprises sections 3, 4, and 5 of this paper.

The advantages of continuous least squares over Galerkin schemes are well known, the bilinear form associated with least squares being symmetric and admitting appropriate solutions to a wider class of boundary value problems. However, certain Galerkin schemes have the advantage that they admit coarser families of approximating functions than does least squares because the associated bilinear form is obtained via integration by parts. Another important feature of continuous least squares is its connection to discrete least squares and collocation under appropriate numerical quadrature schemes. In particular, continuous least squares is discrete least squares under quadrature, which in turn is collocation when the associated matrix is square and nonsingular. This feature, which we discuss in section 4, has been observed by several authors, notably by Russell and Varah [17], Sammon [18], and Raviart [15]. It is also present in classical Galerkin schemes when the associated bilinear form arises from pure orthogonality constraints which are not obtained by integration by parts (see section 6, [13], and [17]).

2. Mathematical Preliminaries

In this section we present some notation and results needed in the sequel. Let $L^2[a, b]$ denote the real Hilbert space of square integrable functions on [a, b] with inner product and norm

$$(u, v) = \int_a^b u(t)v(t)dt \quad \text{and} \quad \|u\|_2 = (u, u)^{1/2}$$

and let $L^\infty[a, b]$ denote the real Banach space of essentially bounded measurable functions on [a, b] with norm

$$\|u\|_\infty = \text{ess.sup.}\{|u(t)|: a \le t \le b\}.$$

The space $H^n[a, b]$ introduced in section 1 is a Banach space under the equivalent norms

$$\|u\|_{H^n[a,b]} = \left(\sum_{j=0}^{n} \|u^{(j)}\|_2^2 \right)^{1/2}$$

and

$$|u|_n = \sum_{j=0}^{n-1} \|u^{(j)}\|_\infty + \|u^{(n)}\|_2 ,$$

the first norm being the usual Sobolev norm.

Under our assumption that L is invertible, the differential operator L is 1-1 and continuous from $\mathcal{D}(L)$ under the Sobolev topology onto $L^2[a, b]$ under the L^2 topology, with its inverse L^{-1} satisfying

(2.1) $$\|L^{-1}f\|_{H^n[a,b]} \leq \|L^{-1}\| \, \|f\|_2 \quad \text{for all } f \in L^2[a, b].$$

Moreover, if $Lu = f$, then $f \in H^k[a, b]$ implies $u \in H^{n+k}[a, b]$ and $f \in C^k[a, b]$ implies $u \in C^{n+k}[a, b]$.

Fix a point $t \in [a, b]$ and consider the linear functional $F: L^2[a, b] \to \mathbb{R}$ defined by

$$F(f) = L^{-1}f(t) \qquad \text{for } f \in L^2[a, b].$$

Since evaluation at a point is a continuous linear functional on $H^n[a, b]$ under its Sobolev structure, it follows that F is a continuous linear functional on $L^2[a, b]$. Therefore, by the Riesz representation theorem there exists a unique function $K(t, \cdot) \in L^2[a, b]$ such that

(2.2) $$L^{-1}f(t) = F(f) = \int_a^b K(t, s)f(s)ds \quad \text{for all } f \in L^2[a, b],$$

and equation (2.2) is valid for all $t \in [a, b]$. The function $K(t, s)$ is just the standard Green's function for L.

Starting from this point, one can proceed to develop all the well-known properties of the Green's function, and indeed, Dunford and Schwartz [8, Chapter XIII] give a very elegant treatment of this operator-theoretic approach to the Green's function. One property we will need is the following (see Theorem XIII.3.8 in [8]): if $\psi_1, \ldots, \psi_n$ and $\omega_1, \ldots, \omega_n$ are C^∞ functions which form bases for the solution spaces of $\tau\psi = 0$ and $\tau^*\omega = 0$, respectively, then there exist unique nxn scalar matrices $\Gamma = [\gamma_{ij}]$ and $\Gamma' = [\gamma'_{ij}]$ such that

$$(2.3)\qquad \begin{aligned} K(t,\ s) &= \sum_{i,j=1}^{n} \gamma_{ij}\psi_i(t)\omega_j(s),\ a \le s < t \le b,\\ K(t,\ s) &= \sum_{i,j=1}^{n} \gamma'_{ij}\psi_i(t)\omega_j(s),\ a \le t < s \le b. \end{aligned}$$

From (2.3) we see immediately that $K(t, s)$ is infinitely differentiable in both variables for $t \neq s$ with

$$(2.4)\qquad \begin{aligned} \frac{\partial^{p+q}K}{\partial t^p \partial s^q}(t,\ s) &= \sum_{i,j=1}^{n} \gamma_{ij}\psi_i^{(p)}(t)\omega_j^{(q)}(s),\ a \le s < t \le b,\\ \frac{\partial^{p+q}K}{\partial t^p \partial s^q}(t,\ s) &= \sum_{i,j=1}^{n} \gamma'_{ij}\psi_i^{(p)}(t)\omega_j^{(q)}(s),\ a \le t < s \le b, \end{aligned}$$

for $p, q = 0, 1, 2, \ldots$, and these partial derivatives are all bounded. The constants γ_{ij} and γ'_{ij} are determined by the jump equations and the boundary conditions satisfied by $K(t, s)$.

<u>Remark</u>. If the differential operator L is not invertible, then one can introduce the <u>generalized inverse</u> $L^\dagger$ and the <u>generalized Green's function</u> $K(t, s)$ of L. The first author has recently completed a study [11] of these concepts, characterizing them in terms of the $2n^{th}$ order differential operators LL^* and L^*L. They may be useful in future work dealing with ill-posed problems.

To construct approximate solutions to (1.1) from a subspace of splines, we begin with a partition Δ: $a = t_0 < t_1 < \ldots < t_N = b$ of $[a, b]$. Let $Sp(m, \Delta, k)$, $m \ge k \ge n - 1$, denote the space of polynomial splines:

$$Sp(m,\ \Delta,\ k) = \{u \in C^k[a,\ b]:\ u \in P_m[I_i],\ 1 \le i \le N\}.$$

Here $I_i = [t_{i-1}, t_i]$ and $P_m[I_i]$ denotes the set of polynomials of degree m on I_i. Let $h = \max\{t_i - t_{i-1}: 1 \le i \le N\}$ denote the maximum mesh grading of Δ, and let $\underline{h} = \min\{t_i - t_{i-1}: 1 \le i \le N\}$ denote the minimum mesh grading of Δ. We will assume throughout that $\Delta = \Delta(N)$ is selected from a family $\mathcal{M}$ of uniformly graded meshes, i.e., $\Delta \in \mathcal{M}$ implies there exists a constant σ independent of Δ such that $h/\underline{h} \le \sigma$.

The partition Δ gives rise to two additional spaces of functions which we will use. Specifically, let

$$H_{\Delta}^{n,k}[a, b] = \{u \in H^n[a, b]: u \in H^{n+k}[I_i], 1 \le i \le N\}$$

with the norm

$$\|u\|_{H_{\Delta}^{n,k}[a,b]} = \left(\sum_{i=1}^{N} \|u\|_{H^{n+k}[I_i]}^2\right)^{1/2},$$

and let

$$C_{\Delta}^{n,k}[a, b] = \{u \in H^n[a, b]: u \in C^{n+k}[I_i], 1 \le i \le N\}$$

with the norm

$$\|u\|_{C_{\Delta}^{n,k}[a,b]} = \max_{0 \le j \le n+k} \|u^{(j)}\|_{\infty}.$$

We also let $H_{\Delta}^{p}[a, b] = H_{\Delta}^{0,p}[a, b]$ and $C_{\Delta}^{p}[a, b] = C_{\Delta}^{0,p}[a, b]$.

To simplify the discussion, we will use the subspace $Sp(2n-1, \Delta, n-1)$ of Hermite splines; the theory easily carries over to subspaces of smoother splines of the same degree, i.e., $k > n - 1$. The <u>cardinal</u> <u>basis</u> $\Phi_N = \{\phi_{ij}: 0 \le i \le N, 0 \le j \le n - 1\}$ for $Sp(2n-1, \Delta, n-1)$ consists of the unique splines from $Sp(2n-1, \Delta, n-1)$ solving the interpolation problems

$$\phi_{ij}^{(k)}(t_\ell) = \delta_{i\ell}\delta_{jk}, \quad 0 \le i, \ell \le N, \ 0 \le j, k \le n - 1.$$

The exact formulas for these basis functions are

$$\phi_{ij}(t) = \begin{cases} \dfrac{(t-t_i)^j}{j!}\left(\dfrac{t-t_{i-1}}{t_i-t_{i-1}}\right)^n \displaystyle\sum_{k=0}^{n-1-j}\binom{n-1+k}{k}\left(\dfrac{t-t_i}{t_{i-1}-t_i}\right)^k, & t_{i-1} \le t \le t_i, \\ \dfrac{(t-t_i)^j}{j!}\left(\dfrac{t-t_{i+1}}{t_i-t_{i+1}}\right)^n \displaystyle\sum_{k=0}^{n-1-j}\binom{n-1+k}{k}\left(\dfrac{t-t_i}{t_{i+1}-t_i}\right)^k, & t_i \le t \le t_{i+1}, \\ 0 \quad \text{otherwise}, \end{cases}$$

where $0 \le i \le N$ and $0 \le j \le n - 1$ (see [4, p. 37]). Each ϕ_{ij} has its support in $[t_{i-1}, t_{i+1}]$, and each $v \in Sp(2n-1, \Delta, n-1)$ has the unique representation

$$v(t) = \sum_{i=0}^{N} \sum_{j=0}^{n-1} v^{(j)}(t_i)\phi_{ij}(t).$$

Our analysis makes use of certain growth rates on the derivatives of members of Φ_N. Specifically, direct computation readily verifies

Lemma 2.1. There exist constants γ depending only on n, a, and b such that

(2.5) $$|\phi_{i\ell}^{(j)}(t)| \leq \gamma h^{\ell-j}, \quad a \leq t \leq b, \text{ t not a knot,}$$

for $0 \leq i \leq N$, $0 \leq \ell \leq n - 1$, and $0 \leq j \leq 2n$, and

(2.6) $$\sum_{i=0}^{N} \sum_{\ell=0}^{n-1} |\phi_{i\ell}^{(j)}(t)| \leq \gamma h^{-j}, \quad a \leq t \leq b,$$

for $0 \leq j \leq n - 1$.

For each $u \in C^{n-1}[a, b]$ it is well known that there exists a unique $\bar{u} \in Sp(2n-1, \Delta, n-1)$ satisfying

(2.7) $$\bar{u}^{(j)}(t_i) = u^{(j)}(t_i) \text{ for } 0 \leq i \leq N,\ 0 \leq j \leq n - 1.$$

The function $\bar{u}$ is the piecewise Hermite interpolate of degree $2n - 1$ to u. In the next two theorems we give optimal estimates for $\|u^{(j)} - \bar{u}^{(j)}\|_p$, $p = 2$ or $p = \infty$; the constants γ depend only on the integers n and k and on the end points a and b. Proofs for these two theorems for most cases can be found in the paper [19] of Swartz and Varga (see Theorem 6.1 and Corollary 6.2). The proofs given here are based on Lemma 2.1 and are very simple.

Theorem 2.2. If $u \in H_\Delta^{n,k}[a, b]$ where $0 \leq k \leq n$, then

(2.8) $$\|u^{(j)} - \bar{u}^{(j)}\|_2 \leq \gamma \|u^{(n+k)}\|_2 h^{n+k-j} \text{ for } 0 \leq j \leq n+k,$$

and

(2.9) $$\|u^{(j)} - \bar{u}^{(j)}\|_\infty \leq \gamma \|u^{(n+k)}\|_2 h^{n+k-j-1/2} \text{ for } 0 \leq j \leq n+k-1.$$

Proof. Fix i with $1 \leq i \leq N$ and let v be the $(n+k-1)$st Taylor polynomial for u about t_{i-1}:

$$v(t) = \sum_{p=0}^{n+k-1} \frac{u^{(p)}(t_{i-1}^+)}{p!}(t - t_{i-1})^p.$$

Clearly

$$\|u^{(n+k)} - v^{(n+k)}\|_{L^2[I_i]} = \|u^{(n+k)}\|_{L^2[I_i]}.$$

Since $u^{(n+k-1)}(t_{i-1}^+) = v^{(n+k-1)}(t_{i-1})$, for each $t \in I_i$ we have

$$|u^{(n+k-1)}(t)-v^{(n+k-1)}(t)| = |\int_{t_{i-1}}^{t} [u^{(n+k)}(s)-v^{(n+k)}(s)]ds|$$

$$\leq \|u^{(n+k)}-v^{(n+k)}\|_{L^2[I_i]}(t_i-t_{i-1})^{1/2}$$

$$\leq \|u^{(n+k)}\|_{L^2[I_i]}h^{1/2},$$

and it follows immediately that

$$\|u^{(n+k-1)} - v^{(n+k-1)}\|_{L^\infty[I_i]} \leq \|u^{(n+k)}\|_{L^2[I_i]}h^{1/2}$$

and

$$\|u^{(n+k-1)} - v^{(n+k-1)}\|_{L^2[I_i]} \leq \|u^{(n+k)}\|_{L^2[I_i]}h.$$

Proceeding by induction we obtain

$$(*) \qquad \|u^{(j)} - v^{(j)}\|_{L^2[I_i]} \leq \|u^{(n+k)}\|_{L^2[I_i]}h^{n+k-j}, \quad 0 \leq j \leq n+k$$

and

$$(**) \qquad \|u^{(j)} - v^{(j)}\|_{L^\infty[I_i]} \leq \|u^{(n+k)}\|_{L^2[I_i]}h^{n+k-j-1/2}, \quad 0 \leq j \leq n+k-1.$$

Next, fix j with $0 \leq j \leq n + k$. Clearly $v \in Sp(2n-1, \Delta, n-1)$, so for any $t \in (t_{i-1}, t_i)$ we have by (**) and Lemma 2.1

$$|v^{(j)}(t) - \bar{u}^{(j)}(t)| = |\sum_{p=i-1}^{i}\sum_{q=0}^{n-1}[v^{(q)}(t_p)-u^{(q)}(t_p)]\phi_{pq}^{(j)}(t)|$$

$$\leq \sum_{p=i}^{i}\sum_{q=0}^{n-1}\|u^{(n+k)}\|_{L^2[I_i]}h^{n+k-q-1/2}\gamma h^{q-j}$$

$$= n\gamma\|u^{(n+k)}\|_{L^2[I_i]}h^{n+k-j-1/2}.$$

Combining this inequality with (*) and (**) yields (2.8) and (2.9). Q.E.D.

A similar argument gives

Theorem 2.3. If $u \in C_{\Delta}^{n,k}[a, b]$ where $0 \le k \le n$, then

(2.10) $$\|u^{(j)} - \bar{u}^{(j)}\|_{\infty} \le \gamma \|u^{(n+k)}\|_{\infty} h^{n+k-j} \quad \text{for } 0 \le j \le n + k.$$

3. Continuous Least Squares

Given a function f in $L^2[a, b]$, in the sequel we let u denote the unique solution of equation (1.1). Also, we let γ, γ_1, ... denote generic constants which are independent of Δ and u. The approximates to u are sought in the space

$$S_N = Sp(2n-1, \Delta, n-1) \cap \mathcal{D}(L).$$

For continuous least squares one determines the unique $\tilde{u} \in S_N$ which satisfies

(3.1) $$\|Lu - L\tilde{u}\|_2 = \min_{v \in S_N} \|Lu - Lv\|_2 ,$$

or equivalently,

(3.2) $$(Lu - L\tilde{u}, Lv) = 0 \qquad \text{for all } v \in S_N.$$

If we let $\mathcal{B}_N = \{\phi_i : 1 \le i \le nN\}$ be a basis for S_N and if we set

$$\tilde{u} = \sum_{i=1}^{nN} \tilde{x}_i \phi_i ,$$

then (3.2) leads to the linear system

(3.3) $$\sum_{j=1}^{nN} (L\phi_i, L\phi_j)\tilde{x}_j = (f, L\phi_i), \qquad 1 \le i \le nN,$$

for the coefficients $\tilde{x}_i$.

We want to establish L^2 and L^{∞} error estimates for the derivatives $u^{(j)} - \tilde{u}^{(j)}$. Let us assume initially that $u \in H_{\Delta}^{n,k}[a, b]$ where $0 \le k \le n$. From Theorem 2.2 one estimate is immediate, namely if $\bar{u}$ is the piecewise Hermite interpolate to u, then $\bar{u} \in S_N$ and

(3.4) $$\|Lu - L\tilde{u}\|_2 \le \|Lu - L\bar{u}\|_2 \le \gamma \|u^{(n+k)}\|_2 h^k.$$

Let $K_j(t, s) = \frac{\partial^j K}{\partial t^j}(t, s)$ for $0 \le j \le n - 1$, where for convenience we frequently write $K_{j,t}(s) \equiv K_j(t, s)$. We know that

$$(3.5) \qquad u^{(j)}(t) - \tilde{u}^{(j)}(t) = \int_a^b K_j(t, s)[Lu(s) - L\tilde{u}(s)]ds$$

for $a \leq t \leq b$ and $0 \leq j \leq n - 1$. For each point $t \in [a, b]$ and for $0 \leq j \leq n - 1$, let $G_{j,t}(s) \equiv G_j(t, s)$ be the function of s in $\mathcal{D}(L)$ which satisfies $LG_{j,t}(s) = K_{j,t}(s)$, i.e.,

$$G_j(t, s) = \int_a^b K(s, \xi)K_j(t, \xi)d\xi, \qquad a \leq t, s \leq b.$$

Inserting the representations (2.3) and (2.4) into the above integral, we see that $G_j(t, s)$ is infinitely differentiable in the two variables t and s for $t \neq s$, and these partial derivatives are each bounded. Also, as a function of s we know that

$$K_{j,t} \in H^{n-1-j}[a,b] \cap C^{\infty}[a, b] \cap C^{\infty}[t, b],$$

and hence,

$$G_{j,t} \in H^{2n-1-j}[a, b] \cap C^{\infty}[a, t] \cap C^{\infty}[t, b].$$

For $t \in [a, b]$ and $0 \leq j \leq n - 1$ let $\phi_{j,t} \in Sp(2n-1, \Delta, n-1)$ be the piecewise Hermite interpolate to $G_{j,t}$. Clearly $\phi_{j,t} \in S_N$, and from (3.2) and (3.5) we obtain

$$(3.6) \qquad u^{(j)}(t) - \tilde{u}^{(j)}(t) = \int_a^b [LG_{j,t}(s) - L\phi_{j,t}(s)][Lu(s) - L\tilde{u}(s)]ds$$

for $a \leq t \leq b$ and $0 \leq j \leq n-1$. Using this result we now outline error estimates for continuous least squares (for details see [12]).

<u>Theorem 3.1</u>. (Superconvergence at Knots) If $u \in H_{\Delta}^{n,k}[a, b]$ where $0 \leq k \leq n$, then

$$(3.7) \qquad |u^{(j)}(t_i) - \tilde{u}^{(j)}(t_i)| \leq \gamma \|u^{(n+k)}\|_2 h^{n+k}$$

for $0 \leq i \leq N$ and $0 \leq j \leq n-1$.

<u>Proof</u>. Since $G_{j,t_i} \in H_{\Delta}^{n,n}[a, b]$, from Theorem 2.2 we have

$$\|LG_{j,t_i} - L\phi_{j,t_i}\|_2 \leq \gamma_1 \|G_{j,t_i}^{(2n)}\|_2 h^n \leq \gamma_2 h^n.$$

Combine this with (3.4) and (3.6). Q.E.D.

Theorem 3.2. If $u \in H_{\Delta}^{n,k}[a, b]$ where $0 \leq k \leq n$, then

(3.8) $$\|u^{(j)} - \tilde{u}^{(j)}\|_2 \leq \gamma \|u^{(n+k)}\|_2 h^{n+k-j}$$

and

(3.9) $$\|u^{(j)} - \tilde{u}^{(j)}\|_\infty \leq \gamma \|u^{(n+k)}\|_2 h^{n+k-j-1/2}$$

for $0 \leq j \leq n-1$.

Proof. Let $\bar{u} \in S_N$ be the piecewise Hermite interpolate to u. Using Theorem 3.1 and Lemma 2.1 we have

$$|\bar{u}^{(j)}(t) - \tilde{u}^{(j)}(t)| = |\sum_{i=0}^{N} \sum_{\ell=0}^{n-1} [u^{(\ell)}(t_i) - \tilde{u}^{(\ell)}(t_i)]\phi_{i\ell}^{(j)}(t)|$$

$$\leq \gamma_1 \|u^{(n+k)}\|_2 h^{n+k}\gamma_2 h^{-j},$$

and hence,

(3.10) $$\|\bar{u}^{(j)} - \tilde{u}^{(j)}\|_p \leq \gamma \|u^{(n+k)}\|_2 h^{n+k-j}$$

for $0 \leq j \leq n-1$ and $p = 2, \infty$. The proof is completed using Theorem 2.2 and (3.10). Q.E.D.

A similar argument using Theorem 2.3 yields

Theorem 3.3. If $u \in C_{\Delta}^{n,k}[a, b]$ where $0 \leq k \leq n$, then

(3.11) $$\|u^{(j)} - \tilde{u}^{(j)}\|_\infty \leq \gamma \|u^{(n+k)}\|_\infty h^{n+k-j}$$

for $0 \leq j \leq n-1$.

Remark. The error estimates (3.8) and (3.11) are also valid for derivatives of order $n \leq j \leq n+k$, while those in (3.9) are valid for $n \leq j \leq n + k - 1$. This is shown in [12] using Theorems 2.2 and 2.3 together with Schmidt's inequality [2] and Markov's inequality [20].

4. Discrete Least Squares - Existence and Uniqueness

The introduction of numerical quadrature transforms the continuous least squares problem to a discrete least squares problem. To set the problem in a specific context, let $\{t_{ij}: 1 \leq i \leq N, 1 \leq j \leq m\}$ be a set of points of [a, b] constrained by the inequalities

$$t_{i-1} \le t_{i1} < t_{i2} < \dots < t_{im} \le t_i,$$

and for i=1,…, N and for $\phi \in C[t_{i-1}, t_i]$ let

(4.1)
$$\int_{t_{i-1}}^{t_i} \phi(t)dt \cong \sum_{j=1}^{m} w_{ij}\phi(t_{ij})$$

be a quadrature rule with positive weights $w_{i1}, w_{i2}, \dots, w_{im}$ which integrates polynomials of degree M exactly on $[t_{i-1}, t_i]$. For $i = 1, \dots, N$ and for each pair of functions $\phi, \psi \in C[t_{i-1}, t_i]$, we introduce the discrete or pseudo inner product and norm

$$\langle\phi, \psi\rangle_i = \sum_{i=1}^{N} w_{ij}\phi(t_{ij})\psi(t_{ij}), \quad |\phi|_i = \langle\phi, \phi\rangle_i^{1/2},$$

while for $\phi, \psi \in C_\Delta^0[a, b]$ we have the analogous discrete inner product and norm

$$\langle\phi, \psi\rangle = \sum_{j=1}^{m} \langle\phi, \psi\rangle_i, \quad |\phi| = \langle\phi, \phi\rangle^{1/2}.$$

Since the quadrature points $t_{(i-1)m}$, t_{i1} and t_{im}, $t_{(i+1)1}$ are allowed to coalesce, care must be taken to use the appropriate left or right hand limits of ϕ in evaluating ϕ at such quadrature points.

Assuming the solution u to equation (1.1) belongs to $C_\Delta^{n,0}[a,b]$, a discrete least squares approximate to u is any $\hat{u} \in S_N$ solving the minimization problem

(4.2)
$$|Lu - L\hat{u}| = \min_{v \in S_N} |Lu - Lv|.$$

This problem always has a solution $\hat{u}$, and $\hat{u}$ satisfies the orthogonality constraint

(4.3)
$$\langle Lu - L\hat{u}, Lv\rangle = 0 \qquad \text{for all } v \in S_N.$$

Moreover, $\hat{u}$ is given by

$$\hat{u} = \sum_{i=1}^{nN} \hat{x}_i\phi_i$$

where the coefficients $\hat{x}_i$ satisfy the linear system

$$(4.4) \qquad \sum_{j=1}^{nN} \langle L\phi_i, L\phi_j \rangle \hat{x}_j = \langle f, L\phi_i \rangle, \quad 1 \le i \le nN.$$

Thus, in this context discrete least squares is equivalent to the introduction of numerical quadrature rules into the continuous least squares algorithm. In the event $m = n$ and $M = 2n - 1$, discrete least squares becomes collocation at the n Gauss points of each subinterval when the $nN \times nN$ matrix $[\langle L\phi_i, L\phi_j \rangle]$ is nonsingular, i.e., discrete least squares is orthogonal collocation in this context (see Theorem 2.2 in [17] and §8 in [18]).

We are interested in establishing the uniqueness of $\hat{u}$ for h sufficiently small. The main ideas of this theory are due to Sammon [18] or follow simply from generalizations of well-known ideas of Douglas and Dupont used in a similar context to study discrete Galerkin schemes (see Lemma 5 and Corollary 6 in [6]).

<u>Theorem 4.1</u>. (Sammon) Let $M \ge 2n - 2$, and let the solution u of (1.1) belong to $C_{\Delta}^{n,0}[a, b]$. Then there exists a constant $\gamma > 0$ such that for h sufficiently small:

$$(4.5) \qquad |L\phi|^2 \ge \gamma \, \|\phi\|^2_{H^n[a,b]} \quad \text{for all } \phi \in S_N.$$

Moreover, for h sufficiently small the discrete least squares approximate $\hat{u}$ to u from S_N is unique.

<u>Proof</u>. See Theorem 7.1 in [18].

5. Convergence Rates for Discrete Least Squares

We now extend the error analysis of section 3 to include discrete least squares approximates $\hat{u}$ to the solution u of equation (1.1). The spirit of the analysis is much the same as that for continuous least squares, including superconvergence at the knots of the partition Δ, which in turn leads to L^2 and L^∞ estimates for $u^{(j)} - \hat{u}^{(j)}$. Sammon

[18] has obtained similar L^2 estimates by a somewhat more complicated analysis, but he does not establish either the superconvergence or the L^∞ estimates. In the event $m = n$ and $M = 2n - 1$, so discrete least squares reduces to collocation at the n Gauss points of each subinterval, such estimates have been established by DeBoor and Swartz [5].

Throughout this section it is assumed that the hypothesis of Theorem 4.1 holds. In particular, $M \geq 2n - 2$ and h is sufficiently small to yield the estimate (4.5) and to guarantee the uniqueness of $\hat{u}$. Also, $\bar{u}$ denotes the piecewise Hermite interpolate to u in S_N.

<u>Lemma 5.1</u>. If $u \in H_\Delta^{n,k}[a, b]$ where $1 \leq k \leq n$, then

$$|Lu - L\bar{u}| \leq \gamma \, \|u^{(n+k)}\|_2 \, h^{k-1/2}, \tag{5.1}$$

and if $u \in C_\Delta^{n,k}[a, b]$ where $0 \leq k \leq n$, then

$$|Lu - L\bar{u}| \leq \gamma \, \|u^{(n+k)}\|_\infty \, h^k. \tag{5.2}$$

<u>Proof</u>. From Theorem 2.2 we have

$$|Lu - L\bar{u}|^2 = \sum_{i=1}^{N} \sum_{j=1}^{m} w_{ij} [Lu(t_{ij}) - L\bar{u}(t_{ij})]^2$$

$$\leq \gamma^2 \, \|u^{(n+k)}\|_2^2 \, h^{2k-1} (b - a).$$

Q.E.D.

<u>Lemma 5.2</u>. If $u \in C_\Delta^{n,0}[a, b]$, then

$$|Lu - L\hat{u}| \leq |Lu - L\bar{u}| \tag{5.3}$$

and

$$\|L\hat{u} - L\bar{u}\|_2 \leq \gamma |Lu - L\bar{u}|. \tag{5.4}$$

<u>Proof</u>. The first estimate is immediate from the definition of $\hat{u}$. For the second estimate we use the continuity of L and (4.5).

Q.E.D.

The analogue of (3.4) is contained in

<u>Lemma 5.3</u>. If $u \in H_\Delta^{n,k}[a, b]$ where $1 \leq k \leq n$, then

(5.5) $$\|Lu - L\hat{u}\|_2 \leq \gamma \|u^{(n+k)}\|_2 h^{k-1/2},$$

and if $u \in C_\Delta^{n,k}[a, b]$ where $0 \leq k \leq n$, then

(5.6) $$\|Lu - L\hat{u}\|_2 \leq \gamma \|u^{(n+k)}\|_\infty h^k.$$

Proof. Apply the triangle inequality, Theorems 2.2 or 2.3, Lemma 5.2, and finally, Lemma 5.1.

Q.E.D.

In terms of "optimal" estimates for $u^{(j)} - \hat{u}^{(j)}$, one might expect rates of convergence of order h^{n+k-j} in case u belongs to $H_\Delta^{n,k}[a, b]$ or $C_\Delta^{n,k}[a, b]$, but unfortunately, we are unable to obtain these types of results. However, if we assume that u has added local smoothness, say u belongs to $H_\Delta^{n,n+k}[a, b]$ or $C_\Delta^{n,n+k}[a, b]$, then we can derive these higher rates of convergence, and in particular, can obtain order h^{2n} for $u - \hat{u}$. This requirement of the extra smoothness on u also appears in [5], [6], and [18].

To establish the higher rates of convergence, we require one additional lemma.

Lemma 5.4. If $u \in H_\Delta^{n,n+k}[a, b]$ where $k \geq 0$, then

(5.7) $$\|Lu - L\hat{u}\|_{H_\Delta^{0,n+k}[a,b]} \leq \gamma \|u\|_{H_\Delta^{n,n+k}[a,b]}.$$

Proof. Using Theorem 2.2 and the Schmidt inequality, we have

$$\|Lu-L\hat{u}\|_{H_\Delta^{0,n+k}[a,b]} \leq \gamma_1 \|u-\bar{u}\|_{H_\Delta^{n,n+k}[a,b]} + \gamma_1 \|\hat{u}-\bar{u}\|_{H_\Delta^{n,n+k}[a,b]}$$

$$\leq \gamma_2 \|u\|_{H_\Delta^{n,n+k}[a,b]} + \gamma_2 h^{1-n} \|\hat{u}-\bar{u}\|_{H^n[a,b]}.$$

But by Lemma 5.2 and Lemma 5.1

$$\|\hat{u} - \bar{u}\|_{H^n[a,b]} \leq \gamma_3 \|L\hat{u} - L\bar{u}\|_2$$
$$\leq \gamma_4 |Lu - L\bar{u}|$$
$$\leq \gamma_5 \|u^{(2n)}\|_2 h^{n-1/2}.$$

Q.E.D.

Lemmas 5.1, 5.2 and 5.4 enable us to establish the following superconvergence result, which is the analogue of Theorem 3.1.

Theorem 5.5. (Superconvergence at Knots) Let $M \geq 2n - 2$, and let $u \in H_\Delta^{n,n+k}[a, b]$ where $0 \leq k \leq n$. Then

$$|u^{(j)}(t_i) - \hat{u}^{(j)}(t_i)| \leq \gamma \|u\|_{H_\Delta^{n,n+k}[a,b]} h^{n+k} \tag{5.8}$$

for $0 \leq i \leq N$ and $0 \leq j \leq n - 1$.

Proof. Fix i, j with $0 \leq i \leq N$, $0 \leq j \leq n-1$, and let $K_{j,t}(s)$, $G_{j,t}(s)$, and $\phi_{j,t}(s)$ be the functions introduced in section 3. We know that

$$u^{(j)}(t_i) - \hat{u}^{(j)}(t_i) = \langle K_{j,t_i}, Lu - L\hat{u}\rangle + E(t_i)$$

where

$$E(t_i) = (K_{j,t_i}, Lu - L\hat{u}) - \langle K_{j,t_i}, Lu - L\hat{u}\rangle.$$

Now $G_{j,t_i} \in C_\Delta^{n,n}[a, b]$ and $u \in H_\Delta^{n,n}[a, b]$, and hence, Lemmas 5.1 and 5.2 yield

$$\begin{aligned} |\langle K_{j,t_i}, Lu - L\hat{u}\rangle| &= |\langle LG_{j,t_i} - L\phi_{j,t_i}, Lu - L\hat{u}\rangle| \\ &\leq |LG_{j,t_i} - L\phi_{j,t_i}|\,|Lu - L\hat{u}| \\ &\leq \gamma_1 \|G_{j,t_i}^{(2n)}\|_\infty h^n \gamma_2 \|u^{(2n)}\|_2 h^{n-1/2} \\ &\leq \gamma_3 \|u\|_{H_\Delta^{n,n+k}[a,b]} h^{2n-1/2}. \end{aligned}$$

To estimate the quadrature error $E(t_i)$, we note that the quadrature rule is exact for polynomials of degree $n + k - 1$, and consequently, by the Peano Kernel Theorem and Lemma 5.4

$$\begin{aligned} |E(t_i)| &\leq \gamma_4 h^{n+k} \sum_{p=1}^{N} \int_{t_{p-1}}^{t_p} \left|\left(\frac{d}{dt}\right)^{n+k} \{K_{j,t_i}(t)[Lu(t) - L\hat{u}(t)]\}\right| dt \\ &\leq \gamma_5 h^{n+k} \|K_{j,t_i}\|_{H_\Delta^{0,n+k}[a,b]} \|Lu - L\hat{u}\|_{H_\Delta^{0,n+k}[a,b]} \\ &\leq \gamma_6 h^{n+k} \|u\|_{H_\Delta^{n,n+k}[a,b]}. \end{aligned}$$

Q.E.D.

A similar proof leads to the following variation of Theorem 5.5.

Theorem 5.6. (Superconvergence at Knots) Let $M \geq 2n - 1$, and let $u \in C_\Delta^{n,n+k}[a, b]$ where $0 \leq k \leq n$. Then

$$|u^{(j)}(t_i) - \hat{u}^{(j)}(t_i)| \leq \gamma \|u\|_{C_\Delta^{n,n+k}[a,b]} h^{n+k} \tag{5.9}$$

for $0 \leq i \leq N$ and $0 \leq j \leq n - 1$.

These superconvergence results provide the basis for our principal error estimates for discrete least squares, which are the analogues of Theorems 3.2 and 3.3 and have similar proofs. We simply state the results.

Theorem 5.7. Let $M \geq 2n - 2$, and let $u \in H_\Delta^{n,n+k}[a, b]$ where $0 \leq k < n$. Then

$$\|u^{(j)} - \hat{u}^{(j)}\|_p \leq \gamma \|u\|_{H_\Delta^{n,n+k}[a,b]} h^{n+k-j} \tag{5.10}$$

for $0 \leq j \leq n + k$ and $p = 2, \infty$.

Theorem 5.8. Let $M \geq 2n - 1$, and let $u \in C_\Delta^{n,n+k}[a,b]$ where $0 \leq k \leq n$. Then

$$\|u^{(j)} - \hat{u}^{(j)}\|_p \leq \gamma \|u\|_{C_\Delta^{n,n+k}[a,b]} h^{n+k-j} \tag{5.11}$$

for $0 \leq j \leq n + k$ and $p = 2, \infty$.

6. Operators Which Factor and Galerkin Schemes

The previous sections demonstrate the relationship between collocation, discrete least squares, and continuous least squares schemes for solving (1.1). It is interesting to ask whether there is a class of differential operators for which continuous least squares and Galerkin schemes are equivalent. In general, the answer is no. To clarify this statement we introduce the factorization of a differential operator.

A function $u^{\#} \in S_N$ is said to be a <u>nonpermissive</u> or <u>strong Galerkin approximate</u> to the solution u of (1.1) provided

$$(Lu - Lu^{\#}, v) = 0 \quad \text{for all } v \in S_N. \tag{6.1}$$

Calculations involving (6.1) require that S_N be a subspace of $\mathcal{D}(L) \subset H^n[a, b]$. It is in this sense that (6.1) is "nonpermissive" and that the approximate solution $u^{\#}$ is "strong" due to its regularity matching that of the solution of (1.1).

Suppose n = 2m and the operator L factors as $L = TT^*$ where T is an mth order differential operator with $\mathcal{D}(T)$ and $\mathcal{D}(T^*)$ appropriate subspaces of $H^m[a, b]$ (see [9]). If $u \in \mathcal{D}(L)$ satisfies (1.1), then

$$(TT^*u, v) = (f, v) \quad \text{for all } v \in L^2[a, b],$$

and hence,

$$(T^*u, T^*v) = (f, v) \quad \text{for all } v \in \mathcal{D}(T^*). \tag{6.2}$$

Conversely, if $u \in \mathcal{D}(T^*)$ and (6.2) holds for all $v \in \mathcal{D}(T^*)$, then clearly $T^*u \in \mathcal{D}(T)$ and $TT^*u = f$, i.e., $u \in \mathcal{D}(L)$ and u is a solution of (1.1). Permissive or weak Galerkin schemes arise in this context. Indeed, letting $\bar{S}_N$ be a finite-dimensional subspace of $\mathcal{D}(T^*) \subset H^m[a,b]$, a function $u^+ \in \bar{S}_N$ is said to be a <u>permissive</u> or <u>weak Galerkin approximate solution</u> to (1.1) provided

$$(T^*u^+, T^*v) = (f, v) \quad \text{for all } v \in \bar{S}_N. \tag{6.3}$$

The scheme is "permissive" in that u^+ is allowed to come from approximating subspaces whose regularity is less than that of strong solutions. Integrating the right hand side of (6.3) by parts leads to

$$(T^*u - T^*u^+, T^*v) = 0 \quad \text{for all } v \in \bar{S}_N. \tag{6.4}$$

Equation (6.4) can be interpreted as a least squares problem, and it is in this sense that Galerkin is equivalent to least squares. From (6.1) it is clear that strong Galerkin approximate solutions are weak Galerkin approximate solutions but not the converse.

In most applied problems $L=TT^*+A$ where A is a differential operator of order less than or equal to that of T^* and $\mathcal{D}(A) \supset \mathcal{D}(T^*)$. For such operators there is usually no connection between Galerkin and least squares schemes.

REFERENCES

[1] U. Ascher, Discrete least squares approximations for ordinary differential equations, MRC Technical Summary Report #1654 (1976).

[2] R. Bellman, A note on an inequality of E. Schmidt, Bull. Amer. Math. Soc., 50 (1944), pp. 734-736.

[3] J. H. Bramble and A. H. Schatz, Least squares methods for $2m^{th}$ order elliptic boundary-value problems, Math. Comput., 25(1971), pp. 1-32.

[4] P. J. Davis, Interpolation and Approximation, Dover Publications, New York, 1975.

[5] C. DeBoor and B. Swartz, Collocation at Gaussian points, SIAM J. Numer. Anal., 10 (1973), pp. 582-606.

[6] J. Douglas, Jr., and T. Dupont, Galerkin approximations for the two-point boundary problem using continuous, piecewise polynomial spaces, Numer. Math., 22 (1974), pp. 99-109.

[7] J. Douglas, Jr., T. Dupont, and L. Wahlbin, Optimal L_∞ error estimates for Galerkin approximations to solutions of two-point boundary value problems, Math. Comput., 29 (1975), pp. 475-483.

[8] N. Dunford and J. T. Schwartz, Linear Operators. I, II, Pure and Appl. Math., vol. 7, Interscience, New York, 1958, 1963.

[9] R. Kannan and J. Locker, Nonlinear boundary value problems and operators TT*, J. Differential Equations, to appear.

[10] E. B. Karpilovskaya, A method of collocation for integro-differential equations with biharmonic principal part, U.S.S.R. Computational Math. and Math. Phys., 10 (1970), pp. 240-260.

[11] J. Locker, The generalized Green's function for an nth order linear differential operator, Trans. Amer. Math. Soc., to appear.

[12] J. Locker and P. M. Prenter, Optimal L^2 and L^∞ error estimates for continuous and discrete least squares methods for boundary value problems, to appear.

[13] P. M. Prenter, Splines and Variational Methods, John Wiley & Sons, New York, 1975.

[14] P. M. Prenter and R. D. Russell, Orthogonal collocation for elliptic partial differential equations, SIAM J. Numer. Anal., 13 (1976), to appear.

[15] P. A. Raviart, The use of numerical integration in finite element methods for solving parabolic equations, Topics in Numerical Analysis (Proc. Roy. Irish Acad. Conf., Dublin, 1972), J. Miller, ed., Academic Press, London, 1973, pp. 233-264.

[16] R. D. Russell and L. F. Shampine, A collocation method for boundary value problems, Numer. Math., 19 (1972), pp. 1-28.

[17] R. D. Russell and J. M. Varah, A comparison of global methods for linear two-point boundary value problems, Math. Comput., 29(1975), pp. 1-13.

[18] P. Sammon, The discrete least squares method, Master's Thesis University of British Columbia. See also: The discrete least squares method, Math. Comput., 31 (1977), pp. 60-65.

[19] B. K. Swartz and R. S. Varga, Error bounds for spline and L-spline interpolation, J. Approx. Theory, 6 (1972), pp. 6-49.

[20] J. Todd (ed.), A Survey of Numerical Analysis, McGraw-Hill, New York, 1962.

[21] G. Vainikko, The convergence of the collocation method for non-linear differential equations, U.S.S.R. Computational Math. and Math. Phys., 6 (1966), pp. 35-42.

[22] M. F. Wheeler, An optimal L_∞ error estimate for Galerkin approximations to solutions of two-point boundary value problems, SIAM J. Numer. Anal., 10 (1973), pp. 914-917.

Averaging to Improve Convergence of Iterative Processes

W. Robert Mann
University of North Carolina
Chapel Hill, N. C. 27514

It has been widely known for nearly a century that mean value methods play a fundamental role in associating a sum with an infinite series. Féjèr's major discovery that arithmetic means solve, in a large sense, the convergence problem for Fourier series furnished strong impetus for the rapid development of what we call summability theory.

In 1953 it was discovered [13] that these same averaging methods could be applied to the construction of generalized iteration procedures which sometimes succeed where ordinary iteration fails. Enough work has now been done with this idea to enable us to evaluate its importance, and to contrast the limitations of averaging methods in iteration with their wider applicability in summing series. It is the purpose of this article to give such a survey. We shall begin by showing how averaging methods can be combined with iteration, and then describe some of the discoveries which delineate the applicability of the idea.

A setting sufficiently general to reveal the essential features is that where E is a compact convex set in a Banach space, X, and f is a continuous self-mapping of E. If ordinary iteration fails we try the following method of approximating a fixed point of f. We begin as usual with a first guess x_1 and take x_2 to be $f(x_1)$. But from here on we use 2-step procedure to go from the nth approximation to the $(n + 1)$ st. First we compute an average of the first n approximations

(1) $$v_n = a_{n1} x_1 + a_{n2} x_2 + \ldots + a_{nn} x_n,$$

and then we take x_{n+1} to be the image of this average rather than the image of x_n, i.e.

(2) $$x_{n+1} = f(v_n)$$

To make sure that v_n will be in the domain of f, the mean used in

(1) is required to be an *internal* mean i.e. a mean with the property that the average of a set of vectors belongs to the convex hull of the set being averaged. This restricts us to weighted arithmetic means where the coefficients in (1) satisfy the two conditions of being non-negative and having one as their sum. These weights could of course be chosen to depend on the first n generalized iterates $x_1, x_2, \ldots, x_n$, but this would presuppose a greater ability than we usually have to derive from these iterates information as to the location of the fixed point. Therefore we shall assume these coefficients to be scalar constants.

A complication in the procedure just described is that it generates two sequences -- the iterates $\{x_n\}_1^\infty$ and their averages $\{v_n\}_1^\infty$. Without some further restriction on the averaging process, not much can be said about the relation of these two sequences to each other or to the fixed points of f. To see how to simplify the situation, think of the averaging process in terms of the following infinite lower triangular matrix, A.

$$A = \begin{pmatrix} 1 & 0 & 0 \ldots & & & \\ a_{21} & a_{22} & 0 \ldots & & & \\ a_{31} & a_{32} & a_{33} \ldots & & & \\ & & & & & \\ a_{n1} & a_{n2} \ldots & & & a_{nn} & 0 \ldots \\ & & & & & \end{pmatrix}$$

where we have already imposed the conditions that all entries are non-negative and all row sums are one. The nth row determines the mean of the first n generalized iterates in (1). If we now impose the further condition that A be *regular* in the summability sense, it becomes easy to prove [13] that if either of the two sequences $\{x_n\}_1^\infty$ and $\{v_n\}_1^\infty$ converges then the other converges to the same limit, and that their common limit is fixed point of f. A condition necessary and sufficient to imply that the matrix A, as described above, be regular in that the columns form null sequences. Henceforth this will be assumed. In fact all of our matrices from now on will be, like A, infinite lower triangular matrices in which

$$a_{ij} \geq 0; \quad \sum_{j=1}^{i} a_{ij} = 1; \quad \lim_{i\to\infty} a_{ij} = 0.$$

Such matrices will be called *iteration matrices*. A generalized iteration process is determined by the starting point x_1, the function f and the iteration matrix A. Such a process is therefore represented by (x_1, f, A). That iteration matrix, C, defined by $a_{nk} = \frac{1}{n}$, $k = 1, 2, \ldots n$ will be called the Césaro matrix. Notice that the identity matrix, I, is also an iteration matrix and that (x_1, f, I) is just the ordinary iteration process.

A couple of simple examples in the special case where X is the real line are suggestive. Here the convex compact subset is a compact interval. The involution $f(x) = 5/x$ is a continuous bijection of $[1,5]$ having $\sqrt{5}$ as its unique fixed point. Ordinary iteration is useless here but generalized iteration using the Césàro matrix, i.e. the process $(x_1, 5/x, C)$ seems to converge to $\sqrt{5}$ for all choices of x_1, although rather slowly. We shall return to this example later.

The second example is the use of Newton's method of solve $f(x) = 0$ where $f(x) = \sqrt[3]{x}$. The function to be iterated is $N(x) = x - \frac{f(x)}{f'(x)} = -2x$. Since N is very expansive there is no compact interval (of more than 1 point) which N maps into itself, and ordinary iteration obviously fails. It is quite easy to show however that generalized iteration with the Césàro matrix not only converges but gives the exact solution in just three steps.

In view of these and other simple examples which the reader can easily construct for himself, the following theorem is not surprising.

Theorem 1 If f is a continuous self-mapping of a compact interval $[a, b]$ then for all x_1 belonging to $[a, b]$, the generalized iteration process (x_1, f, C) converges to a fixed point of f.

In case the fixed point is unique, the theorem is easy and was proved in [13]. Elimination of the uniqueness hypothesis was achieved in [8].

As soon as we shift attention from the line to the plane it becomes obvious that no theorem approaching the generality of Theorem 1 can possibly hold. On the other hand it is equally clear that generalized iteration does sometimes succeed in higher dimensional cases where ordinary iteration fails.

The next major step in this field (although it was not immediately recognized as such) was made by Krasnoselskii [12] in which he proved the following result.

<u>Theorem 2</u> Let f be a completely continuous non-expansive self-mapping of a closed convex subset E of a Banach space into itself. If the Banach space is uniformly convex, then for every starting point $x_1 \in E$, the sequence

$$x_{n+1} = \frac{1}{2}(x_n + fx_n) \tag{3}$$

converges to a fixed point of f.

Krasnoselskii stated that his result was related to those of [13] involving iteration matrices, but if he was aware of exactly what the relationship was he didn't reveal it. It was some years later when Dotson [2] made the relationship explicitly clear by pointing out that the Krasnoselskii iteration process is the same as (x_1, f, K) where K may be called the Krasnoselskii matrix

$$K = \left(\begin{array}{lllll} 1 & 0 & 0 & 0 & \\ 1/2 & 1/2 & 0 & 0 & \\ 1/4 & 1/4 & 1/2 & 0 & \\ 1/8 & 1/8 & 1/4 & 1/2 & \ldots \\ \ldots & \ldots & \ldots & \ldots & \\ \ldots & \ldots & \ldots & \ldots & \end{array}\right.$$

defined by $a_{n,n} = 1/2$ for $n > 1$ and $a_{n+1,k} = 1/2\, a_{n,k}$ if $k < n + 1$. Recall that by Theorem 1, $(x_1, 5/x, C)$ converges to $\sqrt{5}$. Krasnoselskii's theorem, interpreted in the light of Dotson's observation, shows that $(x_1, 5/x, K)$ also converges to $\sqrt{5}$ and is in fact just the familar divide-and-average square root method long associated with the name of Archytas [1].

In 1969 this theorem of Krasnoselskii was extended by Reinermann [15] who generalized (3) to

$$(4) \qquad v_{n+1} = (1 - c_n)\, v_n + c_n f(v_n),$$

where $c_1 = 1$, $0 < c_n \leq 1$, and $\sum_1^\infty c_n$ diverges. Reinermann proved that if one also assumes that $\lim_{n\to\infty} c_n = 0$, then for continuous self-mappings of a compact interval having only one fixed point, the sequence $\{v_n\}_1^\infty$ converges to that fixed point. Like Krasnoselskii, he then went on to get some results for non-expansive mappings in uniformly convex Banach spaces. We shall show a little later that Reinermann's procedure, (4), can also be represented in terms of a process (x_1, f, A) for suitable matrices A.

Uniform convexity of the space X is now known not to be as important in iteration as it once seemed. In 1966, Edelstein [6] showed that Krasnoselskii's theorem could still be derived when uniform convexity is replaced by the weaker hypothesis of strict convexity. Finally, in 1976, Ishekawa showed, in a theorem to be stated later, that not even strict convexity is necessary.

Unfortunately, the non-expansivity hypothesis is much harder to relax Since iteration of strict contractions on complete matric spaces is such a simple process, an extension which goes only from "strictly contractive" to "non-expansive" may well appear disappointing. Actually, a few results have been obtained for Banach space mappings which don't quite have to be non-expansive, and for linear topological spces without norms [4]. However, to delineate the role of averaging in iteration we do not have to follow these developments.

From the beginning it seemed clear that the Césàro means used in Theorem 1 could not possibly be best for iteration purposes. If an iteration is succeeding at all then the more recent iterates are better than the earlier ones and ought to be weighed more heavily. In other words, a more promising iteration matrix would have the mass shifted toward the diagonal rather than evenly distributed. The question of which iteration matrix is best for a particular problem is still unanswered, but a property which seems to be of great importance in all iteration matrices was discovered independently by Outlaw and Groetsch [17], Reinermann [15] and Dotson [4]. It is that

(5) $$a_{n+1,k} = (1 - a_{n+1,\, n+1})\, a_{n,k} \quad k = 1, 2, \ldots, n.$$

This condition, which we shall call stability, emerged in different ways in the papers of the above authors and had more or less the aspect of an ad hoc hypothesis. In retrospect it can now be seen as a very natural requirement whose importance ought to have been foreseen from the beginning. It should perhaps be included in the definition of "average" as that term is used in iteration. This can be seen as follows.

In order that the averaging work together with the iteration to improve convergence, it is important that the averaging process not displace the average far form the fixed point once the average is close to the fixed point. If v_n is near the fixed point, its image $x_{n+1} = f(v_n)$ must also lie near the fixed point, by the continuity of f. We wish to make sure that in this situation v_{n+1} will also remain near the fixed point. Since the averaging process is also continuous, the natural way to achieve this is to admit only those averages satisfying the condition that if $\bar{x}$ is the average of the n-term sequence $x_1, x_2, \ldots, x_n$ then $\bar{x}$ is also the average of the (n+1)-term sequence $x_1, x_2, \ldots, x_n, \bar{x}$. Means with this property are said to be *stable*. Once this is agreed to, the question becomes that of finding a necessary and sufficient condition on the iteration matrix A to guarantee that the average which it determines has this property. It is an easy exercise, which can well be left to the reader, to show that such a necessary and sufficient condition is simply (5).

Those familiar with the Nagumo-Kolmogoroff theory of means [14], [11] extended by de Finetti [7] and summarized by Hille [9] will recognize the property implied by

(5) as a weakened form of what de Finetti calls the associativity axiom. By a finite mean on a set S we mean a sequence of functions $\{M_n\}$ such that for each natural number n, M_n associates with each n-term sequence $s_1, s_2, \ldots, s_n$ in S a member $M_n(s_1, s_2, \ldots, s_n)$ of S. The only property essential to all means is that if $s_1 = s_2 = \ldots = s_n = s$, then $M_n(s_1, s_2, \ldots, s_n) = s$. In practice there are various other properties which are usually found in means. For example, finite means on vector spaces usually have the property that the mean of a sequence of vectors belongs to their convex hull. Such means are said to be *internal*, and we have already required that all the means which we shall use be internal. Associativity is a very strong consistency condition holding among the members of the sequence $\{M_n\}_1^\infty$. It It can be stated as follows: Let $i_1, i_2, \ldots, i_k$ be any k-element subsequence of the first n natural numbers, and let $(s_1, s_2, \ldots, s_n)$ be any n-element sequence in S. If $M_k(s_{i_1}, s_{i_2}, \ldots s_{i_k}) = \bar{s}$, then $M_n(s_1, s_2, \ldots, s_n) = M_n(t_1, t_2, \ldots, t_n)$ where $t_k = \bar{s}$ if $k = i_j$ for some j ; $t_k = s_k$ otherwise.

It is not hard to prove that the arithmetic mean, the geometric mean and the harmonic mean are both stable an associative, and it is easy to prove that associativity implies stability. However there are stable means which are not associative. An example of this is the antiharmonic mean

$$AH(x_1, x_2, \ldots x_n) = \frac{x_1^2 + x_2^2 + \ldots + x_n^2}{x_1 + x_2 + \ldots\ldots + x_n}$$

which is used in computing the reduced length of a compound pendulum. It is easy to show that means defined by a stable iteration matrix do have a limited kind of associativity, namely

$$M_n(x_1, x_2, \ldots, x_k, x_{k+1}, \ldots x_n) = M_n(\mu, \mu, \ldots, \mu, x_{k+1}, \ldots x_n)$$

where $\mu = M_k(x_1, x_2, \ldots x_k)$.

Consider now a generalized iteration process (x_1, f, A) where A is an iteration matrix satisfying the stability condition (5). By (1) and (2) we have

$$
\begin{aligned}
v_{n+1} &= a_{n+1,1}x_1 + \dots + a_{n+1,n}x_n + a_{n+1,n+1}x_{n+1} \\
&= (1-a_{n+1,n+1})\, a_{n,1}x_1 + \dots + (1-a_{n+1,n+1})\, a_{n,n}x_n + a_{n+1,n+1}f(v_n) \\
&= (1-a_{n+1,n+1})\, v_n + a_{n+1,n+1}f(v_n),
\end{aligned}
$$

which is exactly the sequence (4) which was introduced by Reinermann. What this implies is that if one uses a stable iteration matrix, then the resulting generalized iteration process is so simple that it does not require the use of any matrix at all -- the process reduces to Reinermann's sequence generating formula (4) in which $c_n = a_{nn}$. Reinermann's condition that $\sum_1^\infty c_n$ diverge has an easy iterpretation in terms of the iteration matrix R representing his method.

$$
R = \begin{pmatrix}
1 & & & \\
c_2' & c_2 & & \\
c_2'c_3' & c_2c_3' & c_3 & \\
c_2'c_3'c_4' & c_2c_3'c_4' & c_3c_4' & c_4
\end{pmatrix}
$$

$$
a_{n1} = c_1 \prod_{k=2}^{4} c_k' \;,\; a_{nk} = c_k \prod_{i=k+1}^{n} c_i' \;,\; a_{nn} = c_n
$$

where $c_i' = 1 - c_i$. Regularity in the sense of summability requires that the columns form null sequences, which here requires that $\prod_{i=k+1}^{n} (1-c_i)$ tend to zero as n approaches infinity. If we require that $0 \le c_i < 1$, then it is well known that a necessary and sufficient condition that $\prod_{i=k+1}^{n} (1-c_i)$ approach zero is that $\sum_1^\infty c_n$ diverge.

In 1976 Ishikawa [10] greatly extended the work of Krasnoselskii and Reinermann. He considered f to be a nonexpansive mapping from a closed subset E of a Banach space X to a compact subset D of X. His hypotheses on the diagonal sequence $\{c_n\}_1^\infty$ were that $0 \leq c_n \leq b < 1$ and that $\sum_1^\infty c_n = \infty$. He was then able to prove, without any convexity hypothesis on X, that if there exists a v_1 such that the Reinermann sequence (4) lies in D then f has a fixed point in D, and $\{v_n\}_1^\infty$ converges to a fixed point of f.

In conclusion it should be pointed out that although stable means now seem to be the right ones to use in iteration, non-stable means seem better suited for use in some other fields. Outlaw [16] and Dotson [2] for example have made use of the non-stable means defined by

$$\begin{pmatrix} 1 & & & \\ 1/2 & 1/2 & & \\ 1/3 & 0 & 2/3 & \\ 1/4 & 0 & 0 & 3/4 \\ & & & \end{pmatrix}$$

to describe several well known ergodic theorems, and Dotson's papers [3], [5] using means in an application of ergodic theory did not reveal any relevance of the idea of stability in the means naturally arising there.

Finally, it must be mentioned that although what we have called interation matrices always define internal means the concept of stability in no way depends on the means being internal. If we drop the requirement that the iteration matrices be non-negative (but keep, of course, the requriement that the rows sum to 1) the matrices will still define means on vector spaces, but the means will no longer have to be internal. Such means offer some hope of improving iterative methods in cases where the iterates approach the fixed point from one side, so to speak. If, for example, there is some hyperplane which separates the iterates from the fixed point, then averaging by internal means could be expected to retard rather than to accelarate convergence. A wise choice of external means seems to be what is needed here. Their use would be quite analogous to the well known method of over-relaxation in the solution of linear algebraic systems. We have stressed the internal means here for two reasons -- one being the obvious advantage they offer where there is a necessity for staying within a given convex set. But more important is the common occurence of problems calling for iteration where internal means are obviously indicated. In nonlinear diffusion theory, foı example, the iteration frequently takes place in a partially ordered Banach space with the function having the property of being order reversing. Internal means here are empirically observed to improve convergence even though theorems to this effect are hard to prove because the averaging process usually scrambles chains i.e. transforms simply ordered subsets into subsets which are no longer simply ordered.

REFERENCES

1. C. B. BOYER, A history of mathematics, John Wiley & Sons, Inc. 1968 page 78.

2. W. G. DOTSON, JR., Ph.D. thesis, University of North Carolina at Chapel Hill (1968).

3. W. G. DOTSON, JR., An application of ergodic theory to the solution of linear functional equations in Banach spaces, Bull. Amer. Math. Soc. 75 (1969) 347-352.

4. W. G. DOTSON, JR., On the Mann iterative process, Trans. Amer. Math. Soc. 149 (1970) 65-73.

5. W. G. DOTSON, JR., Mean ergodic theorems and iterative solution of linear functional equations, J. Math. Anal. and Appl. 34 (1971) 141-150.

6. M. EDELSTEIN, A remark on a theorem of M. A. Krasnoselskii, Amer. Math. Monthly 73 (1966).

7. B. de FINETTI, Sul concetto di media, Giorn. Inst. Italiano 2 (1931) 369-396.

8. R. L. FRANKS and R. P. MARZEC, A theorem on mean value iterations, Proc. Amer. Math. Soc. 30 (1971) 324-326.

9. E. HILLE, Methods in classical and functional analysis, Addison-Wesley Publishing Company 1972.

10. S. ISHEKAWA, Fixed points and iteration of a nonexpansive mapping in a Banach space, Proc. Amer. Math. Soc. 59 , no. 1, 1976.

11. A. N. KOLMOGOROFF, Sur la notion de la moyenne, Rend. Acad. Lincei 12 1930 388-391.

12. M. A. KRASNOSELSKII, Two remarks about the method of successive approximations, Uspekhi Matematicheskikh Nauk Vol. 10 1955 no. 1 (63) 123-127.

13. W. R. MANN, Mean value methods in iteration, Proc. Amer. Math. Soc. 4 (1953) 506-510.

14. M. NAGUMO, Ueber eine klasse der mittelwerte, Jap. J. Math. 1 1930 71-79.

15. J. REINERMANN, Ueber Toeplitzche Iterationsverfahren und einize ihre Anwendungen in der konstructive Fixpanktheorie, Studia Math. 32 (1969) 209-227.

16. C. OUTLAW, Ph.D. thesis (unpublished) University of North Carolina at Chapel Hill (1966).

17. C. OUTLAW and C. W. GROETSCH, Averaging iteration in a Banach space, Bull. Amer. Math. Soc. 75 (1969) 430-432.

On the Perturbation Theory for Generalized Inverse Operators in Banach Spaces

M. Z. Nashed

Department of Mathematics
and Institute for Mathematical Sciences
University of Delaware
Newark, Delaware 19711

Introduction

One of the most important chapters in applications of functional analysis is the perturbation theory for linear operators. The classic and beautiful book by Kato [1] is the best evidence that this author knows of the "power" of functional analysis. Various aspects of perturbation theory for linear operators (bounded or unbounded) may be found in [1], Gohberg and Krein [2] and Golberg [3]. Aspects of perturbation theory for matrices, linear equations, and eigenvalue problems are thoroughly developed in the classic books by Householder [4] and Wilkinson [5].

The perturbation theory for generalized inverses of matrices and linear operators has many subtle points which have no counterpart in the perturbation theory for inverses. The recent papers [6], [7] address many aspects of perturbation theory for generalized inverses and contain some new results. Stewart [7] surveys perturbation theory for the Moore-Penrose generalized inverse, for the orthogonal projection onto the column space of a matrix, and for the linear least squares problem. The emphasis in [7] is on some fundamental contributions by Wedin [18], [19], In [6; Sections 3 and 4] the author develops perturbations and continuity properties of generalized inverses of linear operators in Banach and Hilbert spaces, but the development also sheds more light on the case of matrix generalized inverses. Perturbation and computation of linear least squares problems are studied in the books by Lawson and Hanson [8] and Stewart [9]. The extensive bibliography [10] surveys the relevant literature and provides some annotations.

In the present note, we address two questions in perturbation theory for generalized inverses of linear operators which do not seem to have been studied before even in the case of matrices. The first question is related to effects on the generalized inverse of perturbations in the complementary spaces to the null space and the range (or the closure of the range) of a linear operator. The second question concerns the stability of inner inverses and outer inverses.

The generalized inverse $A^{\dagger}$ depends not only on A but also on the complements to the subspaces mentioned earlier (or equivalently on the projectors P and Q onto the null space and range of A, respectively, along chosen complements). The Moore-Penrose inverse involves orthogonal complements. We show that for linear (but not necessarily bounded) operators with bounded generalized inverses, $A^{\dagger}$ depends continuously on the projectors. The notion of the gap between two closed subspaces is relevant to the measure of "smallness" of perturbations of the complementary subspaces. In the case of linear operators in Hilbert spaces, and in particular for matrices, perturbations of the complementary orthogonal subspaces may be viewed as perturbations of the inner products. Thus the results include the effects of perturbations of the inner products (say due to computational error, or to the use of wrong inner products) on the Moore-Penrose inverse.

The second aspect of generalized inverse perturbation theory that we consider concerns the stability of inner inverses $(ABA = A)$ and of outer inverses $(BAB = B)$. We prove a stability property for the latter and indicate the instability for the former.

1. Generalized Inverses of Linear Operators in Banach Spaces.

We recall that a subspace M of a Banach space X is said to have a topological complement if there exists a subspace N such that $X = M \oplus N$ (topological direct sum). A subspace M has a topological complement if and only if there exists a continuous linear idemponent mapping (hereafter called a projector) P of X onto M. A projector P on X induces a decomposition of X into two topological complements PX and $(I-P)X$.

Let X and Y be Banach spaces and let A be a linear operator with domain $D(A)$ in X and range $R(A)$ in Y. We assume that $D(A)$ is dense in X, and that the null space of A, $N(A)$, has a topological complement M. Let P denote the projector of X onto $N(A)$ along M, so $X = R(P) \oplus N(P) = N(A) \oplus M$, $R(P) = N$, $R(P) = N(A)$, and it follows easily that

$$X = N(A) + \overline{M \cap D(A)} . \tag{1.1}$$

Operators with the property (1.1) form a subset of the set of so-called "domain-decomposable" operators (see [11]). We call $M \cap D(A) = N(P) \cap D(A)$ the carrier of A with respect to P, and denote it by C_P. If A is a closed linear operator with dense domain in a Hilbert space H, or if A is a bounded linear operator on H, then $N(A)$ is a closed subspace and hence has a topological complement (in particular an orthogonal complement), so (1.1) holds. However, this is not necessarily the case in Banach spaces since an infinite-dimensional closed subspace does not necessarily have a topological complement.

The following theorem is a special case of the general theory of generalized inverse operators announced in [10] and fully developed in [11], and is easy to prove directly.

Theorem 1.1. Let X and Y be Banach spaces and let $A : \mathcal{D}(A) \subset X \to Y$ be a linear operator with dense domain. Assume that $\mathcal{N}(A)$ has a topological complement M and let P be the projector of X onto $\mathcal{N}(A)$ along M. Assume that $\overline{\mathcal{R}(A)}$, the closure of the range of A, has a topological complement S; and let Q denote the projector of Y onto $\overline{\mathcal{R}(A)}$ along S. Then there exists a unique linear operator $A^{\dagger} = A^{\dagger}_{P,Q}$ which satisfies the following relations:

$$(1.2)\qquad \begin{cases} \mathcal{D}(A^{\dagger}) = \mathcal{R}(A) + \mathcal{N}(Q), & \\ \mathcal{R}(A^{\dagger}) = \mathcal{C}_P(A) & , \\ A^{\dagger}A = I-P & \text{on} \quad \mathcal{D}(A), \\ AA^{\dagger} = Q & \text{on} \quad \mathcal{D}(A^{\dagger}). \end{cases}$$

The operator $A^{\dagger}_{P,Q}$ is called the generalized inverse of A with respect to the projectors P and Q. From (1.2) it follows immediately that $AA^{\dagger}A = A$ and $A^{\dagger}AA^{\dagger} = A^{\dagger}$ on $\mathcal{D}(A)$ and $\mathcal{D}(A^{\dagger})$ respectively, and $\mathcal{N}(A^{\dagger}) = \mathcal{N}(Q)$. The generalized inverse $A^{\dagger}$ is also characterized as the unique solution B of the equations

$$(1.3)\qquad \begin{cases} BABx = x & \text{for} \quad x \in \mathcal{R}(A)+\mathcal{N}(Q) =: \mathcal{D}(B), \\ BA = I-P & \text{on} \quad \mathcal{D}(A), \\ AB = Q & \text{on} \quad \mathcal{D}(B)\,. \end{cases}$$

The operator $A^{\dagger}$ depends on A and the complements M, S (equivalently the projectors P, Q). A mapping $A^{\dagger}(A,P,Q)$ is thus induced as A varies over the class of all linear operators which satisfies the assumptions of Theorem 1.1. The purpose of this paper is to study the dependence of $A^{\dagger}$ on A, P and Q and to develop an associated perturbation theory.

We also consider stability properties of inner and outer inverses. Recall that an inner inverse of a linear operator $A : X \to Y$ is a linear operator $B : \mathcal{D}(B) \subset Y \to X$ such that $ABA = A$ on $\mathcal{D}(A)$. While every linear operator has an inner inverse, a bounded (or closed) linear operator need not necessarily have any bounded (or closed) inner inverse. Also if B is a bounded inner inverse of A, then for any linear operator C with $R(C) \subset N(A)$, the operator B + C is an inner inverse of A; hence unless A is invertible, A has infinitely many unbounded inner inverses. A bounded (or closed densely defined) linear operator has a bounded inner inverse if and only if $N(A)$ and $R(A)$ are closed complemented subspaces of X and Y, respectively.

Similarly an <u>outer inverse</u> for A is a linear operator $B : D(B) \subset Y \to X$ such that BAB = B on $D(B)$. Every operator has a nonzero outer inverse. We can deduce immediately that for any outer inverse B, the maps BA and AB are idempotent, $R(BA) = R(B)$, $R(AB) \subset R(A)$, $N(A) \subset N(BA)$, and $N(B) = N(AB)$, and B induces the <u>algebraic</u> direct sum decompositions:

$$(1.4) \qquad \begin{cases} X = R(B) \dotplus N(BA) \\ Y = N(B) \dotplus R(AB) \end{cases}.$$

Clearly B is an outer inverse for A if and only if A is an inner inverse for B. Thus we have the following theorem on the existence of bounded outer inverses and the corresponding dual theorem for bounded inner inverses. We let $L(X,Y)$ denote the set of all bounded linear operators on X into Y, and $C(X,Y)$ the space of all closed linear operators with dense domain in X.

<u>Theorem 1.2</u>. Let X and Y be Banach spaces and let $A \in L(X,Y)$ or $A \in C(X,Y)$. A necessary and sufficient condition for $B \in L(Y,X)$, $B \neq 0$, to be an outer inverse for A is that $R(B)$ and $N(B)$ be closed complemented subspaces of X and Y respectively. If $B \in L(Y,X)$ is an outer inverse for A, then BA is a (continuous) projector of Y onto $N(B)$ and I-AB is a (continuous) projector of Y on $N(B)$, and the algebraic decompositions (1.4) are also

topological decompositions:

(1.5)
$$\begin{cases} X = R(B) \oplus N(BA) \\ Y = N(B) \oplus R(AB) \end{cases}$$

Corollary 1.3. If $N(A)$ and $R(A)$ are closed complemented subspaces of X and Y, respectively, then A has a bounded inner inverse and a bounded outer inverse. Furthermore A in this case exists and is bounded with $D(A^{\dagger}) = Y$.

2. Continuous Dependence of $A^{\dagger}_{P,Q}$ on the Projectors P and Q.

We first consider the case when A is a linear operator with a bounded generalized inverse. Let

(2.1)
$$X = N(A) \oplus M, \quad Y = R(A) \oplus S$$

and also,

(2.2)
$$X = N(A) \oplus M_n, \quad Y = R(A) \oplus S_n, \quad n = 1,2,\ldots$$

Let P denote the projector of X onto $N(A)$ along M, and let P_n denote the projector of X onto $N(A)$ along M_n. Similarly, let Q be the projector of Y onto $R(A)$ along S, and Q_n be the projector of Y onto $R(A)$ along S_n. In this section we consider the behavior of $A^{\dagger}_{P_n,Q_n} - A^{\dagger}_{P,Q}$.

We first note that $A^{\dagger}_{P,Q}$ and $A^{\dagger}_{P_n,Q_n}$ are bounded and defined on all of Y. Furthermore

(2.3)
$$A^{\dagger}_{P_n,Q_n} = (I+P-P_n)A^{\dagger}_{P,Q}(I-Q+Q_n) .$$

This relation is deduced in [11]. A crucial step for this deduction is a lemma which characterizes the set of all linear idempotents whose range is a given subspace of a vector space. On the other hand, one may verify the validity of (2.3) directly. To this end, it follows immediately from (2.3) that $A^{\dagger}_{P_n,Q_n} = (I-P_n)A^{\dagger}_{P,Q}Q_n$. Denote the right-hand side of this expression by F. By simple manipulations, it is easy to verify that $FAF = F$, $FA = I-P_n$ and $AF = Q_n$; so by uniqueness of the generalized inverse (Theorem 1.1), (2.3) is valid. We shall abbreviate $A^{\dagger}_{P_n,Q_n}$ by $A^{\dagger}_n$, and $A^{\dagger}_{P,Q}$ by $A^{\dagger}$. It follows from (2.3) that for any $y \in Y$,

$$A^{\dagger}_n y - A^{\dagger} y = (P-P_n)A^{\dagger}y + (P-P_n)A^{\dagger}(Q_n-Q)y + A^{\dagger}(Q_n-Q)y \ . \tag{2.4}$$

Thus if $||P_n x - Px|| \to 0$ and $||Q_n y - Qy|| \to 0$ as $n \to \infty$, then $||A^{\dagger}_n y - A^{\dagger} y|| \to 0$ for each $y \in Y$. It follows from the Banach-Steinhaus Theorem that the sequence $\{A^{\dagger}_n\}$ is uniformly bounded in this case. From (2.2) we have the error estimate

$$||A^{\dagger}_n - A^{\dagger}|| \leq (\alpha+\beta+\alpha\beta)||A^{\dagger}||, \tag{2.5}$$

where

$$\alpha := ||P-P_n|| \quad \text{and} \quad \beta := ||Q-Q_n|| \ . \tag{2.6}$$

In particular, if P_n and Q_n converge uniformly to P and Q respectively, then $A^{\dagger}_n$ converges to A in the uniform operator topology. Thus we have established the continuous dependence of $A^{\dagger}_{P,Q}$ on the projections P,Q for bounded generalized inverses in Banach space. Formally, we have:

<u>Theorem 2.1</u>. Under the assumptions (2.1) and (2.2), if $P_n \to P$ and $Q_n \to Q$ (pointwise, uniformly) for all $x \in X$ and $y \in Y$, then $A^{\dagger}_{P_nQ_n} \to A^{\dagger}_{P,Q}$ (pointwise, uniformly), and the error bound (2.5) holds.

For the general case of a linear operator with unbounded generalized inverse (the setting of Theorem 1.1), the domains of $A^{\dagger}_{P_n,Q_n}$ and $A^{\dagger}_{P,Q}$ are dense in Y and will be different in general. In fact, $D(A^{\dagger}_n) = D(A^{\dagger})$ for all n if and only if $R(Q_n-Q) \subset R(A)$ for all n

(see [11], p. 27). Also since $R(A)$ is not closed, for any $y_o \notin \overline{R(A)}$, we can select Q and Q_n so that $y_o \in D(A^\dagger)$ and $y_o \notin D(A_n^\dagger)$; see [11], p. 28. Thus one cannot expect a result similar to Theorem 2.1 to hold. However, some weaker error estimates are possible. Suppose that instead of (2.1) and (2.2) we have the decompositions required by Theorem 1.1, namely,

$$X = N(A) \oplus M, \quad Y = \overline{R(A)} \oplus S \quad , \tag{2.7}$$

$$X = N(A) \oplus M_n, \quad Y = \overline{R(A)} \oplus S_n, \tag{2.8}$$

and let P, Q, P_n and Q_n be the respective associated projectors. In this case $D(A^\dagger) = R(A) \dotplus S$ and $D(A_n^\dagger) = R(A) \dotplus S_n$. For $y \in D(A^\dagger)$ we write $y = y_r + y_s$, where $y_r \in R(A)$ and $y_s \in S$. Similarly, for $y_n \in D(A_n^\dagger)$, we write $y_n = y_{nr} + y_{ns}$, where $y_{nr} \in R(A)$ and $y_{ns} \in S_n$. We then have

$$\begin{aligned} ||A_n^\dagger y_n - A^\dagger y|| &= ||A_n^\dagger y_{nr} - A^\dagger y_r|| \\ &\leq ||A_n^\dagger y_{nr} - A^\dagger y_{nr}|| + ||A^\dagger (Q_n - Q) y|| \ . \end{aligned} \tag{2.9}$$

For the first term on the right-hand side of the inequality, we can still use (2.3), which is valid for all $y \in R(A)$. Thus using (2.4) with y replaced by y_{nr}

$$\begin{aligned} ||A_n^\dagger y_n - A^\dagger y|| \leq{} & ||(P-P_n)A^\dagger y_{nr}|| + ||(P-P_n)A^\dagger (Q_n-Q) y_{nr}|| \\ & + ||A^\dagger (Q_n-Q) y_{nr}|| + ||A^\dagger (Q_n-Q) y|| \ . \end{aligned}$$

The second and third terms vanish. If $(P-P_n)A^\dagger Q_n y_n$ and $A^\dagger (Q_n-Q) y$ approach zero as $y_n \to y$, then $A_n^\dagger y_n \to A^\dagger y$.

One important class of unbounded generalized inverses can be easily resolved, however, under the assumptions (2.7) and (2.8). This is the case when A is a closed densely defined operator. Then $A^\dagger$ is also a closed operator with dense domain $D(A^\dagger) = R(A) \dotplus S$. Using the graph norm $||y||_G := ||A^\dagger y|| + ||y||$ for $y \in D(A^\dagger)$, A becomes a bounded operator and then the previous theory applies.

3. Estimates using the Notion of the Gap Between Two Subspaces.

Estimates for $\alpha = ||P-P_n||$ and $\beta = ||Q-Q_n||$ in (2.5) can be given using the notion of a gap between two subspaces.

Let M and N be two closed subspaces of a Banach space X. Define

$$\delta(M,N) = \sup\{\mathrm{dist}(x,N) : x \in M, \quad ||x|| = 1\} \tag{3.1}$$

if $M \neq \{0\}$, and define $\delta(\{0\},N) = 0$ for any N. In constrast, as seen from (3.1), $\delta(M,\{0\}) = 1$ for any $M \neq \{0\}$. In general, $\delta(M,N) \neq \delta(N,M)$. The gap between M and N (see Kato [1]) is defined by

$$g(M,N) = \max\{\delta(M,N), \quad \delta(N,M)\} . \tag{3.2}$$

It follows directly from the definition that $0 \leq g(M,N) \leq 1$, and $g(M,N) = 0$ if and only if $M = N$. $g(M,N) < 1$ implies $\dim M = \dim N$ ([1], p. 200).

Let N,M and M' be closed subspaces of X such that

$$X = N \oplus M = N \oplus M' ,$$

and let P and P' be the projectors onto N along M and M', respectively. If $||P'||g(M,M') < 1$, then it is not hard to show ([6], [16]) that

$$||P-P'|| \leq \frac{||I-P'|| \, ||P'|| g(M,M')}{1-||P'|| g(M,M')} .$$

If X is a Hilbert space, then $\operatorname{dist}(x,N)$ is attained for any closed subspace N of X, and

$$\inf\{||x-u|| : u \in N\} = ||x-P_N x||,$$

where P_N is the orthogonal projector of X onto N. Thus

$$\begin{aligned} \delta(M,N) &= \sup\{||x-P_N x|| : x \in M, ||x|| = 1\} \\ &= \sup\{||P_M v-P_N P_M v|| : v \in X, ||v|| = 1\}. \\ &= ||(I-P_N)P_M||. \end{aligned}$$

Therefore,

$$g(M,N) = \max\{||(I-P_N)P_M||, ||(I-P_M)P_N||\}. \tag{3.3}$$

Now let X be a Hilbert space with respect to each of two equivalent inner product $(\cdot,\cdot)$ and $[\cdot,\cdot]$. Let N be a closed subspace of X and let M and M' be the orthogonal complements of N with respect to these inner products. Let P and P' be the orthogonal projectors of X onto N along M and M' respectively. Then from (3.3),

$$g(M,M') = ||P-P'||. \tag{3.4}$$

4. Stability of Outer Inverses

It is well known that the dependence of the Moore-Penrose generalized inverse $A^\dagger$ on A is discontinuous in general. A necessary and sufficient condition that

$$\lim_{B\to A} B^\dagger = A^\dagger$$

is that rank (B) = rank (A) as B approaches A. Penrose [15] was the first to observe this requirement. Several different proofs together with explicit perturbation bounds are now given in the literature ([19], [17], [6], [7]). It follows from the formula

$$A^\dagger_{P,Q} = (I-P)A^\dagger Q$$

that this discontinuous dependence prevails also for other types of generalized inverses obtained relative to nonorthogonal projectors, e.g. the generalized inverse with specified range and null space, weighted generalized inverses, etc.

When we consider various perturbation theories for generalized inverse operators, it is necessary to make precise what is meant by a "small" perturbation. The search for a natural notion of "smallness" is the pivotal point of perturbation theory. Clearly the smallness of the norm $||B-A||$ is not adequate for perturbation theory of generalized inverses. Wedin [18], [19] was the first to give a comprehensive treatment of perturbation theory for pseudo-inverses of matrices, introducing the notion of A and B "in the acute case", which is called "acute perturbation" in [7]. In terms of the concept of the gap, this means $g(R(A),R(B)) < 1$ and $g(R(A^*),R(B^*)) < 1$.

If B is an acute perturbation of A, then $\operatorname{rank}(A) = \operatorname{rank}(B)$. If $\operatorname{rank}(A) = \operatorname{rank}(B)$ and $||B-A||_2 < 1/||A^{+}||_2$, then B is an acute perturbation of A.

A totally different approach to perturbation theory for generalized inverses was developed in [17] for matrices and for operators on Banach spaces. It includes as special cases some of the results based on acute perturbations (see, e.g., Corollary 1, p. 6).

All approaches to generalized inverse perturbation seem to be only concerned with the Moore-Penrose inverse for matrices or for linear operators on Hilbert spaces, with the exception of [6], [17] which treat the generalized inverse $A^{+}_{P,Q}$ in a Banach space setting. Perturbation and stability properties for inner inverses and for outer inverses do not seem to have been considered in the literature. The approach in [17] makes it possible to infer some of these properties. However, we shall take here a more direct approach to prove that the set of all $A \in L(X,Y)$ with bounded outer inverses is open in the uniform operator topology. This is not true for the set of all bounded linear operators with bounded inner inverses.

Theorem 4.1. Let $A \in L(X,Y)$ and let $B \in L(Y,X)$ be an outer inverse of A, i.e. $BAB = B$. Let $A \in L(Y,X)$ be such that $||B(\tilde{A}-A)|| < 1$. Then $\tilde{B} := [I+B(\tilde{A}-A)]^{-1}B$ is a bounded outer inverse to $\tilde{A}$ with $N(\tilde{B}) = N(B)$ and $R(\tilde{B}) = R(B)$. Moreover,

$$(4.1) \qquad ||\tilde{B}-B|| \leq \frac{||\tilde{A}-A||\cdot||B||^2}{1-||B(\tilde{A}-A)||}$$

Proof. Let $C := I+B(\tilde{A}-A)$. Since $||B(\tilde{A}-A)|| < 1$, it follows from Banach's lemma that C has a bounded inverse. Since $BAB = B$, we have $BAC = BAB\tilde{A} = B\tilde{A}$. Thus

$$BA = B\tilde{A}C \tag{4.2}$$

Let $\tilde{B} := C^{-1}B$. Then $\tilde{B}\tilde{A}\tilde{B} = C^{-1}B\tilde{A}C^{-1}B = C^{-1}BAB = C^{-1}B = \tilde{B}$, using (4.2) in the second term. Clearly $\tilde{B}$ is bounded and $N(\tilde{B}) = N(B)$.

Now we prove that $R(\tilde{B}) = R(B)$. Let $x \in R(B)$, i.e., $BAx = x$. Then $\tilde{B}\tilde{A}x = C^{-1}B\tilde{A}x = C^{-1}B\tilde{A}BAx$. But $CB = [I+B(\tilde{A}-A)]x = B\tilde{A}B$, so $B = C^{-1}B\tilde{A}B$. Thus $\tilde{B}\tilde{A}x = C^{-1}B\tilde{A}BAx = BAx = x$. This proves that $x \in R(\tilde{B})$. The reverse inclusion $R(\tilde{B}) \subset R(B)$ is established by summetry. Let $\tilde{C} = I+\tilde{B}(A-\tilde{A})$. Then $C\tilde{C} = C[I+C^{-1}B(A-\tilde{A})] = C+B(A-\tilde{A}) = I+B(\tilde{A}-A)+B(A-\tilde{A}) = I$. Now since C^{-1} exists, it follows that $\tilde{C} = C^{-1}$ and $\tilde{C}^{-1} = C$. This proves that if $\tilde{B}\tilde{A}\tilde{B} = \tilde{B}$, then $BAB = B$ and $R(\tilde{B}) \subset R(B)$. Finally the bound (4.1) follows from $\tilde{B} = C^{-1}B = \tilde{C}B = [I+\tilde{B}(A-\tilde{A})]B$ and the known estimate: $||(I+L)^{-1}|| \le \frac{1}{1-||L||}$ if $||L|| < 1$.

We remark that outer inverses of matrices and linear operators have useful applications in iterative methods (e.g., Newton's method) for nonlinear operator equations with singular Fréchet derivative and in nonlinear least-squares problems (see, e.g. [14]).

Theorem 4.1 suggests that the inherent instability (or the discontinuous dependence of) $A^{\dagger}$ (on A) resides in the relation $AA^{\dagger}A = A$. Inner inverses do not enjoy the stability property (stated in Theorem 4.1) that outer inverses have. In fact inner inverses are quite unstable. To dramatize this instability, we recall the technical fact if A is a closed linear operator with closed range in Banach space and if the dimension of the null space and the co-dimension of the range of A are both infinite, then there always exists a compact operator K of infinite rank, so that no matter how small $||K||$, the operator $A+K$ has a nonclosed range. Thus, whereas the operator A has a bounded inner inverse, the perturbed operator $A+K$ has no bounded inner inverse no matter how small K is in norm!

References

[1] T. Kato, Perturbation Theory for Linear Operators, Springer-Verlag, New York, 1966.

[2] I. C. Gohberg and M. G. Krein, Introduction to the Theory of Non-selfadjoint Operators, American Mathematical Society, Providence, R. I., 1969.

[3] S. Goldberg, Unbounded Linear Operators, McGraw-Hill Book Co., New York, 1966.

[4] A. S. Householder, The Theory of Matrices in Numerical Analysis, Dover, New York, 1964.

[5] J. H. Wilkinson, The Algebraic Eigenvalue Problem, Oxford University Press, London, 1965.

[6] M. Z. Nashed, Perturbations and approximations for generalized inverses and linear operator equatios, in [12], pp. 325-396.

[7] G. W. Stewart, On the perturbation of pseudo-inverses, projections and linear least squares problems, SIAM Review 19 (1977), pp. 634-662.

[8] C. L. Lawson and R. J. Hanson, Solving Linear Least Squares Problems, Prentice-Hall, Englewood Cliffs, N.J., 1974.

[9] G. W. Stewart, Introduction to Matrix Computations, Academic Press, New York, 1973.

[10] M. Z. Nashed and G. F. Votruba, A unified approach to generalized inverses of linear operators: I. Algebraic, topological and projectional properties, Bull. Amer. Math. Soc., 80(1974), pp. 825-830; II. Extremal and proximinal properties, ibid., 80 (1974), pp. 831-835.

[11] M. Z. Nashed and G. F. Votruba, A unified operator theory of generalized inverses, in [12], pp. 1-109.

[12] M. Z. Nashed, editor, Generalized Inverses and Applications, Academic Press, New York, 1976.

[13] M. Z. Nashed and L. B. Rall, Annotated bibliography on generalized inverses and applications, in [12], pp. 771-1041; especially Section M.

[14] A. Ben-Israel and T. N. E. Greville, Generalized Inverses: Theory and Applications, Wiley-Interscience, New York, 1974.

[15] R. Penrose, A generalized inverse for matrices, Proc. Cambridge Philos. Soc., 51 (1955), pp. 506-513.

[16] J. E. Osborn, Spectral approximation for compact operators, Math. Comp., 29 (1975), pp. 712-725.

[17] R. H. Moore and M. Z. Nashed, Approximations to generalized inverses of linear operators, SIAM J. Appl. Math., 27 (1974), pp. 1-16.

[18] P. Å. Wedin, On pseudo-inverses of perturbed matrices, Lund University Computer Science Tech. Rept., Lund, Sweden, 1969.

[19] P. Å. Wedin, Perturbation theory for pseudo-inverses, BIT, 13 (1973), pp. 217-232.

BOUNDARY VALUE PROBLEMS FOR SYSTEMS OF NONLINEAR PARTIAL DIFFERENTIAL EQUATIONS

J. W. Neuberger, Emory University

A Hilbert space lemma from [3] is used to give a method for solving boundary value problems for nonlinear systems. The lemma may be applied in various function space settings to yield L_2 solutions to systems of nonlinear partial differential equations (see [3]) but emphasis here is on a discrete setting which yields a readily programable numerical method for such problems. Basic to the discussion is the consideration of a family of Dirichlet spaces and their associated Laplacians.

1. A Hilbert space lemma. Suppose H is a real Hilbert space and $B(H)$ is the space of continuous linear transformations from H to H. Denote by $B_1(H)$ the subset of $B(H)$ consisting of all symmetric elements of $B(H)$ which have numerical range in $[0,1]$. If $T \in B_1(H)$, then $T^{1/2}$ denotes the unique member of $B_1(H)$ whose square is T.

Lemma H. Suppose $W \in H$, P is an orthogonal projection on H and L is a strongly continuous function from H to $B_1(H)$. Suppose also that $Q_0 = P$ and

$$Q_{n+1} = Q_n^{1/2} L(Q_n^{1/2} W) Q_n^{1/2}, \quad n = 0, 1, \ldots .$$

Then $\{Q_n^{1/2} W\}_{n=1}^{\infty}$ converges to an element $Z \in H$ so that $PZ = Z$ and $L(Z)Z = Z$.

For completeness, a proof is given in Section 6. The author is grateful to Alan Lambert for information concerning square roots of symmetric nonnegative transformations. Appreciation is also expressed to Tom Manteuffel for helpful information on computation and to Philip Walker for his comments.

2. Some potential theory on finite grids. Suppose m is a positive integer, Ω is a bounded open subset of R^m and G_0 is a finite rectangular grid with uniform spacing between adjacent points δ. Assume that G_0 intersects Ω in such a way that if $p \in G \equiv G_0 \cap \Omega$, then for $i = 1, \ldots, m$, either $p + \delta e_i$ or $p - \delta e_i$ is in G where $e_1, \ldots, e_m$ denotes the standard basis of R^m.

Denote by K the vector space of all real-valued functions u on G, denote by K^+ the nonnegative members of K and denote by K^{++} the positive members of K. Take $\|u\| = (\Sigma_{r\in G}(u(r))^2)^{1/2}$, $u \in K$. For $i = 1, \ldots, m$, define $D_i : K \to K$ so that if $u \in K$ and $p \in G$,

$$(D_i u)(p) = \begin{cases} (1/2\delta)(u(p+\delta e_i) - u(p-\delta e_i)) & \text{if } p+\delta e_i,\ p-\delta e_i \in G \\ (1/\delta)(u(p+\delta e_i) - u(p)) & \text{if } p+\delta e_i \in G,\ p-\delta e_i \notin G \\ (1/\delta)(u(p) - u(p-\delta e_i)) & \text{if } p+\delta e_i \in G,\ p-\delta e_i \in G \end{cases}$$

Take $D_0 u = u$, $u \in K$.

Denote by M_0 the collection of all $\alpha = (\alpha_0, \alpha_1, \ldots, \alpha_m)$, $\alpha_i \in K^{++}$, $i = 0, 1, \ldots, m$. For such α, denote by K_α the space whose points are those of K but for $u \in K_\alpha$

$$\|u\|_\alpha = (\Sigma_{i=0}^m \Sigma_{r\in G}((D_i u)(r))^2 \alpha_i(r))^{1/2}.$$

It may be verified that K_α is a Dirichlet space in the sense of Beurling and Deny [2]. Denote inner products in K and K_α by $\langle\,,\rangle$ and $\langle\,,\rangle_\alpha$ respectively.

Denote by Δ_α the Laplacian for K_α, that is

$$\langle u,v\rangle_\alpha = \langle \Delta_\alpha u, v\rangle, \quad u,\ v \in K.$$

Theorem 2 gives an explicit expression for Δ_α. It follows from [2] that Δ_α^{-1} is positive in the sense that if $u \in K^+$, then $\Delta_\alpha^{-1} u \in K^+$.

Suppose that G' is a proper subset of G. If $\lambda \geq 1$, define $\alpha(\lambda) = (\alpha_{0,\lambda}, \alpha_{1,\lambda}, \ldots, \alpha_{m,\lambda}) \in M_0$ so that

$$\alpha_{0,\lambda}(p) = \lambda, \quad p \in G'$$

$$\alpha_{0,\lambda}(p) = 1, \quad p \in G\backslash G'$$

$$\alpha_{i,\lambda}(p) = 1, \quad p \in G, \quad i = 1, \ldots, m.$$

Define $P' : K \to K$ so that $(P'u)(p) = u(p)$, $p \in G'$, $(P'u)(p) = 0$, $p \in G\backslash G'$.

<u>Lemma to Theorem 1</u>. If $\lambda \geq 1$, then $\Delta_{\alpha(\lambda)} = (\lambda - 1)P' + \Delta_{\alpha(1)}$.

<u>Theorem 1</u>. The following hold:

(i) $J \equiv \lim_{\lambda\to\infty} \Delta_{\alpha(\lambda)}^{-1}$ exists

(ii) $JP' = P'J = 0$

(iii) $J|_{R(I-P')} = C^{-1}$ where $C \equiv (I-P')\Delta_{\alpha(1)}|_{R(I-P')}$.

3. Embedding of Dirichlet spaces. For $\alpha = (\alpha_0, \alpha_1, \ldots, \alpha_m) \in M_0$, denote by H_α the vector space whose points are those of K^{m+1} but if $z = (z_0, z_1, \ldots, z_m) \in H_\alpha$, then $|||z|||_\alpha = (\Sigma_{i=0}^m \Sigma_{r\in G}(z_i(r))^2\alpha(r))^{1/2}$. Define $D : K \to K^{m+1}$ by

$$Du = (D_0 u, D_1 u, \ldots, D_m u), \quad u \in K.$$

Denote by H'_α the subspace of H_α whose points are $R(D)$. Observe that $\|u\|_\alpha = |||Du|||_\alpha$, $u \in K_\alpha$ so that K_α and H'_α are equivalent spaces. Denote inner product in H_α by $(\, , \,)_\alpha$.

We adopt the following notation: If F is a linear transformation from K to K^{m+1} and $\alpha \in M_0$, then $F^{*\alpha}$ denotes the transformation from H_α to K so that

$$(Fz,x)_\alpha = \langle z, F^{*\alpha}x\rangle \quad \text{for} \quad z \in K, \ x \in H_\alpha .$$

Similarly if $\alpha \in M_0$ and E is a linear transformation from H_α to K, then $E^{*\alpha}$ denotes the linear transformation from K to H_α so that

$$\langle Ex, z\rangle = (x, E^{*\alpha}z)_\alpha \quad \text{for} \quad z \in K, \ x \in H_\alpha .$$

Theorem 2. If $\alpha \in M_0$, then $\Delta_\alpha = D^{*\alpha}D$.

For $\alpha \in M_0$, P_α denotes the orthogonal projection of H_α onto H'_α.

Theorem 3. If $\alpha \in M_0$, then $P_\alpha = D\Delta_\alpha^{-1}D^{*\alpha}$.

If F is a linear transformation from $K^{m+1} \times K^{m+1}$ to $K \times K$ and $\gamma = (\alpha, \beta) \in M_0 \times M_0$, then $F^{*\gamma}$ denotes the transformation from $K \times K$ to $H_\alpha \times H_\beta$ so that $\langle F\binom{x}{y}, \binom{u}{v}\rangle = \langle \binom{x}{y}, F^{*\gamma}\binom{u}{v}\rangle$ for $\binom{x}{y} \in H_\alpha \times H_\beta$, $\binom{u}{v} \in K \times K$ where the first inner product is in $K \times K$ and the second is in $H_\alpha \times H_\beta$.

4. Boundary value problems. Here we consider boundary value problems for systems of two nonlinear difference equations. The considerations may be extended without difficulty to higher order systems and of course specialized to the case of a single equation. The discrete system considered may arise from finite difference approximation to boundary value problems for nonlinear systems of partial differential equations.

Denote $K^{m+1} \times K^{m+1}$ by H and $K \times K$ by M. If $\binom{w}{z} \in M$, write

$$D\binom{w}{z} \quad \text{for} \quad \binom{Dw}{Dz}.$$

Suppose A is a continuous function from M to $L(H,M)$ and $\binom{f}{g} \in M$. We seek $Z = \binom{w}{z} \in M$ so that the following boundary value problem is satisfied:

$$(*)\quad \begin{cases} A(Z)DZ = 0 \\ w(p) = f(p), \quad p \in G' \\ z(p) = g(p), \quad p \in G'' \end{cases} .$$

We describe an iteration procedure for $(*)$ and give a sufficient condition for this process to converge to a solution.

Pick an increasing divergent number sequence $\{\lambda_i\}_{i=1}^{\infty}$ with $\lambda_1 > 1$. In terms of this sequence define $\alpha[i] = \alpha(\lambda_i) \in M_0$, $i = 1, 2, \ldots$ for G' as in section 2 and similarly define $\beta[i] = \beta(\lambda_i)$, $i = 1, 2, \ldots$ for G''.

Suppose i is a positive integer. Denote $(\alpha[i],\beta[i])$ by $\gamma[i]$, denote $H_{\alpha[i]} \times H_{\beta[i]}$ by H_i and denote by P_i the orthogonal projection on H_i so that

$$P_i\binom{x}{y} = \begin{pmatrix} D\Delta^{-1}_{\alpha[i]}D^{*\alpha[i]}x \\ D\Delta^{-1}_{\beta[i]}D^{*\beta[i]}y \end{pmatrix}, \quad \binom{x}{y} \in H_i .$$

Define $\Delta_{\gamma[i]}: M \to M$ so that

$$\Delta_{\gamma[i]}\binom{x}{y} = \begin{pmatrix} \Delta_{\alpha[i]}x \\ \Delta_{\beta[i]}y \end{pmatrix}, \quad \binom{x}{y} \in M .$$

Note that

$$\Delta^{-1}_{\gamma[i]}\binom{x}{y} = \begin{pmatrix} \Delta^{-1}_{\alpha[i]}x \\ \Delta^{-1}_{\beta[i]}y \end{pmatrix} \quad \text{and}$$

$$D^{*\gamma[i]}\binom{x}{y} = \begin{pmatrix} D^{*\alpha[i]}x \\ D^{*\beta[i]}y \end{pmatrix}, \quad \binom{x}{y} \in M .$$

Choose $W = \binom{r}{s} \in M$ satisfying the boundary conditions of $(*)$, i.e.,

$$r(p) = f(p), \quad p \in G'$$
$$s(p) = g(p), \quad p \in G'' .$$

If $(x,y) = ((x_0, x_1, \ldots, x_m), (y_0, y_1, \ldots, y_m)) \in H$, then define

$$\pi\binom{x}{y} = \binom{x_0}{y_0} .$$

Denote by H_0 the kernel of π and for $Y \in M$, denote by $A_0(Y)$ the restriction of $A(Y)$ to this kernel. Condition B denotes the proposition that for all $Y \in M$,

$$(A_0(Y)A_0(Y)^*)^{-1} \text{ exists.}$$

Theorem 4. Suppose i is a positive integer and condition B holds. For each $z \in H$, define

$$L_i(Z) = I - A(\pi Z)^{*\gamma[i]}[A(\pi Z)A(\pi Z)^{*\gamma[i]}]^{-1}A(\pi Z).$$

Suppose also that $W \in M$ and define $Q_{0,i} = P_i$,

$$Q_{n+1,i} = Q_{n,i}^{1/2}L_i(Q_{n,i}^{1/2}DW)Q_{n,i}^{1/2}, \quad n = 0, 1, \ldots .$$

Then $\{Q_{n,i}\}_{n=0}^{\infty}$ and $\{Q_{n,i}^{1/2}\}_{n=0}^{\infty}$ converge.

If T is a linear transformation on a finite dimensional vector space, all eigenvalues of T are ≥ 0 and T has a square root, then $T^{1/2}$ denotes the unique square root of T which has all eigenvalues ≥ 0. Note that this is consistent with the previous definition of $T^{1/2}$ for the case T is nonnegative and symmetric.

Theorem 5. Suppose condition B holds and $Q_{n,i}$, $n = 0, 1, \ldots, i = 1, 2, \ldots$ are as in Theorem 4. Define $R_{n,i} = \pi Q_{n,i}D$, $n = 0, 1, \ldots, i = 1, 2, \ldots$. Then $\{R_{n,i}\}_{n=0}^{\infty}$ and $\{R_{n,i}^{1/2}\}_{n=0}^{\infty}$ converge, $i = 1, 2, \ldots$.

The following is the main result of this paper:

Theorem 6. Suppose condition B is satisfied and $R_{n,i}$, $n = 0, 1, \ldots, i = 1, 2, \ldots$ are as in Theorem 5. Suppose also that $\{R_{n,i}^{1/2}\}_{n,i=1}^{\infty}$ is bounded and that the convergence of $\{R_{n,i}^{1/2}\}_{n=0}^{\infty}$ is uniform for $i = 1, 2, \ldots$. Then there is a solution to (*). Moreover a solution may be computed in either of the following ways:

(1) Define $Z_i = \lim_{n\to\infty} R_{n,i}^{1/2}W$, $i = 1, 2, \ldots$. Then $\{Z_i\}_{i=1}^{\infty}$ converges to Z satisfying (*).

(2) Define $Y_n = \lim_{i\to\infty} R_{n,i}^{1/2}W$, $n = 0, 1, \ldots$. Then Z of (1) is also the limit of $\{Y_n\}_{n=0}^{\infty}$.

The process in this section may loosely be described as being for Dirichlet-type conditions (i.e., the specifying of a solution at given points). It is achieved by using sequences $\{\alpha[i]\}_{i=1}^{\infty}$ in M_0 with $\alpha[i]_j(p) = 1$, $j = 1, \ldots, m$, $i = 1, 2, \ldots, p \in G$ and $\alpha[i]_0$ weighting the boundary points more and more as $i \to \infty$. To introduce Neumann-type conditions, components $\alpha[i]_j$, $j = 1, \ldots, m$

should be similarly weighted with $\alpha[i]_0$ constant at 1. Except for a few computer runs, this is uncharted territory but it is surely tractable and appears to hold considerable promise.

The conditions of Theorem 6 may be tested (but of course not verified conclusively) by computation. If for a given initial estimate the conditions of the theorem seem to fail, points may be deleted from G' or G'' and the resulting conditions tested again. One may continue this deletion of points and retesting until suitable boundary conditions are found. Conversely, if the conditions of Theorem 6 hold for some set of boundary conditions, additional points may be added to G' or G'' until the conditions of the theorem fail. In this way one hopes to determine maximal sets G' and G'' (relative to a given initial choice W). Maximal G', G'' for a given W seem to correspond to uniqueness of solution.

Some additional background for this work is briefly described in [3]. It is hoped that this work will find application in the area of numerical solution of conservation equations.

5. Systems of partial differential equations. We give here a brief discussion of an underlying function space developement for the numerical process in sections 2, 3, 4. It is a combination of ideas from these sections and ideas from [3].

Suppose each of m and k is a positive integer and Ω is an open (possibly unbounded) subset of R^m. Denote by K the space of square integrable (relative to Lebesgue measure) R^k-valued functions on Ω. Denote K^{m+1} by H. Denote by B a strongly continuous function from K to $L(H,K)$ so that

$$B(x)B(x)^* = I, \quad x \in K.$$

Define $L : H \to L(H,H)$ by

$$L(x) = I - B(\pi x)^* B(\pi x), \quad x \in H$$

where π projects elements of K^{m+1} onto their first components.

Denote by H_0' those members of H which are of the form

$$Du \equiv (u, D_1 u, \ldots, D_m u), \quad u \in C^{(1)}(\Omega) \cap K,$$

where D_i denotes differentiation in the ith place of u, $i = 1, \ldots, m$. Denote by H' the closure of H_0' in H. This is just a way of introducing L_2 generalized derivatives (cf [1]).

We may write many nonlinear systems, for example those commonly written

$$u_t + f(u)_x = 0$$

where f has continuous first partial derivatives, in the form

$$B(u)Du = 0 .$$

An example illustrates this. For the equation

$$u_t + uu_x = 0$$

on a region Ω of R^2, choose $m = 2$, $k = 1$ and

$$B(u)\begin{pmatrix} p \\ q \\ r \end{pmatrix} = (1+u^2)^{-1/2}(q+ur), \quad u, p, q, r \in K .$$

Clearly then the equation $u_t + uu_x = 0$ is equivalent to $B(u)Du = 0$.

The iteration of Lemma H can then start with any member $W \in H$ and leads to a solution u to $B(u)Du = 0$. Actually a variety of functional nonlinear partial differential systems may be included in this theory but very few deductions have been made in this direction (see however [4] for some results for linear inhomogeneous systems).

The above procedure makes no reference to boundary conditions; it picks out, in a sense, a solution 'nearest' to the initial estimate - at least if $W = Du$ for some $u \in K$. It seems clear however that a study of boundary conditions may be made following the numerical case of sections 2, 3, 4. Replace members $\alpha \in M_0$ by families of measures on Ω. For a given initial estimate, apply Lemma H infinitely often using a sequence of measures which converge to singular measures concentrated on appropriate portions of the 'boundary' on which conditions are to be specified.

Consider a system of two partial differential equations on Ω with boundary conditions specified for the first unknown on an arc Γ_1 and for the second unknown on an arc Γ_2. For G_δ a finite rectangular grid (with mesh δ) approximating Ω, denote by G_δ', G_δ'' subsets of G_δ approximating Γ_1 and Γ_2 respectively. Choose a $C^{(1)}$ function W on Ω satisfying the boundary conditions. Denote by W_δ the E_2-valued function on G_δ which agrees with W on G. Use W_δ to start the iteration of Theorems 4, 5, 6. Fairly extensive computer investigations (mainly with a single nonlinear partial differential equation) suggest that if there is a solution to the boundary value problem then iterates of the numerical process give good approximation to the solution as $\delta \to 0$ even though for fixed δ, the limit of the process on G_δ may not be a good approximation to the solution. A detailed error analysis is being attempted.

6. Proofs. Proof of Lemma H. Note that $Q_0 \in B_1(H)$. Since $R(L) \subset B_1(H)$ one has by induction that each of $Q_0, Q_1, \ldots \in B_1(H)$. For each nonnegative integer n and $x \in H$,

$$\langle Q_{n+1}x,x\rangle = \langle Q_n^{1/2}L(Q_n^{1/2}W)Q_n^{1/2}x,x\rangle = \langle L(Q_n^{1/2}W)Q_n^{1/2}x,Q_n^{1/2}x\rangle \le \langle Q_n^{1/2}x,Q_n^{1/2}x\rangle$$

$= \langle Q_n x,x\rangle$ so that $Q_n \ge Q_{n+1}$. Hence $\{Q_n\}_{n=0}^{\infty}$ converges strongly on H to $Q \in B_1(H)$ and $\{Q_n^{1/2}\}_{n=0}^{\infty}$ converges strongly to $Q^{1/2}$. In particular, $\{Q_n^{1/2}W\}_{n=0}^{\infty}$ converges to an element Z of H and so $\{L(Q_n^{1/2}W)\}_{n=0}^{\infty}$ converges strongly to $L(Z)$. Hence $\{L(Q_n^{1/2}W)Q_n^{1/2}W\}_{n=0}^{\infty}$ converges to $L(Z)Z$. Since $PQ_0 = P$ it follows by induction that $PQ_n = Q_n$ and hence that $PQ_n^{1/2} = Q_n^{1/2}$, $n = 1, 2, \ldots$ and so $PQ^{1/2} = Q^{1/2}$. In particular $PZ = PQ^{1/2}W = Q^{1/2}W = Z$. Since for each non-negative integer n and $x \in H$, $\langle Q_{n+1}x,x\rangle = \langle L(Q_n^{1/2}W)Q_n^{1/2}x,Q_n^{1/2}x\rangle$, it follows that $\langle Qx,x\rangle = \langle L(Z)Q^{1/2}x,Q^{1/2}x\rangle$ and hence $\langle (I-L(Z))Q^{1/2}x,Q^{1/2}x\rangle = 0$. Therefore $(I-L(Z))Q^{1/2}x = 0$ since $I-L(Z) \in B_1(H)$. In particular $(I-L(Z))Z = 0$, i.e., $L(Z)Z = Z$.

Proof of Theorem 2. If $x, y \in K$, $\alpha \in M_0$ then $\langle \Delta_\alpha x,y\rangle = \langle x,y\rangle_\alpha = (Dx,Dy)_\alpha = \langle D^{*\alpha}Dx,y\rangle$. Hence $\Delta_\alpha = D^{*\alpha}D$.

Proof of Theorem 3. Suppose $\alpha = (\alpha_0, \alpha_1, \ldots, \alpha_m) \in M_0$ and $x \in K$. Then $\langle \Delta_\alpha x,x\rangle = (Dx,Dx)_\alpha = \Sigma_{i=0}^m \Sigma_{r\in G}((D_i x)(r))^2\alpha_i(r) \ge \Sigma_{r\in G}(x(r))^2 > 0$ unless $x = 0$. Therefore $\Delta_\alpha \ge I$ and so Δ_α^{-1} exists. Clearly Δ_α is symmetric. Define $S = D\Delta_\alpha^{-1}D^{*\alpha}$. Then $S^2 = S$, $S^{*\alpha} = S$ and $R(S) = R(D)$. This gives that $S = P_\alpha$ and so the proof is complete.

Proof of lemma to Theorem 1. Suppose $\lambda \ge 1$ and $x, y \in K$. Then $\langle Dx,Dy\rangle_{\alpha(\lambda)} = \Sigma_{i=0}^m \Sigma_{r\in G}((D_i x)(r))((D_i y)(r))\alpha(\lambda)_i(r) = (\lambda-1)\Sigma_{r\in G'}x(r)y(r) + \langle Dx,Dy\rangle_{\alpha(1)} =$

$(\lambda - 1)\Sigma_{r\in G}((P'x)(r))y(r)) + \langle Dx, Dy\rangle_{\alpha(1)}$ and so $\langle \Delta_{\alpha(\lambda)}x,y\rangle = \langle D^{*\alpha(\lambda)}Dx,y\rangle$ $= \langle(\lambda-1)P'x,y\rangle + \langle D^{*\alpha(1)}Dx,y\rangle = \langle((\lambda-1)P' + \Delta_{\alpha(1)})x,y\rangle$. Therefore $\Delta_{\alpha(\lambda)} = (\lambda-1)P' + \Delta_{\alpha(1)}$.

Proof of Theorem 1. Note that if $0 < a < c$ and $x \in K$, then $\langle cP' + \Delta_{\alpha(1)})x,x\rangle \geq \langle(aP' + \Delta_{\alpha(1)})x,x\rangle \geq \langle \Delta_{\alpha(1)}x,x\rangle \geq \langle x,x\rangle$, so that $\Delta_{\alpha(c+1)} = cP' + \Delta_{\alpha(1)} \geq aP' + \Delta_{\alpha(1)} = \Delta_{\alpha(a+1)} \geq \Delta_{\alpha(1)} \geq I$ so that $0 \leq (cP' + \Delta_{\alpha(1)})^{-1} \leq (aP' + \Delta_{\alpha(1)})^{-1} \leq I$. Hence $\lim_{\lambda\to\infty}(\lambda P' + \Delta_{\alpha(1)})^{-1}$ exists, i.e., $J \equiv \lim_{\lambda\to\infty}\Delta_\lambda^{-1}$ exists. Note that J is symmetric and $J \geq 0$.

For $x \in K$, $\lambda > 0$, define $y_\lambda = (\lambda P' + \Delta_{\alpha(1)})^{-1}x$. Then $\|y_\lambda\| \leq \|x\|$, $x = \lambda P'y_\lambda + \Delta_{\alpha(1)}y_\lambda$, $\lambda\|P'y_\lambda\| \leq \|x\|(1 + |\Delta_{\alpha(1)}|)$ and hence $\|P'y_\lambda\| \to 0$ as $\lambda \to \infty$, i.e., $P'(\lambda P' + \Delta_{\alpha(1)})^{-1}x \to 0$ as $\lambda \to \infty$. Therefore $P'J = 0$ and consequently $JP' = 0$ since $(P'J)^* = JP'$.

With C defined as in (iii) it is clear that C^{-1} exists since $\Delta_{\alpha(\lambda)}^{-1}$ exists. Suppose $x \in R(I - P')$. Define $y_\lambda = (\lambda P' + \Delta_{\alpha(1)})^{-1}x$. Then $\lambda P'y_\lambda + \Delta_{\alpha(1)}y_\lambda = x$, $\lambda(I-P')P'y_\lambda + (I-P')\Delta_{\alpha(1)}y_\lambda = x$, $(I-P')\Delta_{\alpha(1)}(I-P')y_\lambda + (I-P')\Delta_{\alpha(1)}P'y_\lambda = x$, $y_\lambda = C^{-1}x + C^{-1}(I-P')\Delta_{\alpha(1)}P'y_\lambda$. Hence $Jx = \lim_{\lambda\to\infty}y_\lambda = C^{-1}x$ since $\lim_{\lambda\to\infty}P'y_\lambda = P'Jx = 0$.

Proof of Theorem 4. For i a positive integer, $Z \in H_i$, $L_i(Z)$ as defined is an orthogonal projection on H_i. Furthermore, L_i is a continuous transformation from H_i to $L(H_i,H_i)$. By Lemma H $Q_{n,i}$ and $Q_{n,i}^{1/2}$ converge strongly as $n \to \infty$. Since H_i is finite dimensional, $Q_{n,i}$ and $Q_{n,i}^{1/2}$ converge in the space $L(H_i,H_i)$ as $n \to \infty$.

Proof of Theorem 5. This follows from Theorem 4 since $R_{n,i} = \pi Q_{n,i}D$, $R_{n,i}^{1/2} = \pi Q_{n,i}^{1/2}D$, $n = 0, 1, \ldots,$ $i = 1, 2, \ldots$.

Proof of Theorem 6. Suppose i is a positive integer. Define $B_i(Y) = (A(Y)A(Y)^{*\gamma[i]})^{-1/2}A(Y)$, $y \in M = K \times K$. Then since $Q_{n,i}^{1/2} = P_\gamma Q_{n,i}^{1/2}$ $= D\Delta_{\gamma[i]}^{-1}D^{*\gamma[i]}Q_{n,i}^{1/2}$,

$$\begin{aligned} R_{n+1,i} &= \pi[Q_{n,i} - Q_{n,i}^{1/2}B_i(\pi Q_{n,i}^{1/2}DW)^{*\gamma[i]}B_i(\pi Q_{n,i}^{1/2}DW)Q_{n,i}^{1/2}]D \\ &= R_{n,i} - \pi D\Delta_{\gamma[i]}^{-1} D^{*\gamma[i]}Q_{n,i}^{1/2}B_i(R_{n,i}^{1/2}W)^{*\gamma[i]}B_i(R_{n,i}^{1/2}W)Q_{n,i}^{1/2}D \\ &= R_{n,i} - \Delta_{\gamma[i]}^{-1}(B(R_{n,i}^{1/2}W)DR_{n,i}^{1/2})^*(B_i(R_{n,i}^{1/2}W)DR_{n,i}^{1/2}) . \end{aligned}$$

Note that $R_{0,i} = I = R_{0,i}^{1/2}$ so that $R_{0,i}$, $R_{0,i}^{1/2}$ converge as $i \to \infty$.

Assume now that n is a nonnegative integer and $R_{k,i}$, $R_{k,i}^{1/2}$ converge as $i \to \infty$, $k = 0, 1, \ldots, n$. Denote by $r_0, \ldots, r_n$ the limits of $R_{0,i}^{1/2}, \ldots, R_{n,i}^{1/2}$ respectively as $i \to \infty$. Note that

$$B_i(R_{k,i}^{1/2}W) = (A(R_{k,i}^{1/2}W)A(R_{k,i}^{1/2}W)^{*\gamma[i]})^{-1/2}A(R_{k,i}^{1/2}W) \to (A_1(r_k)A_1(r_k)^*)^{-1/2}A(r_k)$$

as $i \to \infty$, where $A_1(Y) \equiv A(Y)|_{H''}$, $Y \in M$, where H'' is the collection of all points $(((I-P')x_0, x_1, \ldots, x_m), ((I-P'')y_0, y_1, \ldots, y_m))$ for $((x_0, x_1, \ldots, x_m), (y_0, y_1, \ldots, y_m)) \in H_i$. Note that $(A_0(Y)A_0(Y)^*)^{-1}$ and hence $(A_1(Y)A_1(Y)^*)^{-1}$ exists since $A_1(Y)A_1(Y)^* \geq A_0(Y)A_0(Y)^* > 0$, $y \in M$.

Hence $R_{n+1,i}$ converges as $i \to \infty$ since $R_{k,i}$, $B_i(R_{k,i}^{1/2}W)$, $R_{k,i}^{1/2}W$, $k = P, 1, \ldots, n$, all converge as $i \to \infty$. It will now be shown that $R_{n+1,i}^{1/2}$ converges as $i \to \infty$. It may be established that $\Delta_{\gamma[i]}^{1/2}R_{n,i}\Delta_{\gamma[i]}^{-1/2}$ is a symmetric and nonnegative member of $L(M,M)$, $i = 1, 2, \ldots$. But a member of $L(M,M)$ which is similar to a symmetric nonnegative element of $L(M,M)$ itself has all eigenvalues ≥ 0 and also has a unique square root with this property. Since $\{R_{n+1,i}^{1/2}\}_{i=1}^{\infty}$ is bounded (by hypothesis) it has a convergent subsequence. If it were to have two subsequences converging to different limits, each of these limits would be a square root of $R_n \equiv \lim_{i\to\infty} R_{n,i}$ which has all eigenvalues ≥ 0, an impossibility. Hence $R_{n+1,i}^{1/2}$ converges as $i \to \infty$.

Denote $\lim_{n\to\infty} R_{n,i}^{1/2}W$ by Z_i, $i = 1, 2, \ldots$ and denote $\lim_{i\to\infty} R_{n,i}^{1/2}W$ by Y_n, $n = 0, 1, \ldots$. Since by hypothesis $\{R_{n,i}^{1/2}W\}_{n=0}^{\infty}$, $i = 1, 2, \ldots$ converge uniformly, it follows that $\{Y_n\}_{n=0}^{\infty}$ converges. Denote its limit by Z. But the above uniform convergence, together with the convergence of $R_{n,i}^{1/2}W$ to Y_n as $i \to \infty$, $n = 0, 1, \ldots,$ implies that $\{Z_i\}_{i=1}^{\infty}$ converges also to Z.

It remains to be shown that Z satisfies (*). First we show that Z satisfies the difference equation $A(Z)DZ = 0$.

For each positive integer i, $DZ_i = \lim_{n\to\infty} Q_{n,i}^{1/2}DW$ since for each positive integer n, $R_{n,i}^{1/2}W = \pi Q_{n,i}^{1/2}DW$ and so $DR_{n,i}^{1/2}W = Q_{n,i}^{1/2}DW$ and hence $\lim_{n\to\infty} Q_{n,i}^{1/2}DW = DZ_i$. By lemma H, $L_i(DZ_i) = DZ_i$ and so

$$DZ_i - A(Z_i)^{*\gamma[i]}(A(Z_i)A(Z_i)^{*\gamma[i]})^{-1}A(Z_i)DZ_i = DZ_i \,,$$

i.e., $A(Z_i)DZ_i = 0$. Since $Z_i \to Z$ as $i\to\infty$, it follows that

$$A(Z)DZ = 0 \,.$$

We now show that the boundary conditions are satisfied. Suppose n is a positive integer. For each positive integer i,

$$R_{n,i} = I - \Delta_{\gamma[i]}^{-1}\Sigma_{q=0}^{n-1}(B_i(R_{n,i}^{1/2}W)DR_{n,i}^{1/2})^*(B_i(R_{n,i}^{1/2}W)DR_{n,i}^{1/2}) \,.$$

Denote

$$\lim_{i\to\infty}\Sigma_{q=0}^{n-1}(B_i(R_{n,i}^{1/2}W)DR_{n,i}^{1/2})^*(B_i(R_{n,i}^{1/2}W)DR_{n,i}^{1/2}) \quad \text{by} \quad T$$

and denote $\lim_{i\to\infty} R_{n,i}$ by S. Write $T \in L(M,M) = L(K\times K,\ K\times K)$ as a matrix with entries in $L(K,K)$:

$$T = \begin{pmatrix} T_{11} & T_{12} \\ T_{21} & T_{22} \end{pmatrix} \,.$$

Then

$$S = \begin{pmatrix} I & 0 \\ 0 & I \end{pmatrix} - \begin{pmatrix} J & 0 \\ 0 & E \end{pmatrix}\begin{pmatrix} T_{11} & T_{12} \\ T_{21} & T_{22} \end{pmatrix}$$

where $J \equiv \lim_{i\to\infty}\Delta_{\alpha[i]}^{-1}$, $E \equiv \lim_{i\to\infty}\Delta_{\beta[i]}^{-1}$.

Hence

$$S = \begin{pmatrix} I - JT_{11} & -JT_{12} \\ -ET_{21} & I - ET_{22} \end{pmatrix}$$

and so

$$S^* = \begin{pmatrix} I - T_{11}^{*}J & -T_{21}^*E \\ -T_{12}^*J & I - T_{22}^*E \end{pmatrix}$$

since J and E are symmetric.

Take P' defined for G' as in section 2 and similarly take P'' defined for G''. Then

$$S^*\binom{P'x}{P''y} = \binom{P'x}{P''y}, \quad \binom{x}{y} \in M$$

since by Theorem 1, $JP' = 0 = EP''$. Hence

$$(S^{1/2})^*\binom{P'x}{P''y} = (S^*)^{1/2}\binom{P'x}{P''y} = \binom{P'x}{P''y}, \quad \binom{x}{y} \in M.$$

In particular for $\binom{r_n}{s_n} = Y_n$, $\langle\binom{P'r_n}{P''s_n}, \binom{x}{y}\rangle = \langle S^{1/2}\binom{r}{s}, \binom{P'x}{P''y}\rangle$

$= \langle\binom{r}{s}, (S^{1/2})^*\binom{P'x}{P''y}\rangle = \langle\binom{r}{s}, \binom{P'x}{P''y}\rangle = \langle\binom{P'r}{P''s}, \binom{x}{y}\rangle = \langle\binom{P'f}{P''g}, \binom{x}{y}\rangle$, $\binom{x}{y} \in M$,

since $P'r = P'f$, $P''s = P''g$. Hence $P'r_n = P'f$, $P''s_n = P''g$, i.e.,

$$r_n(p) = f(p), \quad p \in G'$$

$$s_n(p) = g(p), \quad p \in G'',$$

i.e., Y_n satisfies the boundary condition part of (*). Since $Z = \lim_{n\to\infty} Y_n$, Z must also satisfy the boundary conditions.

7. Late addition. The following Lemma H' was very recently found by this writer. It appears to lead to an alternative to the main material of this paper. The main advantage of the alternative is that the taking of square roots of matrices (numerically a cumbersome procedure for this writer) may be avoided entirely.

Lemma H'. Suppose that each of H and K is a Hilbert space and that A is a strongly continuous function from H to $L(H,K)$ so that $A(x)A(x)^* = I$ for all $x \in H$. Suppose also that P is an orthogonal projection on H, $W \in H$, $0 < c \leq 1$ and $Q_0 = P$,

$$Q_{n+1} = Q_n - cQ_nA(Q_nW)^*A(Q_nW)Q_n,$$

$n = 0, 1, \ldots$. Then $\{Q_nW\}_{n=1}^{\infty}$ converges to $Z \in H$ so that $PZ = Z$, $A(Z)Z = 0$.

The lemma is established by showing that each of $Q_0, Q_1, \ldots$ is a symmetric nonnegative element of $L(H,H)$ and $Q_{n+1} \leq Q_n$, $n = 0, 1, \ldots$. So $Q_0, Q_1, \ldots$ converges strongly on H. In particular $Q_0W, Q_1W, \ldots$ converges to $Z \in H$. That $PZ = Z$ is established as in the proof of Lemma H. That $(I - A(Z)^*A(Z))Z = Z$ is established in the same way that $L(Z)Z = Z$ was proved in Lemma H. But $(I - A(Z)^*A(Z))Z = Z$ implies that $A(Z)Z = 0$.

REFERENCES

[1] S. Agmon, "Lectures on Elliptic Boundary Value Problems," D. van Nostran Co., 1965.

[2] A. Beurling et J. Deny, Espaces de Dirichlet I. Le Cas Elementaire, Acta Math. 99(1958), 203-224.

[3] J. W. Neuberger, Projection Methods for Linear and Nonlinear Systems of Partial Differential Equations, Springer-Verlag Lecture Notes, 564(1976) 341-349.

[4] ______________, Square Integrable Solutions to Linear Inhomogeneous Systems, J. Diff. Equations (to appear).

On the solvability of nonlinear equations involving abstract and differential operators

W.V. Petryshyn*

Introduction. We continue the study of the solvability of differential equations initiated in [21,24] by means of the A-proper[1] mapping theory. In [22] the author studied the relationship of A-properness to maps of monotone and K-monotone type and the abstract results gotten were used, on the one hand, to deduce as special cases the basic surjectivity results for $T:X \to X^*$ of monotone type obtained earlier by other authors (see [22]) and, on the other, to obtain the approximation-solvability of semilinear general elliptic BV Problems of order 2 in the Sobolev space W_2^2. It is known that the theory of monotone type operators is applicable essentially to differential equations involving operators in divergence form. As is shown in [22] and in the present paper, the theory of maps of A-proper type is applicable to abstract and differential equations which involve a more general class of operators which act from X to Y with Y not necessarily equal to X^*.

The purpose of this paper is two-fold. In Section 1 we extend some abstract theorems in [22] and, in virtue of the recent result of Toland [28], we show how they apply to equations to which neither the theory of monotone type nor condensing type maps is applicable. In the first part of Section 2 we deduce some consequences of the

---

*Supported in part by the NSF Grant MCS 76-06352.

[1]See Section 1 for precise definitions and assertions mentioned in the Introduction.

results in Section 1 for problems where unbounded discontinuous operators are involved. The abstract results are modelled on the situation which arises when one treats semilinear elliptic problems on unbounded domains. The second part treats ordinary and partial differential equations.

We now outline briefly some of our results. Let X and Y be real Banach spaces, $T:X \to Y$ a nonlinear map, $\Gamma=\{X_n,V_n;E_n,W_n\}$ an admissible scheme for

$$(1) \qquad Tx=f \qquad (x\in X,\ f\in Y)$$

and

$$(2) \qquad T_n(X_n)=W_n(f) \qquad (x_n\in X_n, W_nf\in E_n, T_n=W_nT|_{X_n})$$

a sequence of finite dimensional equations associated with (1) by Γ.

In Section 1 we first study (see Theorems 1.1 and 1.2) the approximation-solvability of (1) under conditions that T is A-proper w.r.t. Γ and there is an $r_0>0$ and a suitable map $K:X \to Y^*$ such that

(c) $||Tu||+(Tu,Ku)/||Ku||>0 \ \forall ||u||\geq r_0$ and T satisfies condition (+) given by

(+) If $\{x_j\}$ is any sequence with $Tx_j \to g$ for some $g\in Y$, then $\{x_j\}$ is bounded. The interesting feature of Theorems 1.1 and 1.2 is that unlike, the results in [22,16] and others, Γ is not required to be such that $(W_ng,K_nx)=(g,Kx)$ for $x\in X_n$ when T is bounded. If $Y\neq X^*$, the latter condition involves K in the construction of Γ and this imposes restriction on T for it to be A-proper w.r.t. such schemes. In the second part Theorem 1.2 is used to establish the solvability of (1) when T is not A-proper but is a uniform limit of A-proper maps with T satisfying (c). The relation of our results to those of other authors is indicated in the text. It should be noted, however, that even when $Y=X^*$ and $T:X \to X^*$ is quasimonotone our Theorem 1.3 implies a new result for this general class of maps whose study was initiated in Calvert-Webb [6] and Hess [13], and further studied by Fitzpatrick [9] and others. Theorem 1.4 gives a surjectivity result for a pseudo-A-proper map.

Assuming that $Y=X=H$, where H is a Hilbert space, in the first part of Section 2 we use Theorem 1.1 to study the solvability of

$$Au-Bu+Nu=f \qquad (u\in D(A),\ f\in H), \tag{3}$$

where A is a densely defined, positive definite, self-adjoint operator whose essential spectrum is bounded below, $B:H \to H$ is α-ball-contraction and $N:D(A^{\frac{1}{2}}) \to H$ is such that $N=N_1+N_2$ with N_1 monotone and N_2 compact. Recent results of Toland [28] and Stuart [27] allow us to reduce (3) to an equivalent equation in H to which Theorem 1.1 applies. Eq. (3) is also studied when $A=I$ and B is semicontractive in the sense of Browder [3]. Theorem 2.1 is then applied to the solvability of

$$-(p(x)u')'+q(x)u'-g(x,u)+h(x,u)=f(x) \qquad (x\in(0,\infty),u(0)=0), \tag{4}$$

under certain conditions on p,q q and h. As a second example we use Theorem 1.1 to obtain the solvability in $\mathring{W}_2^m \cap W_2^{2m}$ of

$$\sum_{|\alpha|,|\beta|\leq m} (-1)^{|\beta|} D^\beta (a_{\alpha\beta}(x) D^\alpha u) - N(x,D^\gamma u) = f(x) \qquad (f\in L_2,\ |\gamma|\leq 2m-1), \tag{5}$$

where $Q\subset R^n$ is a bounded domain and $a_{\alpha\beta}$ and N satisfy suitable conditions. As special cases of our Theorem 2.6 we deduce Theorem 1 and 2 of De Figueiredo [11]. Finally, using Theorem 1.3, we give a simple proof of the result due to Wille [32] concerning a weak solvability of ODE of order 2 involving a map $T:\overset{\circ}{W}{}^1_2 \to \overset{\circ}{W}{}^1_2$ which is quasimonotone and such that $||Tu||+(Tu,u)/||u|| \to \infty$ as $||u|| \to \infty$ but T is not coercive.

Section 1. Let $\{X_n\}$ and $\{E_n\}$ be sequences of oriented finite dimensional spaces with $\{X_n\} \subset X$, V_n an inclusion map of X_n into X, and W_n a linear map of Y onto E_n. We use the same symbol $\|\cdot\|$ to denote the norms in X, Y and E_n, and "$\longrightarrow$" and "$\rightharpoonup$" to denote strong and weak convergence respectively.

Definition 1.1. The quadruple of sequences $\Gamma = \{X_n, V_n; E_n, W_n\}$ is said to be an admissible scheme for (X,Y) provided that $\dim X_n = \dim E_n$ for each n, $\operatorname{dist}(x, X_n) \to 0$ as $n \to \infty$ for each x in X, and $\{W_n\}$ is uniformly bounded.

Note that we do not require $\{E_n\}$ to be subspaces of Y nor do we assume that $\{X_n\}$ is nested. For later references we describe three typical schemes.

(a) Injective scheme for (X,X^*). If $E_n = X_n^*$ and $W_n = V_n^*$, then $\Gamma_I = \{X_n, V_n; X_n^*, V_n^*\}$ is admissible for (X,X^*).

(b) Projective scheme for (X,Y). If $E_n = Y_n \subset Y$ and $W_n = Q_n$, where $Q_n: Y \to Y_n$ is a linear projection such that $\|Q_n\| \leq \beta$, then $\Gamma_P = \{X_n, V_n; Y_n, Q_n\}$ is admissible for (X,Y).

(c) Projective scheme for (X,X). If $E_n = X_n$ and $W_n = P_n$, where $P_n: X \to X_n$ is a linear projection such that $\|P_n\| \leq \alpha$ for some $\alpha \geq 1$, then $\Gamma_\alpha = \{X_n, P_n\}$ is admissible for (X,X) and, in fact, Γ_α is projectionally complete in the sense that $P_n(x) \to x$ in X for each x in X.

In case X and Y are Hilbert spaces, then the projections are always assumed to be orthogonal.

Example (a) shows that when X is separable then $\{X,X^*\}$ always has an admissible scheme, while example (c) shows there is an admissible scheme for (X,X) if X has a Schauder basis.

If $Q \subset X$ is bounded we define $\chi(Q)$, the ball-measure of noncompactness of Q, to be $\inf\{r>0 \mid Q$ can be covered by a finite number of balls whose radii $\leq r\}$. Clearly $\chi(Q)=0$ iff $\bar{Q}$ is compact. Given $k \in \mathbb{R}^+$, a continuous map $F:X \to X$ is called k-ball-contractive if $\chi(F(Q)) \leq k\chi(Q)$ for each bounded $Q \subset X$. Then each compact map C is 0-ball-contractive, and if $S:X \to X$ is ℓ-Lipschitzian, then $S+C$ is ℓ-ball-contractive. Fixed point theorems, degree and index theories, mappings, theorems, eigenvalue theorems, and other results have recently been obtained (see [10,19,23,26] and other references cited there).

To extend surjectivity results obtained earlier for strongly monotone maps from reflexive space X to X^* Browder introduced (see [3]) a more general class of maps which he called maps of type (S) and (S_+) and which are defined as follows. $T:X \to X^*$ of of type (S) (of type (S_+)) if $\{x_j\} \subset X$ is any sequence such that $x_j \rightharpoonup x$ in X and $\lim_j (Tx_j, x_j - x) = 0$ $(\overline{\lim}_j (Tx_j, x_j - x) \leq 0)$, then $x_j \to x$ in X. Obviously, maps of type (S_+) form a subclass of maps of type (S) (see also [22]). Extending the surjectivity results for monotone maps, Leray and Lions introduced in [15] maps of pseudo-monotone type, while Brezis introduced in [1] still more general class of maps of type (M) which is defined to be a map $T:X \to X^*$ such that if $x_j \rightharpoonup x$ in X, $Tx_j \rightharpoonup g$ in X^* and $\overline{\lim}(Tx_j, x_j) \leq (g,x)$, then $Tx = g$. Hess [13] and Calvert-Webb [6] have independently introduced a very general class of maps $T:X \to X^*$ which are quasimonotone, i.e., such that if $x_j \rightharpoonup x$ in X, then $\overline{\lim}(Tx_j, x_j - x) \geq 0$. It was shown in [9] that there are maps which are quasimonotone but neither pseudo-monotone nor of type (M) in the sense of Brezis [1].

The importance of the above classes of mappings stems from the fact that certain elliptic operators in divergence form give rise to operator equations in suitable Sobolev spaces involving these various classes of operators. However, if the differential operator is not in divergence form or if the operator T acts from X to Y with $Y \neq X^*$, then one has to redefine the above notions with respect to some suitable linear or nonlinear operator, say, K of X into Y^*. This has been done by the author and others (see [20,21,22] and other references cited there).

The following class of A-proper mappings introduced by the author (see [21]) proved to be useful, since on the one hand, this class includes compact, P -compact, and k-ball-contractive vector fields, mappings of type (S) and (KS) as well as maps of strongly K-monotone type (see [21]) and, on the other hand, the notion of the A-proper mapping proved to be also useful in the constructive solvability of abstract and differential nonlinear equations (see [21,22,24]).

Definition 1.2. $T:X \to Y$ is approximation-proper (A-proper) w.r.t. Γ if $T_n:X_n \to E_n$ is continuous for each n and if $\{x_{n_j} | x_{n_j} \in X_{n_j}\}$ is any bounded sequence such that $||T_{n_j}(x_{n_j})-W_{n_j}g|| \to 0$ as $j \to \infty$ for some g in Y, then there exists an $x \in X$ such that (i): $Tx=g$ and (ii): $x \in \overline{\{x_{n_j}\}}$, where $\overline{\{x_{n_j}\}}$ is the closure of $\{x_{n_j}\}$ in X.

If in Definition 1.2 we drop the requirement (ii), then T is said to be pseudo-A-proper. The latter class of maps has been studied by the author in [20], where it was shown that in addition to A-proper maps it includes weakly continuous maps $T:X \to Y$, maps of type (KM) and others (including monotone, pseudomonotone, and type (M)).

The following examples of A-proper mappings (for others see [3, 21]), which we shall employ in what follows, illustrate the generality of this class; and we also observe that a compact perturbation of an A-proper map remains A-proper.

Example 1. If $F:X \to X$ is k-ball-contractive with $k\varepsilon[0,1)$, (in particular, if $F=S+C$ where C is compact and S is ℓ-Lipschitizian with $\ell\varepsilon(0,1)$), then $I-F$ is A-proper w.r.t. $\Gamma_1=\{X_n,P_n\}$.

In fact one sees that F is P_1-compact, i.e., $\lambda I-F$ is A-proper w.r.t. Γ_1 for each $\lambda\geq 1$ (see [21]). To introduce our next example we recall that $T:X \to Y$ is said to be demicontinuous if $x_j \to x$ in X implies that $Tx_j \rightharpoonup Tx$ in Y; T is of type (KS) if $x_j \rightharpoonup x$ in X and $\lim_j(Tx_j,K(x_j-x))=0$ implies that $x_j \to x$ in X, where K is a suitable map of X into Y^*; T is quasibounded if $\{Tx_j\}$ is bounded whenever $\{x_j\}$ and $\{(Tx_j,Kx_j)\}$ are bounded. The following result was proved in [22] for the case of projective schemes.

Example 2. Suppose X is reflexive, $K\varepsilon L(X,Y^*)$ is injective with $R(K)$ dense in Y^*, K_n is a map of X_n into $D(W_n^*)$ such that $\{K_n(z_n)\}$ is bounded whenever $\{z_n | z_n\varepsilon K_n\}$ is bounded, and

(1.1) $$(W_n g,K_n(x))=(g,Kx) \quad \text{for all } x\varepsilon X_n, \; g\varepsilon Y \text{ and each } n.$$

If $T:X \to Y$ is demicontinuous, quasibounded, and of type (KS), then T is A-proper w.r.t. $\Gamma=\{X_n,V_n;E_n,W_n\}$.

The above assertion is also valid for certain nonlinear maps K. When $Y=X^*$ and for Γ we take $\Gamma_I=\{X_n,V_n;X_n^*,V_n^*\}$, then we may choose $K=I$ and $K_n=I_n$. In this case we see that every quasibounded map $T:X \to X^*$ of type (S) is A-proper w.r.t. Γ_I. Since every monotone map defined on X is quasibounded, firmly monotone (and, in particular, strongly monotone) mappings are A-proper w.r.t. Γ_I.

As the next example, which will prove to be important in our sudy of semilinear elliptic boundary value problems in Section 2, is the following result obtained recently by Toland [28].

Example 3. Suppose H is a real Hilbert space with a nested projectionally complete scheme $\Gamma_1=\{X_n,P_n\}$. If $F:H \to H$ is k-ball-contractive and $N:H \to H$ is continuous, bounded and c-monotone (i.e. $(Nx-Ny, x-y)\geq c||x-y||^2$ $\forall x,y\in H$, $c\geq 0$), then $T=I-F+N$ is A-proper w.r.t. Γ_1 provided $2k-c<1$.

The interesting feature of this class of mappings is that neither the theory of monotone type or k-set-contractive type mappings is applicable to equations involving the above class of mappings. Consequently, the A-proper mapping theory is the only one that can handle equations involving operators of Example 3.

It was also shown in [28] that if F is linear, then one may drop the condition that N is bounded and assume only that $k-c<1$.

It was shown in [31] that if $F\in L(H,H)$ is k-ball-contractive, then for each $\varepsilon>0$ there are maps C_ε, $F_\varepsilon \in L(H,H)$ with C_ε compact and $||L_\varepsilon||\leq k+\varepsilon$ such that $F=F_\varepsilon+C_\varepsilon$. In view of this and Lemma 1 in [22] or Example 2 we have the following improvement of Lemma 2.8 in [28].

Example 4. If $F\in L(H,H)$ is k-ball-contractive and $N:H \to H$ hemicontinuous and c-monotone, then $T=I-F+N$ is A-proper w.r.t. $\Gamma_1=\{X_n,P_n\}$ with $\{X_n\}$ not necessarily nested provided $k-c<1$.

To state our results precisely we need the following

Definition 1.3. Eq. (1) is said to be strongly (feebly) approximation-solvable w.r.t. Γ if there is $n_f \in N$ such that Eq. (2) has a solution $x_n \in X_n$ for each $n \geq n_f$ with the property that $x_n \to x$ in X ($x_{n_j} \to x$ for some subsequence) and $Tx=f$.

Note that the approximation-solvability of (1) implies the solvability of (1) but the converse is not true in general. Our first result is the following theorem concerning the approximation-solvability of Eq. (1), which extends Theorem 1 in [22].

Theorem 1.1. Let $\Gamma=\{X_n,V_n;E_n,W_n\}$ be admissible for (X,Y), K some map of X into Y^* and suppose there is a bounded map $G:X \to Y$ such that

(H1) G is odd, A-proper w.r.t. Γ, and $(Gx,Kx)=||Gx||\,||Kx||>0$ for $0 \neq x \in X$.

(H2) $T_\mu=T+\mu G$ is A-proper w.r.t. Γ for each $\mu>0$.

(H3) $||Tx||+\frac{(Tx,Kx)}{||Kx||}>0$ for all $X-B(0,r_0)$ and some $r_0>0$.

(H4) T satisfies condition (+) or, equivalently, $T^{-1}(Q)=\{x \in X \mid Tx \in Q\}$ is bounded whenever $\bar{Q}$ is compact.

Under the above hypothesis the following assertions are valid:

(A1) If $T:X \to Y$ is A-proper w.r.t. Γ and bounded, then (1) is feebly approximation-solvable for each f in Y. For a given $f \in Y$, (1) is strongly approximation-solvable if it is uniquely solvable.

(A2) If we also assume that there is a map $K_n:X_n \to D(W_n^*)$ such that (1.1) holds, then the assertion of (A1) remains valid for T unbounded provided there exist r_1, γ, $\eta \in \mathbb{R}^+$ such that one of the following two conditions hold:

(1.2) $(Tx,Kx)\geq-\gamma||Kx||$ for $x\epsilon X-B(0,r_1)$.

(1.3) $(Tx,Kx)\geq-\eta||x||\,||Kx||$ for $x\epsilon X-B(0,r_1)$.

We shall deduce Theorem 1 as a corollary of the following slightly more general result which will also prove to be useful in our study of the solvability of (1) in case T is a uniform limit of a suitable sequence of A-proper maps.

Theorem 2. Suppose that all the conditions of Theorem 1 hold except that condition (H4) is replaced by the following

(H4') To each $f\epsilon Y$ there are numbers $r_f\geq\max\{r_0,r_1\}$ and $\alpha_f>0$ such that $||Tx-tf||\geq\alpha_f$ for $x\epsilon\partial B(0,r_f)$ and $t\epsilon[0,1]$.

Then the conclusions of Theorem 1 hold.

Proof. Suppose $f\epsilon Y$ is fixed. Then, by (H4'), there are $r_f\geq\max\{r_0,r_1\}$ and $\alpha_f>0$ such that

(1.4) $||T(x)-tf||\geq\alpha_f$ for all $x\epsilon\partial B(0,r_f)$ and $t\epsilon[0,1]$.

(A1) Suppose first that $T:X\to Y$ is bounded. Consider the homotopy

$$H(x,t)=(1-t)T(x)+tG(x)\qquad (x\epsilon\bar{B}(0,r_f),\ t\epsilon[0,1]).$$

Since T is A-proper and (H2) holds, it follows that $H(\cdot,t)$ is A-proper for each $t\epsilon[0,1]$. Moreover, $H(x,t)\neq0$ for $x\epsilon\partial B(0,r_f)$ and $t\epsilon[0,1]$. If not, then there would exist $x_0\epsilon\partial B(0,r_f)$ and $t_0\epsilon[0,1]$ such that $H(x_0,t_0)=(1-t_0)T(x_0)+t_0G(x_0)=0$. It follows from (H1) that $t_0\neq1$ and from (H3) that $t_0\neq0$. Thus $t_0\epsilon(0,1)$ and, if we set $\mu_0=t_0/(1-t_0)$, we see that $Tx_0+\mu_0Gx_0=0$. Hence $||Tx_0||=\mu_0||Gx_0||$ and $(Tx_0,Kx_0)=-\mu_0(Gx_0,Kx_0)=-\mu_0||Gx_0||\,||Kx_0||$ by (H1). Hence $||Tx_0||+(Tx_0,Gx_0)/||Kx_0||=0$, a contradiction to (H3). Thus, $H(x,t)\neq0$ for $x\epsilon\partial B(0,r_f)$ and $t\epsilon[0,1]$ and, moreover, since T and G are bounded H is continuous in $t\epsilon[0,1]$, uniformly for x in $\bar{B}(0,r_f)$.

Hence, by Theorem 1 in [4], the generalized degree $\mathrm{Deg}(H(\cdot,t),B(0,r_f),0)$ is constant in $t\varepsilon[0,1]$ and therefore $\mathrm{Deg}(T,B(0,r_f),0)=\mathrm{Deg}(G,B(0,r_f),0)$ 0 since G is odd. It follows from this and (1.4) that $0\notin\mathrm{Deg}\,(T-f,B(0,r_f),0)$. Hence, by the definition of Deg., there exists $n_f\varepsilon N$ such that the Brower deg $(T_n-W_nf,B_n(0,r_f),0)\neq 0$ for each $n\geq n_f$. So, by the Brower degree theory, there exist $x_n\varepsilon B_n(0,r_f)$ for each $n\geq n_f$ such that $T_n(x_n)-W_nf=0$. This and the A-properness of T imply the existence of a subsequence $\{x_{n_j}\}$ of $\{x_n\}$ and an $x_0\varepsilon X$ such that $x_{n_j}\to x_0$ in X and $Tx_0=f$. The proof that the entire sequence $\{x_n\}$ converges to x_0 if (1) has at most one solution is obtained in the same way as in [22]. This proves (A1).

(A2). Suppose now that (1.1) and either (1.2) or (1.3) hold but T is not bounded. Then, as in (A1), the homotopy $H(\cdot,t)$ is still A-proper for each $t\varepsilon[0,1]$ and $H(x,t)\neq 0$ for $x\varepsilon\partial B(0,r_f)$ and $t\varepsilon[0,1]$; therefore $\mathrm{Deg}\,(H(\cdot,t),B(0,r_f),0)$ is well-defined for each fixed $t\varepsilon[0,1]$. To show that it is constant in $t\varepsilon[0,1]$, it suffices to show that there is $n_f\varepsilon N$ such that, for each fixed $n\geq n_f$, $H_n(x,t)=(1-t)T_n(x)+tG_n(x)\neq 0$ for $x\varepsilon\partial B_n(0,r_f)$ and $t\varepsilon[0,1]$ for this implies that $\deg\,(H_n(\cdot,t),B_n(0,r_f),0)$ is independent of $t\varepsilon[0,1]$ for each $n\geq n_f$. Our proof will be by contradiction.

Suppose that such an n_f does not exist. Then there would exist an unbounded sequence of integers (which we identify for simplicity with the original sequence $\{n\}$), a sequence $\{t_n\}\subset(0,1)$ with $t_n\to t_0\varepsilon[0,1]$, and $\{x_n|x_n\varepsilon B_n(0,r_f)\}$ such that $(1-t_n)T_n(x_n)+t_nG_n(x_n)=0$ We distinguish between three cases

(i) $t_0=0$: We note that in this case

$$T_n(x_n)=-t_n/(1-t_n)G_n(x_n) \to 0 \quad \text{as} \quad j \to \infty$$

since $\{G_n(x_n)\}$ is bounded. Thus, since T is A-proper, there exist a subsequence, again denoted by $\{x_n\}$, and an x in X such that $x_n \to x$ in X and $Tx=0$ with $||x||=r_f$, in contradiction to (H3).

(ii) $t_0=1$: Since $t_nG_n(x_n)=(t_n-1)T_n(x_n)$ for each n, it follows from this, (H1) and (1.1) that $t_n||Gx_n||\,||Kx_n||=(t_n-1)(Tx_n,Kx_n)$. Since $(t_n-1)<0$ and either $(Tx_n,Kx_n)\geq-\gamma||Kx_n||$ or $(Tx_n,Kx_n)\geq-\eta r_f||Kx_n||$ depending on whether (1.2) or (1.3) holds, it follows that $(1-t_n)(Tx_n,Kx_n)\geq-\gamma_f(1-t_n)||Kx_n||$ where γ_f is either γ or $\gamma\cdot r_f$. Therefore, for each n we have $t_n||Gx_n||\,||Kx_n||\leq\gamma_f(1-t_n)||Kx_n||$. Hence $||Gx_n||\leq\gamma_f(1-t_n)/t_n$. Since $\{||Gx_n||\}$ is bounded, there exists a subsequence, again denoted by $\{||Gx_n||\}$, such that $||Gx_n|| \to c_1$ for some $c_1\geq0$. But $\gamma_f(1-t_n)/t_n \to 0$ and thus $c_1=0$. This implies that $W_nG(x_n)\equiv G_n(x_n) \to 0$ as $n \to \infty$ if either (1.2) or (1.3) holds. This and the A-properness of G imply the existence of a subsequence, again denoted by $\{x_n\}$, and an x in X such that $x_n \to x$ in X and $Gx=0$ with $||x||=r_f$, in contradiction to (H1).

(iii) $0<t_0<1$: Let $\mu_0=t_0/(1-t_0)$, $\mu_n=t_n/(1-t_n)$ and note that $T_n(x_n)+\mu_0G_n(x_n)=(\mu_0-\mu_n)G_n(x_n) \to 0$ since $\mu_n \to \mu_0$ and $\{G_n(x_n)\}$ is bounded. Since, by (H2), $T+\mu_0G$ is A-proper, we may assume that $x_n \to x$ in X and $Tx+\mu_0Gx=0$ with $||x||=r_f$. It follows from this that $||Tx||=\mu_0||Gx||$ and $(Tx,Kx)=-\mu_0(Gx,Kx)=-\mu_0||Gx||\,||Kx||$. Hence $||Tx||+(Tx,Kx)/||Kx||=0$, in contradiction to (H3).

The above discussion shows that Deg $(H(\cdot,t),B(0,r_f),0)$ is also constant in $t\in[0,1]$ when (1.1) and either (1.2) or (1.3) holds. This implies the validity of our assertion in (A2). Q.E.D.

Proof of Theorem 1.1. To prove Theorem 1.1, it suffices to show that (H4) of Theorem 1.1 implies (H4') of Theorem 1.2. Now, if for some f in Y the hypothesis (H4') fails to hold, then we could find sequences $\{t_{n_j}\} \subset [0,1]$ and $\{x_{n_j}\} \subset X$ with $t_{n_j} \to t_0 \varepsilon [0,1]$ and $||x_{n_j}|| \to \infty$ as $j \to \infty$ such that $T(x_{n_j}) - t_{n_j} f \to 0$ as $j \to \infty$. This implies that $T(x_{n_j}) \to t_0 f$ as $j \to \infty$ with $\{x_{n_j}\}$ unbounded, in contradiction to (H4). Q.E.D.

Since (H3) and (H4) of Theorem 1 are implied by the hypothesis

(H5) $||Tx|| + \frac{(Tx,Kx)}{||Kx||} \to \infty$ as $||x|| \to \infty$,

Theorem 1.1 implies the validity of the following useful result.

Proposition 1.1. If (H1) and (H2) of Theorem 1.1 hold, then (A1) and (A2) of Theorem 1.1 remain valid if (H3) and (H4) are replaced by the single hypothesis (H5).

Remark 1.1. It is easy to see that (H4) is implied by

(H6) $||Tx|| \to \infty$ as $||x|| \to \infty$.

Moreover, if (1.2) holds, then (H5) is equivalent to (H6). The latter fact has been used by many authors (see [2 ,17] for references) in their study of operators of monotone type. Condition (1.3) appears to be new and will prove to be useful in the study of semilinear equations with unbounded nonlinear part.

1.2). The interesting feature of (A1) of Theorem 1.1 is that, unlike the results in [22,16] and others, it establishes the solvability of (1) for T acting from X to Y without condition (1.1) which requires Γ to be such that $(W_n g, K_n x) = (g, Kx)$ for $x \varepsilon X_n$ and g in Y. If $Y \neq X^*$ the latter condition essentially involves K in the construction of Γ and this in some cases (as will be seen later) imposes a considerable restriction on T for it to be A-proper with respect to such schemes.

Theorem 1.1 and Proposition 1.1 contain a number of special cases for variou classes of mappings satisfying various growth conditions. Here we state only two such cases since they will be used in Section 2.

Corollary 1.1. If $A:X \to X$ is P_1-compact (and, in particular, ball-condensing) w.r.t. Γ_1 and $T=I-A$ satisfies either (H3)-(H4) or (H5), where K is a section of the normalized duality map J of X into X^* and $G=I$, then (A1) and (A2) of Theorem 1.1 hold for each $f\varepsilon X$.

For bounded A Corollary 1.1 can also be deduced from Proposition 1.4 in [17] and, for A ball-condensing, from Theorem 1.1 in [10]. As in our next corollary we deduce the following result which will prove to be useful in applications.

Corollary 1.2. Suppose $T=G+N$ is A-proper w.r.t. Γ with $G:X \to Y$ bounded and satisfying (H1) of Theorem 1.1.

If $N:X \to Y$ is bounded, $G+tN$ is A-proper w.r.t. Γ for each $t\varepsilon(0,1)$ and $||Gx||-||Nx|| \to \infty$ as $||x|| \to \infty$, then the conclusion (A1) of Theorem 1.1 holds. If (1.1) also holds, then instead of boundedness it suffices to assume that N satisfies the condition $\limsup_{||x||\to\infty} \{||Nx||/||x||\}<\infty$.

We complete this part of Section 1 with the following special case of Theorem 2 in [24] which we shall use in Section 2 and which differs from Corollary 1.2.

Proposition 1.2. *Suppose* T, $L:X \to Y$ *are A-proper w.r.t.* Γ *with* $T=L+F$, *where* L *is odd and 1-homogeneous (i.e.,* $L(tx)=tL(x)\,\forall x\in X,\ t>0$) *and* $\|Fx\|/\|x\| \to 0$ *as* $\|x\| \to \infty$. *Then* $Tx=f$ *is feebly approximation-solvable for each* f *in* Y *provided* $x=0$ *whenever* $Lx=0$.

Using Theorem 1.2, it is now easy to establish a surjectivity theorem for (1) when T is not A-proper but is a uniform limit of a suitable sequence of A-proper mappings.

Theorem 1.3. *Suppose that the maps* T, $G:X \to Y$ *and* $K:X \to Y^*$ *satisfy the hypothesis* (H1)-(H4) *of Theorem 1.1. Then:*

(A1) *If* T *is bounded and* $T(B)$ *is closed in* Y *for each closed ball* B *in* X, *then* (1) *is solvable for each* f *in* Y.

(A2) *If we also assume that* (1.1) *holds and* T *satiifies either* (1.2) *or* (1.3) *of Theorem 1.1, then in* (A1) *we may drop the condition that* T *is bounded.*

Proof. First note that since for nonzero x in X we have

$$\|T_\mu x\|+\frac{(T\ x,Kx)}{\|Kx\|} \geq \|Tx\|-\mu\|Gx\|+\frac{(Tx,Kx)}{\|Kx\|}+\mu\|Gx\|=\|Tx\|+\frac{(Tx,Kx)}{\|Kx\|},$$

it follows from this and (H3) that for each $\mu>0$

$$(1.5)\qquad \|T_\mu x\|+\frac{(T\ x,Kx)}{\|Kx\|}>0 \quad \text{for} \quad \|x\|\geq r_0 .$$

It was shown above that, for each f in Y, (H4) implies the existence of constants $r_f\geq\max\{r_0,r_1\}$ and $\alpha_f>0$ such that (1.4) holds. In view of this and the boundedness of G, we can choose $\mu_f>0$ such that

$$(1.6)\quad \|T_\mu x-tf\|\geq 2^{-1}\alpha_f \quad \text{for} \quad x\in\partial B(0,r_f),\ t\in[0,1),\ \mu\in(0,r_f).$$

Now, obviously T_μ is bounded when T is bounded and T_μ satisfies either (1.2) or (1.3) for $\mu>0$ depending on T. Consequently, for each fixed $\mu\varepsilon(0,\mu_f)$, the operator $T_\mu=T+\mu G$ satisfies all the conditions of Theorem 1.2. Hence applying Theorem 1.2 to the equation $T_{\mu_k}(x)\equiv Tx+\mu_k Gx=f$, where $\mu_k\varepsilon(0,\mu_f)$ is such that $\mu_k\to 0$ as $k\to\infty$, in each of the two cases we find an element $x_k\varepsilon B(0,r_f)$ such that $T_{x_k}+\mu_k G_{x_k}=f$ for each k. Since $\mu_k\to 0$ as $k\to\infty$ and $\{Gx_k\}$ is bounded, it follows that $Tx_k\to f$ in Y as $k\to\infty$. This and the condition that $T(B)$ is closed in Y for each closed ball B in X imply the existence of an element $x\varepsilon B(0,r_f)$ such that $Tx=f$ for a given $f\varepsilon Y$. Since f was fixed but arbitrary, it follows that $T(X)=Y$. Q.E.D.

We note that (A2) of Theorem 1.3 is related to Theorem 2.6 in [16], where the conditions are such that $T+\mu G$ is K-coercive which is not the case in (A2).

Theorem 1.3 includes a number of special cases depending on the choice of Y and K. We illustrate this by the following new result for quasimonotone map $T:X\to X^*$ where X is separable and reflexive which, without loss of generality, we may assume to be such that X and X^* are locally uniformly convex.

Corollary 1.3. *Suppose that* $T:X\to X^*$ *is quasibounded, demicontinuous, quasimonotone,* $T(B)$ *is closed for each closed ball* B *in* X *and either* (d1) *or* (d2) *holds:*

(d1) T *satisfies condition* (+) *and there is* $r>0$ *such that*

$$||Tx||+(Tx,x)/||x||>0 \quad \text{for } ||x||\geq r_0 \tag{1.7}$$

(d2) $||Tx||+(Tx,x)/||x||\to\infty$ *as* $||x||\to\infty$.

Then $T(X)=X^*$ *provided that either* T *is bounded or for some* $r_1>0$ *and* $\eta,\gamma\in\mathbb{R}^+$ *either* $(Tx,x)\geq-\gamma||x||$ *or* $(Tx,x)\geq-\eta||x||^2$ *for* $||x||\geq r_1$.

Proof. If in Theorem 1.3 we set $Y=X^*$, $\Gamma=\Gamma_I$, $K=I$, $K_n=I_n$ and $G=J$, where $J:X\to X^*$ is the normalized duality map, then J is odd, continuous and A-proper w.r.t. Γ_I and $T+\mu J$ is A-proper by the special case of Lemma 2 in [22]. Hence, for the above choice of the operators and spaces, (H1)-(H4) of Theorem 1.3 or (H5) of Proposition 1.1 hold and so the conclusion of Corollary 1.3 follows. Q.E.D.

For the case when X is separable and T quasibounded, Corollary 1.3 extends the basic theorem in [9]. Indeed, if (+) holds, then there is $r_2>0$ such that $||Tx||>0$ for $||x||\geq r_2$. This and the condition $(Tx,x)\geq 0$ for $||x||\geq r_3$ imply (1.7) with $r_0=\max\{r_2,r_3\}$. The surjectivity result in [6] for T bounded and coercive follows from Corollary 1.3 (d2).

We shall see in Section 2 that there are differential operators which satisfy (d2) but are not coercive.

First results under the assumption (d2) together with $(Tx,x)\geq-\eta||x||$ for all $x\in X$ have been obtained in [2]. Condition (d2) has been used in [14,16] and other authors (see [16]) for other classes of maps.

We complete this section with the outline of the proof of the following.

Theorem 1.4. Suppose $T,G:X\to Y$ *and* $K:X\to Y^*$ *satisfy* (H1) *and* (H2) *of Theorem* (1.1) *and* (H5) *of Proposition 1.1.*

(A1) *If* T *is pseudo-A-proper and bounded, then* $T(X)=Y$.

(A2) *If* (1.1) *holds and either* (1.2) *or* (1.3) *of Theorem 1.1 also holds, then in* (A1) *we may drop the condition that* T *is bounded.*

Proof. Since the hypotheses on T are invariant under the change $T_f=T-f$, it suffices to prove Theorem 1.4 for $Tx=0$. Note that, by (H5), there is $r_0>0$ such that

$$(1.8) \qquad ||Tx||+(Tx,Kx)/||Kx||>0 \quad \text{for} \quad ||x||\geq r_0.$$

Consider the homotopy $H_n:\bar{B}_n(0,r_0)\times[0,1]\to E_n$ given by $H_n(x,t)=(1-t)T_n(x)+tG_n(x)$. We claim that there is $n_0\geq 1$ such that $H_n(x,t)\neq 0$ for $x\in\partial B_n(0,r_0)$, $t\in[0,1]$ and $n\geq n_0$. Indeed, if this were not the case, then we could find a sequence integers, which we again denote by $\{n\}$, a sequence $\{t_n\}\subset(0,1)$ with $t_n\to t_0\in[0,1]$ and $x_n\in\partial B_n(0,r_0)$ such that $(1-t_n)T_n(x_n)+t_nG_n(x_n)=0$. We have three cases.

(i) $t_0=0$: In this case we see that $T_n(x_n)=-t_n/(1-t_n)G_n(x_n)\to 0$. Hence, since T is pseudo-A-proper, there exists $x\in X$ such that $Tx=0$. We have thus found a desired element and so may exclude the case $t_0=0$ from further discussion.

The case $t_0=1$ and $t_0\in(0,1)$ are excluded in the same way as in the proof of (A2) of Theorem 1.2.

It follows from this, the finite dimensional degree theory, and the oddness of G_n, that $\deg(T_n,T_n(0,r_0),0)\neq 0$ for $n\geq n_0$. Hence there is $x_n\in B_n(0,r_0)$ such that $T_n(x_n)=0$ for $n\geq n_0$. This and the pseudo-A-properness of T imply the existence of $x\in X$ such that $Tx=0$. Q.E.D.

Section 2. In the first part of this section we deduce some consequences of the results in Section 1 for problems where unbounded discontinuous operators are involved. The abstract results below are modelled on situations which arize when one attempts to treat semilinear elliptic problems on unbounded domains (see [28] and others cited there, where global bifurcation phenomenon is treated for such operators). We add in passing that our first four surjectivity theorems are obtained for the class of mappings to which neither the theory of monotone nor condensing type mappings is applicable. The second part of the section is devoted to some further applications.

(P1) First we consider the equation of the following form

$$Au-Bu+Nu=f \qquad (u\in D(A),\ f\in H), \tag{2.1}$$

where H is a real separable Hilbert space and the operators A, B and C are assumed to satisfy the following hypotheses:

(a1) A is a densely defined, positive definite, self-adjoint linear operator whose essential spectrum $\sigma_e(A)$ is bounded below, i.e., there is a number $\gamma>0$ such that, for each $\varepsilon>0$, $\sigma(A)\cap(-\infty,\gamma-\varepsilon)$ consists of a nonempty set of isolated eigenvalues, each of finite multiplicity, with

$$\lambda_0<\lambda_1<\lambda_2<,\ldots,<\gamma.$$

(a2) $B:H\to H$ is an odd α-ball-contractive map such that $Au-Bu\neq 0$ whenever $0\neq u\in D(A)$.

(a3) $N:H_0\equiv D(A^{\frac{1}{2}})\to H$ is a map such that $N=N_1+N_2$ where $N_2:H_0\to H$ is compact and $N_1:H_0\to H$ is monotone (i.e., $(N_1u-N_1v,u-v)\geq 0$ $\forall u,v\in H_0$), where H_0 is the completion of $D(A)$ in the metric $[u,v]=(Au,v)$ and $||u||_0=(Au,u)^{\frac{1}{2}}$ for all $u,v\in D(A)$.

The space H_0 is continuously imbedded into H, $H_0=D(A^{\frac{1}{2}})$ and $||u||_0=||A^{\frac{1}{2}}u||$ for all u in H_0. It is known (see [8 ,27]) that, in virtue of (a1), the operator A has a bounded inverse A^{-1} defined on H, $A^{-1}:H\to H$ is a bounded, linear self-adjoint and positive operator such that A^{-1} is γ^{-1}-ball-contractive and its square root $S\equiv A^{-\frac{1}{2}}:H\to H$ is $\gamma^{\frac{1}{2}}$-ball-contractive. In view of this and the result in [27], it follows that $\tilde{B}\equiv SBS:H\to H$ is k-ball-contractive with $k=\alpha\gamma^{-1}$. Further, it follows from (a2) that $L\equiv I-\tilde{B}:H\to H$ is odd and $Lv\neq 0$ when $0\neq v\in H$. Now, it is also known that $A^{\frac{1}{2}}$, considered as a map from H_0 to H, is a linear homeomorphism. This and (a3) imply that $F\equiv SNS:H\to H$ is such that $F=F_1+F_2$ with $F_2=SN_2S$ compact and $F_1=SN_1S$ monotone.

Remark 2.1. It is easy to see that $u\in D(A)$ is a solution of (2.1) if and only if $v=A^{\frac{1}{2}}u$ and v is a solution of

$$Lv+Fv=S(f) \qquad (v\in H,\ S(f)\in H). \tag{2.2}$$

We are now in the situation where we can apply to (2.1), or equivalently to (2.2), some of the results of Section 1. To accomplish this we set $Y=X=H$, $\Gamma=\Gamma_1=\{X_n,P_n\}$, $T=L+F$, $G=L$ and choose $K=L$. We distinguish between two cases.

(i) In addition to (a2) and (a3) we assume that $k=\alpha\gamma^{-1}<2^{-1}$, $N_1:H_0\to H$ is bounded and continuous and $\{X_n\}$ is nested.

It follows from the preceding discussion and the additional assumption (i) that F_1 is bounded, continuous and monotone and $\tilde{B}$ is k-ball-contractive with $k<\frac{1}{2}$. Hence $L+tF$ is A-proper w.r.t. Γ_1 for each $t\geq 0$ as was shown in Example 3. This implies that $T+\mu L$ is A-proper for each $\mu>0$. In view of this, our choice L for G satisfies (H1) and (H2) of Theorem 1.1 and thus the following general result is valid for Eq. (2.1).

Theorem 2.1. If conditions (a1)-(a3) and (i) hold, then (2.1) has a solution $u\varepsilon D(A)$ for each f in H provided

$$(2.3)\qquad ||Tv||+\frac{(Tv,Lv)}{||Lv||}\to\infty \quad \text{as } ||v||\to\infty \quad (v\varepsilon H).$$

In general it is difficult to verify (2.3). However, since (2.3) is implied by the condition

$$(2.4)\qquad ||Lv||-||Fv||\to\infty \quad \text{as } ||v||\to\infty \quad (v\varepsilon H),$$

it follows that if we assume additionally that B is 1-homogeneous, then (2.4) is implied by the following verifiable hypothesis

$$(2.5)\qquad \limsup_{||u||_0\to\infty}\{||Nu||/||u||_0\}<m||S||^{-1} \qquad (u\varepsilon H_0),$$

where $||S||$ is the norm of $S:H\to H$ and $m=\inf\{||Ly||:||y||=1,\ y\varepsilon H\}$. First note that, since $Lv\neq 0$ for $v\neq 0$ and $L:H\to H$ is continuous and A-proper and thus proper by Proposition 1.1c in [21], it follows that $m>0$. Since $Sv\varepsilon H_0$ for each v in H with $||Sv||_0=||A^{\frac{1}{2}}Su||=||v||$ and $||Fu||\leq||S||\,||N(Su)||$ we see that $||Sv||_0\to\infty$ as $||v||\to\infty$ and

$$(2.6)\qquad ||Lv||-||Fu||\geq||Sv||_0\{m-||S||\,||N(Sv)||/||Sv||_0\} \text{ for all } v \text{ in } H.$$

The inequality (2.6) and the assumption (2.5) imply (2.4) and thus we deduce from Theorem 2.1 the following corollary.

Corollary 2.1. If A, B and N satisfy the conditions (a1), (a2), (a3), (i) and (2.5), then (2.1) has a solution $u\varepsilon D(A)$ for each f in H.

If in (a2) it is assumed that B is linear, then in virtue of Example 4 and Proposition 1.2 we obtain the following result without the condition that $N_1:H_0 \to H$ is bounded and continuous and that $\{X_n\}$ is nested.

Theorem 2.2. Suppose that A, B and N satisfy conditions (a1), (a2), and (a3) with B linear. Suppose further that

(ii) $k=\alpha\gamma^{-1}<1$ and $N_1:H_0 \to H$ is demicontinuous.

Then (2.1) has a solution $u\in D(A)$ for each f in H provided

$$\frac{||Nu||}{||u||_0} \to 0 \text{ as } ||u||_0 \to \infty \qquad (u\in H_0). \tag{2.7}$$

Proof. Since $L\in L(H,H)$ is injective and A-proper and $L+F:H \to H$ is also A-proper, in virtue of Proposition 1.2, it suffices to show that (2.7) implies that

$$||Fv||/||v|| \to 0 \text{ as } ||v|| \to \infty \qquad (v\in H). \tag{2.8}$$

Now, since $||Fv||\leq||S||\,||N(Sv)||$ for each v in H, to prove (2.8) it suffices to show that $||N(Sv)||/||v|| \to 0$ as $||v|| \to \infty$. But, as above, we note that since $u=Sv \in H_0$ and $|u|_0=||A^{\frac{1}{2}}Sv||=||v||$ it follows from (2.7) that

$$\frac{||N(Sv)||}{||v||} = \frac{||Nu||}{||u||_0} \to 0 \text{ as } ||v||=||u||_0 \to \infty.$$

Q.E.D.

Remark 2.2. If we know that, for some f in H, (2.1) has at most one solution, then (2.1) has a unique solution $u_0\in D(A)$ by Theorems 2.1 and 2.2 (or Corollary 2.1) and in this case u_0 can actually be obtained as a strong limit in H_0-norm of a sequence of Galerkin approximates $u_n\in X_n$ if we choose $\{X_n\}\subset D(A)$ such that $\{Y_n\}=\{A^{\frac{1}{2}}X_n\}$ and the orthogonal projections $P_n:H \to Y_n$ form a projectionally complete scheme $\{X_n, P_n\}$ for (H,H) and determine u_n as a solution of the system of the nonlinear algebraic equations

(2.8) $$(Au_n - Bu_n + Nu_n, u) = (f,u) \quad \text{for all} \quad u \in X_n .$$

It is easy to see that $u_n \in X_n$ is a solution of (2.8) if and only if $v_n = A^{\frac{1}{2}} u_n \in Y_n$ is a solution of

(2.9) $$P_n L v_n + P_n F v_n = P_n S f .$$

Assuming uniqueness for a given f in H, it follows from the proof of Propositions 1.1 and 1.2 that there exists $r_f \in N$ such that (2.9) has a solution $v_n \in Y_n$ for each $n \geq n_f$, $v_n \to v_0$ in H and $Lv_0 + Fv_0 = Sf$. But then $u_0 = Su_0$, $v_n = Su_n$ and $Su_n \to Su_0$ in H. Hence $u_n = A^{-\frac{1}{2}}(v_n) \to A^{-\frac{1}{2}}(v_0) \equiv u_0$ in H_0 since $A^{-\frac{1}{2}}$ is a continuous map from H onto H_0.

The above discussion implies the validity of the following

Theorem 2.3. *Suppose* $\{X_n\} \subset D(A)$ *is such that* $\{A^{\frac{1}{2}} X_n, P_n\}$ *is projectionally complete for* (H,H). *If in Theorems 2.1 or 2.2 we also assume that* $A-B+N : D(A)\ H \to H$ *is injective, then to each* f *in* H *there corresponds* $r_f \in N$ *such that (2.8) has a solution* $u_n \in X_n$ *for each* $n \geq n_f$, $u_n \to u_0$ *in* H_0, $u_0 \in D(A)$ *and* u_0 *is the unique solution of (2.1).*

Remark 2.3. It is especially easy to apply Theorems 2.2 and 2.3 to equations of the form

(2.10) $$Au - \lambda u + Nu = f \qquad (u \in D(A),\ f \in H,\ \lambda \in \mathbb{R})$$

provided λ is not an eigenvalue of A and $|\lambda|\gamma^{-1} < 1$.

(P2) Following [3] we call $F:X \to X$ <u>semicontractive with constant $k>0$</u> if there is a continuous map $V:X\times X \to X$ such that $F(x)=V(x,x)$ for $x\in X$, $||V(x,z)-V(y,z)||\le k||x-y||$ for $x,y,z\in X$, and for each z in X the map $V(z,\cdot):X \to X$ is compact. To indicate further examples of equations to which only the results of Section 1 are applicable we consider the solvability problem for the following equation:

$$u-Fu+Nu+Cu=f \qquad (u,f\in H), \tag{2.11}$$

where F, C, $N:H \to H$ satisfy the following conditions

(b1) F is semicontractive with $k<2^{-1}$ and $||V(0,u)||/||u|| \to 0$ as $||u|| \to \infty$.

(b2) N is bounded, continuous, monotone and

$$|N| = \limsup_{||u||\to\infty} \{||Nu||/||u||\} < 1-k.$$

(b3) C is compact and either $(Cu,u)\ge 0$ $\forall u\in H$ or $||Cu||/||u|| \to 0$ as $||u|| \to \infty$.

<u>Theorem 2.4.</u> <u>Suppose</u> $\Gamma_1=\{X_n,P_n\}$ <u>is complete for</u> H <u>with</u> $\{X_n\}$ <u>nested and conditions</u> (b1)-(b3) <u>hold. Then</u> (2.11) <u>has a solution</u> $u\in H$ <u>for each</u> f <u>in</u> H. <u>If</u> $T\equiv I-F+N+C$ <u>is also injective, then</u> u <u>can be constructed by the Galerkin method.</u>

<u>Proof.</u> It was shown in [30] that when F is semicontractive with $k\in(0,1)$, then F is ball-condensing. The result was improved in [23], where it was shown that, under the same conditions, F is actually k-ball-contractive. In view of this and (b1)-(b3), Lemma 2.7 in [28] implies that T is A-proper w.r.t. Γ_1. Now it follows from (b2) that to each $\varepsilon>0$ there exist an $r>0$ such that $||Nu||\le(|N|+\varepsilon)||u||$ for $||u||\ge r$. Since $||Fu||\le k||u||+||V(0,u)||+2||F(0)||$ for $u\in H$,

it follows from the above that for all u in H we have

$$\frac{(Tu,u)}{||u||} = ||u|| - \frac{(Fu,u)}{||u||} + \frac{(Nu,u)}{||u||} + \frac{(Cu,u)}{||u||} \geq \tag{2.12}$$
$$||u||\{1-k-\frac{||V(0,u)||}{||u||} - \frac{2||F(0)||}{||u||} - \frac{||Nu||}{||u||} + \frac{(Cu,u)}{||u||}.$$

Suppose first that $(Cu,u)\geq 0$ for all u in H. Then it follows from (2.12) and (b1)-(b2) that there exists an $r_1>r$ such that

$$\frac{(Tu,u)}{||u||} \geq ||u||\{1-k-|N|-4\varepsilon\} \quad \text{for} \quad ||u||\geq r_1.$$

Since $\varepsilon>0$ was arbitrary and $1-k-|N|>0$, it follows that $(Tu,u)/||u|| \to \infty$ as $||u|| \to \infty$. The same can be easily shown to be the case when $||Cu||/||u|| \to 0$ as $||u|| \to \infty$. Thus Theorem 2.4 follows from Proposition 1.1. Q.E.D.

In this part of Section 2 we show how the results obtained above are applicable to nonlinear ordinary and partial differential equations involving functions defined on bounded and unbounded domains. To define the problems we first introduce some notation and definitions (see [2 ,12,21]).

Let Q be a domain in R^n, $n\geq 1$, with sufficiently smooth boundary ∂Q. For any fixed $p\varepsilon(1,\infty)$, let $L_p=L_p(Q)$ denote the usual Banach space of real-valued "functions" $u(x)$ on Q with norm $||u||_p$. We use the standard notation for the derivatives $D^\alpha=(\partial/\partial x_n)^{\alpha_n}\dots(\partial/\partial x_n)^{\alpha_n}$, where $\alpha=(\alpha_1,\dots,\alpha_n)$ is a multiindex of nonnegative integers with order of D^α being written as $|\alpha|=\alpha_1+\dots+\alpha_n$. If $m\varepsilon N$, we consider the Sobolev space $W_p^m(Q)=\{u:D^\alpha u\varepsilon L_p(Q)$ for all $|\alpha|\leq m\}$. W_p^m is a separable reflexive Banach space with respect to the norm

$$||u||_{m,p} = (\sum_{|\alpha|\le m} ||D^{\alpha}u||_p^b)^{1/p}.$$

In case $p=2$ we get the Hilbert space W_2^m. Let $C_c^\infty(Q)$ be the family of infinitely differentiable functions with compact support in Q considered as a subset of W_p^m and let $\mathring{W}_p^m$ be the completion in W_p^m of $C_c^\infty(Q)$. Let $<u,v>=\int_Q uvdx$ denote the natural pairing between $u\varepsilon L_p$ and $v\varepsilon L_q$ with $q=p(p-1)^{-1}$.

(P3) Letting $Q=(0,\infty)$, $L_2=L_2(0,\infty)$ and $Du=u(x)'$, we first consider the one-dimensional problem of the form

(I) $-(p(x)u')'+q(x)u-g(x,u)+h(x,u)=f(x) \quad (x\varepsilon(0,\infty),u(0)=0),$

where $f\varepsilon L_2$ and the functions p,q, q and h satisfy the following conditions:

(c1) $p:[0,\infty)\to\mathbb{R}$ is continuous, $p\varepsilon C^1(0,\infty)$, p' is bounded, and $0<a_0\le p(x)\le b_0$ for all $x\ge 0$.

(c2) $q:[0,\infty)\to\mathbb{R}$ is continuous with $\liminf_{x\to\infty} q(x)=\gamma>0$ and $0<a_1\le q(x)\le b_1$ for all $x\ge 0$.

(c3) $g:[0,\infty)\times\mathbb{R}\to\mathbb{R}$ is continuous, $|g(x,\xi_1)-g(x,\xi_2)|\le\alpha|\xi_1-\xi_2|$ for all $\xi_1,\xi_2\varepsilon\mathbb{R}$ and some $\alpha>0$, $g(x,0)\varepsilon L_2$ and $g(x,-\xi)=-g(x,\xi)$ and $g(x,t\xi)=tg(x,\xi)$ for all $x\ge 0$ and $\xi\varepsilon\mathbb{R}$ and $t>0$.

(c4) $h(x,u)=h_1(x,u)+c(x)h_2(x,u)$ with h_1,h_2 and c such that:

(i) $h_1:[0,\infty)\times\mathbb{R}\to\mathbb{R}$ is continuous, $(h_1(x,\xi_1)-h_1(x,\xi_2))(\xi_1-\xi_2)\ge 0$ and $|h_1(x,\xi)|\le a(x)+\beta|\xi|^\delta$ for all $x\ge 0$; all $\xi_1,\xi_2,\xi\varepsilon\mathbb{R}$; some $a(x)\varepsilon L_2$, $\beta>0$ and $\delta\varepsilon(0,1]$.

(ii) $h_2:(0,\infty)\times\mathbb{R}\to\mathbb{R}$ is continuous, $|h_2(x,\xi)|\le\eta$ for all $x\ge 0$ and $\xi\varepsilon\mathbb{R}$ and $c(x)$ is continuous, $c(x)\varepsilon L_2$ and $c(x)\to 0$ as $x\to\infty$.

We shall now show that, in view of the conditions (c1)-(c4), Eq. (I) can be replaced by the corresponding operator equation to which Corollary 2.1 and Theorem 2.3 are applicable.

Let $A(u)(x)=-(p(x)u')'+q(x)u(x)$ for $u\varepsilon D(A)=C_c^\infty(Q)$. In view of the results in [8 ,27], conditions (c1) and (c2) imply that A has a unique self-adjoint extension A in L_2 with $D(A)=\overset{\circ}{W}{}_2^1\cap W_2^2$, A^{-1} exists and is a bounded self-adjoint and positive operator from all of H into H, A^{-1} is γ^{-1}-set-contraction, $S=A^{\frac{1}{2}}:H\to H$ is $\gamma^{-\frac{1}{2}}$-set-contraction, and $A^{\frac{1}{2}}$ is a homeomorphism of $D(A^{\frac{1}{2}})=\overset{\circ}{W}{}_2^1$ onto L_2. It is known (see [7 ,27]) that if $u\varepsilon\overset{\circ}{W}{}_2^1$, then u is continuous on $[0,\infty)$, $u(x)\to 0$ as $x\to\infty$, $\max_{x\geq 0}|u(x)|\leq||u||_{1,2}$ and $||u||_p\leq||u||_{1.2}$ for all $p\geq 2$. Moreover, the norm $||u||_0=||A^{\frac{1}{2}}u||_2$ for all $u\varepsilon\overset{\circ}{W}{}_2^1$ is equivalent with the norm in $\overset{\circ}{W}{}_2^1$. We denote $\overset{\circ}{W}{}_2^1$ by H_0 if $||\cdot||_0$ is used.

In order to discuss the nonlinear terms in I, let $B(u)(x)=g(x,u(x))$ $N_1(u)(x)=h_1(x,u(x))$, and $N_2(u)(x)=c(x)h_2(x,u(x))$ for $x\geq 0$ and $u:[0,\infty)\to\mathbb{R}$. It follows from (c3) and the results in [29] that $B:L_2\to L_2$ is continuous, odd, 1-homogeneous and $||Bu-Bv||_2\leq\alpha||u-v||$ for all $u,v\varepsilon L_2$. The last relation also shows that B is α-ball-contraction. The same arguments show that if (i) of (c4) holds, then N_1 is a continuous, bounded and monotone map of L_2 into L_2 such that $||N_1u||_2\leq||a||_2+\beta||u||_2^\delta$ for all u in L_2. Moreover, since $\overset{\circ}{W}{}_2^1$ (or H_0) is continuously imbedded into L_2, it follows that N_1 is bounded, continuous and monotone as a mapping from $\overset{\circ}{W}{}_2^1$ to L_2. Finally, condition (ii) of (c4) imply that if $u\varepsilon L_2$, then $||N_2u||_2\leq\eta||c||_2$ so that N_2 has a bounded range and is continuous. Moreover, considered as a map from $\overset{\circ}{W}{}_2^1$ to L_2,

N_2 is compact. It is obvious that $N_2:\overset{\circ}{W}{}^1_2 \to L_2$ is continuous. Thus to establish the claim, it remains to show that if $\{u_j\}$ is a bounded sequence in $\overset{\circ}{W}{}^1_2$, then $\{N_2(u_j)\}$ contains a subsequence which converges in L_2. Since, by the imbedding theorem, $N_2:\overset{\circ}{W}{}^1_2(0,n) \to L_2(0,n)$ is compact for each integer $n\varepsilon N$, N_2 has a bounded range and $\lim\limits_{x\to\infty} c(x)=0$, a diagonalization process similar to that used in [5] shows that one can extract a subsequence $\{u_i\}$ of $\{u_j\}$ so that $\{N_2(u_i)\}$ converges in $L_2(0,\infty)$. Note also that $||N_2u||_2/||u||_2 \to 0$ as $||u||_2 \to \infty$.

In view of Corollary 2.1, the above discussion implies the validity of the following theorem for Eq. (I).

Theorem 2.5. *Suppose that the functions* p, q, g *and* h *satisfy the conditions* (c1)-(c4). *Suppose further that* $Au-g(x,u)\neq 0$ *for* $u\varepsilon D(A)$ *with* $u\neq 0$ *and* $\alpha\gamma^{-1}<\frac{1}{2}$. *Then the equation*

$$Au-Bu+N_1u+N_2u=f \qquad (u\varepsilon D(A),\ f\varepsilon L_2) \tag{2.13}$$

has a solution $u\varepsilon D(A)=W^2_2 \cap \overset{\circ}{W}{}^1_2$ *for each* $f\varepsilon L_2$ *provided* $\beta<m||S||^{-1}\eta_1$ *in case* $\delta=1$, *where* $\eta_1=\min\{\sqrt{a_0},\sqrt{a_1}\}$ *and* $m=\inf\{||Ly||_2:||y||_2=1;\ y\varepsilon L_2\}$ *with* $L=I-SBS$ *and no additional provision is necessary if* $\delta\varepsilon(0,1)$.

Remark 2.4. If $N_1(u)=\lambda u$, $\lambda\varepsilon \mathbb{R}$, then it follows from Theorem 2.3 and Remark 2.3, that (2.13) has a solution $u\varepsilon D(A)$ for each f in L_2 provided that $Au-\lambda u=0$ has no nontrivial solutions in $D(A)$ and $|\lambda|\gamma^{-1}$ 1.

(P4) Let $Q\ \mathbb{R}^n$ be a bounded domain with ∂Q of class C^{2m} and let

$$(2.14)\qquad Lu=\sum_{|\alpha|,|\beta|\le m}(-1)^{|\beta|}D^{\beta}(a_{\alpha\beta}(x)D^{\alpha}u)\qquad (x\varepsilon Q),$$

where the following assumptions are made:

(d1) $a_{\alpha\beta}(x)\varepsilon C^{|\beta|}(\bar{Q})$ for all $0\le|\alpha|,|\beta|\le m$

(d2) L is uniformly and strongly elliptic, i.e., there is $\eta_0>0$ with

$$(2.15)\qquad \sum_{|\alpha|=|\beta|=m}a_{\alpha\beta}(x)\xi^{\alpha}\xi^{\beta}\ge\eta_0|\xi|^{2m}\qquad \forall\xi\varepsilon\mathbb{R}^n,\ \ x\varepsilon\bar{Q}.$$

The Dirichlet form associated with L is given by

$$(2.16)\qquad a(u,v)=\sum_{|\alpha|,|\beta|\le m}\int_Q a_{\alpha\beta}(x)D^{\alpha}uD^{\beta}v\,dx.$$

which, in view of (d1), is a bounded bilinear form on $\mathring{W}_2^m$. By Gårding inequality, there are constants $c_0>0$ and k_0 such that

$$(2.17)\qquad a(u,u)\ge c_0||u||_{m,2}^2-k_0||u||_{0,2}^2\qquad \text{for } u\varepsilon\mathring{W}_2^m .$$

It is known (see [12]) that when (d1)-(d2) hold, then the Fredholm alternative implies that one and only one of the following possibilities is true:

(i) For each f in L_2 there exists a unique $u\varepsilon\mathring{W}_2^m$ such that $a(u,v)=\langle f,v\rangle$ for all v in $\mathring{W}_2^m$, or

(ii) there is $0\neq u\varepsilon\mathring{W}_2^m$ such that $a(u,v)=0$ for all v in $\mathring{W}_2^m$.

Furthermore, when (i) holds and ∂Q is of class c^{2m}, the solution operator $A:L_2\to\mathring{W}_2^m$, defined by $Af=u$, is a linear homeomorphism of L_2 onto $H_0\equiv\mathring{W}_2^m\cap W_2^{2m}$. Thus, if we define $G:H_0\to L_2$ by $G=A^{-1}$, then there exists a constant $\eta_1>0$ such that $||Gu||_{0,2}\ge\eta_1||u||_{2,2m}$ for all $u\varepsilon H_0$.

Semilinear Dirichlet BV Problem. Let $N(x,\eta)$ be a function defined on $Q\times\mathbb{R}^M$, where M is the number of all derivatives D^α for $0\leq|\alpha|\leq 2m-1$, such that

(e1) $N: Q\times\mathbb{R}^M \to \mathbb{R}$ satisfies the Caratheodary's conditions

(e2) There are constants $b_\alpha\geq 0$ and $b(x)\in L_2$ such that

(2.18) $$|N(x,\eta)|\leq \sum_{|\alpha|\leq 2m-1} b_\alpha|\eta_\alpha|+b \quad \text{for } \eta\in\mathbb{R}^M \text{ and } x\in Q(\text{a.e.}).$$

It follows from [12] (see also [11]) that, in view of (e1)-(e2), the Nemytskii operator $Nu=N(x,D^\gamma u(x))$, $|\gamma|\leq 2m-1$, is bounded and continuous from W_2^{2m-1} to L_2. Since W_2^{2m} is compactly imbedded in W_2^{2m-1} it follows that $N: W_2^{2m}\to L_2$ is compact.

The generalized Dirichlet BV Problem for

(2.19) $$Lu+N(x,D^\gamma u)=f \qquad (f\in L_2, |\gamma|\leq 2m-1)$$

consists in finding $u\in H_0\equiv \overset{\circ}{W}{}_2^m\cap W_2^{2m}$ such that

(2.20) $$a(u,v)+\langle Nu,v\rangle=\langle f,v\rangle \quad \text{for all } v \text{ in } \overset{\circ}{W}{}_2^m.$$

We now use Proposition 1.1 to establish the following result for the BV Problem (2.20) which extends Theorem 2 (and also Theorem 1) of [11].

Theorem 2.6. Suppose that ∂Q is of class C^{2m}, L satisfies (d1)-(d2) and part (i) of the Fredholm alternative holds. Assume that the function $N(x,\eta)$ satisfies (e1)-(e2) and $T=G+N$ is such that

(2.21) $$||Tu||+(Tu,Gu)/||Gu||_{2,0} \to \infty \text{ as } ||u||_{2,2m}\to\infty.$$

Then, for each f in L_2, Eq. (2.20) has a solution u in

$\overset{\circ}{W}{}_2^m \cap W_2^{2m}$.

Proof. In view of the preceding discussion, the problem of solving (2.20) is equivalent to finding $u \varepsilon H_0$ such that

(2.22) $$Tu=f.$$

Now, let $\{X_n\} \subset H_0$ be a sequence of finite dimensional spaces such that $\text{dist }(u,X_n)=\inf(\{||u-v||_{2,2m}: v\varepsilon X_n\} \to 0$ as $n \to \infty$ for each $u\varepsilon H$, let $Y_n=G(X_n) \subset L_2$ for each n, and let $P_n: H_0 \to X_n$ and $Q_n: L_2 \to Y_n$, be orthogonal projections. Then $\Gamma_p=\{X_n,P_n;Y_n,Q_n\}$ is projectionally complete for (H_0,L_2) and $G:H_0 \to L_2$ is A-proper w.r.t. Γ_p, odd and $||Gu||_{0,2} \to \infty$ as $||u||_{2,2m} \to \infty$. Since $N:H_0 \to L_2$ is compact, it follows that $T=G+N$ is A-proper and $T+\mu G$ is A-proper w.r.t. Γ_p for each $\mu>0$. If in Proposition 1.1 we set $X=H_0$, $Y=L_2$, $\Gamma=\Gamma_p$ and $G=K$, we see that conditions (H1), (H2) and (H5) hold and so Theorem 2.6 follows from (A1) of Proposition 1.1.

Q.E.D.

Remark 2.5. It is obvious that (2.21) is implied by

(2.23) $$||Gu||_{2,0}-||Nu||_{2,0} \to \infty \quad \text{as} \quad ||u||_{2,2m} \to \infty. \quad (u\varepsilon H_0).$$

Condition (2.23) is in turn implied by the hypothesis.

(2.24) $$\limsup_{||u||_{2,2m}\to\infty} \{||Nu||_{2,0}/||u||_{2,2m}\}<\eta_1,$$

where $\eta_1 \varepsilon \mathbb{R}^+$ is the largest for which $||Gu||_{2,0} \geq \eta_1 ||u||_{2,2m}$ for $u\varepsilon H_0$. Note that although the nature of G is not always known in advance, in many cases it is possible to estimate the value of η_1 since it is essentially equal to $||A||^{-1}_{L_2 \to H_0}$. If G is known explicitly, which is the case for example when $k_0=0$, then one actually construct $u_n \varepsilon X_n$ such that $u_n \to u$ in W_2^{2m}-norm if u is the only solution of (2.20). That is, in this case one has the convergence of the Galerkin method in W_2^{2m}.

The condition (2.23) holds if the hypothesis (F-3) used by De Figueiredo in his Theorem 2 [11] holds, where

(F-3): There are constants $0 \le c' \le ||A||^{-1}_{L_2 \to L_2} \equiv \alpha^{-1}$, $0 \le k < 1$, $c_\alpha \ge 0$ for $|\alpha| \le 2m-1$ and $b(x) \varepsilon L_2$ such that

$$|N(x,\eta)| \le c'|\eta_0| + \sum_{|\alpha| \le 2m-1} c_\alpha |\eta_\alpha|^k + b(x) \quad \text{for } \eta \varepsilon \mathbb{R}^M, \ x \varepsilon Q(\text{a.e.}).$$

It was shown in [11] that, in view of (F-3),

$$(2.25) \qquad ||Nu||_{2,0} \le c'\alpha ||Gu||_{2,0} + c''||u||^k_{2,2m} + ||b||_{2,0} \qquad (u \varepsilon H_0),$$

where c'' depends on k and c_α's. Since $c'\alpha < 1$ it follows from (2.25) that $||Gu||_{2,0} - ||Nu||_{2,0} \ge (1-c'\alpha)||Gu||_{2,0} - c''||u||^k_{2,2m} - ||b||_{2,0} \ge$

$$(1-c'\alpha)\eta_1 ||u||_{2,2m} - c''||u||^k_{2,2m} - ||b||_{2,0}$$

for all $u \varepsilon H_0$. The last inequality implies (2.23) and thus Theorem 2 (and Theorem 1) in [11] follow from Theorem 2.6 and Remark 2.5.

(P5) As our final example we consider the following result which is originally due to Wille [32]. We present it here to indicate the generality of our results and also because its proof, based on Theorem 1.3, seems to be simpler. Consider the BV Problem:

$$Tu = -(\text{arctg}\,u')' - u''(x) - \frac{1}{\pi} \int_0^\pi \Big\{\frac{(x-t)^4}{a\pi^4} u(t) - 2[\cos(nt)(k \sin(kx) - n\sin(nx))] - \cos(kt)(n \sin(nx) + k \sin(kx)]u'(t)\Big\}dt, \quad u(0) = u(\pi) = 0,$$

where n and k are positive integers with $n \ne k$. Let $\bar{Q} = [0,\pi]$, $W_2^1 = W_2^1([0,\pi])$ and $\mathring{W}_2^1 = \mathring{W}_2^1([0,\pi])$. Let $A,B,C: \mathring{W}_2^1 \to \mathring{W}_2^1$ be defined by

$$(Au,v)=\int_0^\pi \operatorname{arctg}(u')v\,dx, \qquad (Bu,v)=\int_0^\pi \{u'(x)-\int_0^\pi k(x,t)u'vdt\}dx$$

with $K(x,t)=\frac{2}{\pi}\int_0^x [\cos(nt)(k\sin(ks)-n\sin(ns)-\cos(kt)(n\sin(ns)+ k\sin(ks)]ds, (Cu,v)=-\frac{1}{9\pi^5}\int_0^\pi\int_0^\pi (x-t)^4 u(t)v(x)dtdx$.

Clearly, C and B are linear, C is compact and $(Bu,u)\geq 0$ for $u\in \overset{\circ}{W}{}_2^1$ (see [32]). Further A is nonlinear and monotone since

$$(Au-Av,u-v)=\int_0^\pi (\operatorname{arctg}u'-\operatorname{arctg}v')(u'-v')dx\geq 0 \quad \forall u,v\in \overset{\circ}{W}{}_2^1.$$

Thus, $T=A+C+B:\overset{\circ}{W}{}_2^1 \to \overset{\circ}{W}{}_2^1$ is quasimonotone, demicontinuous and bounded. It was shown in [32] that T is not coercive but, $||Bu||\geq\frac{1}{2}||u||$, $||Au||\leq 2^{-\frac{1}{2}}\pi^{3/2}$ and $||Cu||\leq 9^{-1}||u||$ for all u in $\overset{\circ}{W}{}_2^1$.

Since $(Bu,u)\geq 0$ and $(Au,u)\geq 0$ for u in $\overset{\circ}{W}{}_2^1$, it follows from the above estimates that for all u in $\overset{\circ}{W}{}_2^1$ we have

$$||Tu||+\frac{(Tu,u)}{||u||}\geq ||Tu||+\frac{(Cu,u)}{||u||}\geq ||Bu||-||Au||-2||Cu||\geq$$
$$\frac{1}{2}||u||-\frac{\pi^{3/2}}{2}-\frac{2}{9}||u|| \to \infty \quad \text{as} \quad ||u|| \to \infty.$$

To apply Theorem 1.3 to the weak solvability of (II), i.e., to

$$(2.26) \qquad Tu=f_w, \text{ where } w_f\in \overset{\circ}{W}{}_2^1 \text{ is such that } \langle f,v\rangle=(w_f,v) \quad v\in \overset{\circ}{W}{}_2^1,$$

we set $Y=X=\overset{\circ}{W}{}_2^1$, $G=K=I$, $\Gamma_1=\{X_n,P_n\}$, and observe that $T+\mu I$ is A-proper with respect to Γ_1 for each $\mu>0$ and $T(B)$ is closed for each closed ball B in $\overset{\circ}{W}{}_2^1$. To prove the last assertion, let $\{u_j\}\subset B$ be such that $Tu_j \to g$ in $\overset{\circ}{W}{}_2^1$ and, without loss of generality assume that $u_j \rightharpoonup u_0$ in $\overset{\circ}{W}{}_2^1$. Then $Cu_j \to Cu_0$ and so $(A+B)(u_j) \to g-Cu_0$ in $\overset{\circ}{W}{}_2^1$. Since $((A+B)(u)-(A+B)(u_j),u-u_j)\geq 0$ for all $u\in \overset{\circ}{W}{}_2^1$ and each j, the passage to the limit in the above inequality shows that $((A+B)(u)-(g-Cu_0),u-u_0)\geq 0$ for all $u\in\overset{\circ}{W}{}_2^1$. This implies that $(A+B)(u_0)=g-C(u_o)$ (see [18]), i.e. $T(u_0)=g$. Thus, by the special case of Theorem 1.3, Eq. (2.26) has a solution for each f_w in $\overset{\circ}{W}{}_2^1$ which is the result obtained in [32] by other and more complicated methods. Q.E.D.

Acknowledgement. The writer is indebted to Professor Stuart Antman for having made Toland's report [28] available to him.

References

(1) H. Brezis, Équations et in équations non-lineares dans les espaces en dualité, Ann. Inst. Fourier, Grenoble, 18 (1968), 115-175.

(2) F.E. Browder, Existence theory for boundary value problems for quasilinear elliptic systems with strongly nonlinear lower order terms, Proc. Symp. in Pure Math. 23 (1973), 269-286.

(3) ____________, Nonlinear operators and nonlinear equations of evolution in Banach spaces, Proc. Symp. in Pure Math, AMS, Vol. 18, Part 2 (1976).

(4) F.E. Browder and W.V. Petryshyn, Approximation methods and the generalized topological degree for nonlinear maps in Banach spaces, J. Functional Anal., 3 (1969), 217-245.

(5) K.J. Brown, Some operator equations with an infinite number of solutions, Quart. J. Math (Oxford) (2), 25 (1974), 195-212.

(6) B. Calvert and J.R.L. Webb, An existence theorem for quasimonotone operators, Accad. Naz. Dei Lincei, Ser. 8, 50 (1971), 362-368.

(7) E.N. Dancer, Boundary-value problems for ordinary differential equations on infinite intervals, Proc. London Math. Soc. 30 (1975), 76-94.

(8) N. Dunford and J.T. Schwartz, Linear Operators, Part II; Interscience, New York, 1963.

(9) P.M. Fitzpatrick, Surjectivity results for nonlinear mappings from a Banach space to its dual, Math. Ann. 204 (1973), 177-188.

(10) P.M. Fitzpatrick and W.V. Petryshyn, Positive eigenvalues for nonlinear multivalued noncompact operators with applications to differential operators, J. Differential Equations, 22 (1976), 428-441.

(11) D.G. De Figueiredo, The Dirichlet problem for nonlinear elliptic equations: A Hilbert space approach,

(12) A. Friedman, Partial Differential Equations, Holt, Rinehart and Winston, New York, 1969.

(13) P. Hess, On nonlinear mappings of monotone type homotopic to odd operators, J. Functional Anal., 11 (1972), 138-167.

(14) P.M. Hess, Théorème d'existence pour des perturbations d'opérateurs maximaux monotones, C.R. Acad. Sc. Paris, t.275 (1972), Ser. A, 1171-1173.

(15) J. Leray and J.-L. Lions, Quelques résultats de Višik sur les problemes elliptiques non linéaires par les méthodes de Minty-Browder, Bull. Soc. Math. Frances 93 (1965), 97-107.

(16) P.S. Milojevic and W.V. Petryshyn, Continuation and surjectivity theorems for uniform limits of A-proper mappings with applications, J. Math. Anal. Appl. (to appear).

(17) ______________, Continuation theorems and the approximation-solvability of equations involving multivalued A-proper mappings, J. Math. Anal. Appl. (to appear).

(18) G. Minty, Monotone (nonlinear) operators in Hilbert space, Duke Math. J. 29 (1962), 341-346.

(19) R.D. Nussbaum, The fixed point index for local condensing maps, Ann. Math. Pura Appl. (4) 89 (1971), 217-258.

(20) W.V. Petryshyn, On nonlinear equations involving pseudo-A-proper mappings and their uniform limits with applications, J. Math. Anal. Appl. 38 (1972), 672-720.

(21) ______________, On the approximation-solvability of equations involving A-proper and pseudo-A-proper mappings, Bull. Amer. Math. Soc., 81 (1975), 223-312.

(22) ______________, On the relationship of A-properness to mappings of monotone type with applications to elliptic equations, Fixed Point Theory and its Application (ed. S. Swaminathan), Academic Press, N.Y.), 1976, 149-174.

(23) ______________, Fixed point theorems for various classes of 1-set-contractive and 1-ball-contractive mappings in Banach spaces, Trans. Amer. Math. Soc. 182 (1973), 323-352.

(24) ______________, Fredholm alternative for nonlinear A-proper mappings with applications to nonlinear elliptic boundary value problems, J. Functional Anal. 18 (1975), 288-317.

(25) W.V. Petryshyn and P.M. Fitzpatrick, On 1-set and 1-ball contractions with applications to perturbation problems for nonlinear bijective maps and linear Fredholm maps, Bull. UMI, (4) 7 (1973), 102-124.

(26) B.N. Sadovskii, Ultimately compact and condensing mappings, Uspehi Mat. Nauk 27 (1972), 81-146.

(27) C.A. Stuart, Some bifurcation theory for k-set-contractions, Proc. London Math. Soc. 27 (1973), 531-550.

(28) J.F. Toland, Global bifurcation theory via Galerkin method, University of Essex, Fluid Mechanics Reas. Inst. Report. No. 69 (1976).

(29) M.M. Vainberg, Variational Methods for the Study of Nonlinear Operators, Holden-Day, San Francisco, 1964.

(30) J.R.L. Webb, Mapping and fixed point theorems for nonlinear operators in Banach spaces, Proc. London Math. Soc. (3) 20 (1970), 451-468.

(31) J.R.L. Webb, On a characterization of k-set-contractions, Accad. Naz. Dei Lincei, Ser. 8, 50 (1971), 358-361.

(32) F. Wille, Monotone Operatoren mit Störungen, Arch. Rat. Mech. Anal. 46 (1971), 369-388.

Department of Mathematics
Rutgers University, New Brunswick, N. J.

APPENDIX

After this paper had been written, the author received from J.F. Toland a revised version of his paper [28] published in Nonlinear Analysis, Theory, Methods and Applications, Vol. 1, No. 3 (1977), 305-317. In the Appendix to this paper J.R.L. Webb gives a simpler proof of an improved version of Toland's Lemma 2.7 (i.e. Example 3) in the form:

Proposition A. If $A:H \to H$ is a continuous c-monotone and $B:H \to H$ is k-ball-contraction, then $I-B+A$ is A-proper provided $k-c<1$.

It turns out that Webb's proof can be used to prove the following extension of his result which will improve some of our results in Sections 1 and 2.

Proposition B. Let $\Gamma_1=\{X_n,P_n\}$ be a nested projectionally complete scheme for the Banach space X. Let $B:X \to X$ be k-ball-contractive and $A:X \to X$ continuous and c-accretive, i.e.,

(i) $(Ax-Ay,J(x-y))\geq c||x-y||^2$ for all $x,y\in X$ and some $c\geq 0$, where $J:X \to 2^{X^*}$ is the normalized duality map given by $Jx=\{w\in X^*:(w,x)=||x||^2,||w||=||x||\}$. Then $T=I-B+A$ is A-proper w.r.t. Γ_1 provided $k-c<1$. If B is ball-condensing, $I-B+A$ is A-proper for any $c\geq 0$.

Proof. As in [28] we first show that if $u_n \varepsilon X_n$, then $\chi(\{u_n+P_nAx_n\}) \geq (1+c)\chi(\{u_n\})$. Now since $||P_n||=1$ it is easy to see that $P_nJ(u) \subset J(x)$ for $u\varepsilon X_n$ and $||f||=||P_n^*f||=||u||$ whenever $f\varepsilon J(u)$. It follows from this and (i) that for every $z\varepsilon X_n$ and each f in $J(u_n-z)$ we have $(u_n+P_nAu_n-(z+Az),P_n^*f) \geq (1+c)||u_n-z||^2$.

Hence $||(u_n+P_nAu_n)-(z+Az)|| \geq (1+c)||u_n-z||$ (unless $||u_n-z||$). Now suppose that $\{u_n+P_nAu_n\}$ is covered by finitely many balls $B(v_j,r)$, $1 \leq j \leq \ell$. Since $I+A$ is continuous and $(1+c)$-accretive, it is an homeomorphism by Corollary 3 of Deimling (Manuscripta Math. 13 (1974, 365-374). Hence there are $w_j \varepsilon X$ such that $w_j+Aw_j=v_j$. By continuity of $I+A$ and completeness of $\{X_n\}$, there is $z_j \varepsilon X_{n_j}$ such that $||z_j+Az_j-v_j||<\varepsilon$ for $1 \leq j \leq \ell$. Then for those n such that $\{u_n+P_nAu_n\}$ belong to $B(v_j,r)$ and exceed n_j we have $(1+c)||u_n-z_j|| \leq ||(u_n+P_nAu_n)-(z_j+Az_j)|| \leq ||(u_n+P_nAu_n)-v_j||+\varepsilon \leq r+\varepsilon$. Thus $\{u_n | n \geq \max n_j\}$ is covered by balls $B(z_j,(r+\varepsilon)/(1+c))$. This shows that $\chi(\{u_n\}) \leq (1/(1+c))\chi(\{u_n+P_nAu_n\})$ since ε is arbitrary. To complete the proof, let $\{u_n | u_n \varepsilon X_n\}$ be a bounded sequence such that $u_n-P_nBu_n+P_nAu_n \equiv g_n \to g$ for some g in X. Then $u_n+P_nAu_n=g_n+P_nBu_n$ and so $(1+c)\chi(\{u_n\}) \leq \chi(\{u_n+P_nAu_n\}) \leq \chi(\{g_n+P_nBu_n\}) \leq \chi(\{P_nBu_n\}) \leq k\chi(\{u_n\})$. Since $k<1+c$, it follows that $\chi(\{u_n\})=0$, i.e., $\{u_n\}$ has a convergent subsequence and so the A-properness of T follows from this and the continuity of A and B. The proof of the last assertion is obvious.

Remarks: (1) In virtue of Proposition A, the assertion of Example (3) is valid under the weaker condition: $k-c<1$. Similarly, Theorem 2.1 and Corollary 2.1 are valid under the weaker condition: $k=\alpha\gamma^{-1}<1$.

(2) In virtue of Proposition B, the assertion of Theorem 2.4 is valid for F, N and C acting in a Banach space X with condition (b1) weakened to: $k<1$.

(3) Theorem 2.5 is valid under the weaker condition: $\alpha\gamma^{-1}<1$.

PERTURBATION METHODS FOR THE SOLUTION OF LINEAR PROBLEMS

L. B. Rall

Mathematics Research Center
University of Wisconsin-Madison

Dedicated to Professor Arvid T. Lonseth on his 65th Birthday

Abstract. Linear problems of central interest in numerical analysis are the solution of linear equations, the construction of the inverse or a generalized inverse of a linear operator, finding the eigenvalues and eigenvectors of a linear operator, and linear programming. A survey is made of methods which apply if the data of a solved linear problem is perturbed by operators and vectors of small norm (analytic perturbation), or by operators of finite rank and vectors belonging to a finite-dimensional subspace (algebraic perturbation). Perturbation methods may be used to extend the theory of linear problems, to estimate errors due to inaccurate data and computation, and to solve perturbed problems with economy of effort.

1. Linear problems.

In the abstract framework of functional analysis, a *linear* problem is one which can be formulated in terms of linear spaces and operators [38 , Chapter I]. Naturally, many problems of theoretical and practical interest in numerical analysis belong to this general class. Among these problems, some are important enough to be the subjects of extensive investigations, and also appear in the daily workload of most computing centers devoted to general scientific computation. Of these significant problems, the ones singled out for discussion here are: (a) solution of linear equations, (b) inversion of linear operators, (c) finding the eigen-eigenvalues and eigenvectors of a linear operator, and (d) linear programming. These problems will now be defined in appropriate generality.

a. Solution of linear equations.

Let X, Y denote complete normed linear spaces over a common scalar field Λ. In most applications, one has $\Lambda = R$, the real numbers, or $\Lambda = C$, the complex numbers. The notation $L(X,Y)$ will be used for the set of continuous linear operators from X into Y. Given an operator $A \in L(X,Y)$ and a vector $y \in Y$ as *data*, the problem is to find a *solution* $x \in X$ of the *linear equation*

$$Ax = y . \tag{1.1}$$

For practical as well as abstract treatment of this problem, it is important to be in possession of a *theory* of equation (1.1), which provides information as to which of the following alternatives holds:

Sponsored by the United States Army under Contract No.: DAAG29-75-C-0024.

(1.1a) $\begin{cases} \text{(i)} & \text{For each } y \in Y, \text{ equation (1.1) has a unique solution } x \in X; \\ & \text{or} \\ \text{(ii)} & \text{for some } y \in Y, \text{ equation (1.1) has no solution or several solutions.} \end{cases}$

The choice between (i) existence and uniqueness or (ii) nonexistence or nonuniqueness or solutions exhausts the logical possibilities, and thus the *alternative structure* (1.1a) will be characteristic of the theory of any equation, linear or nonlinear. In case (i), the operator A is said to be *nonsingular;* otherwise (case (ii)), it is called a *singular* operator.

b. Inversion of linear operators.

This problem is closely related to the solution of linear equations. In the nonsingular case (i), equation (1.1) defines the linear (right) *inverse operator* A^{-1} which gives the unique solution x as

(1.2) $$x = A^{-1}y .$$

In many applications, one has $Y = X$ and A^{-1} satisfies

(1.3) $$A^{-1}A = AA^{-1} = I ,$$

where I denotes the *identity operator* in X, that is, $Ix = x$ for all $x \in X$.

In the singular case (ii), the alternatives are nonexistence or nonuniqueness of solutions x of (1.1). The inverse operator A^{-1} of A does not exist in this case, but one may seek a *generalized inverse* $A^{\dagger}$ of A which has some properties which are desirable for the application at hand. For example, in connection with the problem of solving the linear equation (1.1), one might want

(1.4) $$x = A^{\dagger}y$$

to be a solution if the equation is *consistent,* and thus is satisfied by one or more elements of X. It turns out that this is equivalent to the condition that $A^{\dagger}$ satisfies the operator equation

(1) $$AA^{\dagger}A = A.$$

Any operator $A^{\dagger}$ for which (1) holds will be called an *inner inverse* of A [25, pp. 7-11]. The more formal term {1}-*inverse* of A [3, pp. 7-8] has also been applied specifically to operators $A^{\dagger}$ satisfying condition (1).

If equation (1.1) is consistent and $A^{\dagger}$ is an inner inverse of A, then all solutions x may be represented in the form

(1.5) $$x = A^{\dagger}y + (I - A^{\dagger}A)z$$

for $z \in X$. With z arbitrary, formula (1.5) is called the general solution of equation (1.1), as in the elementary theory of linear differential equations.

Generalized inverses may also be useful in case equation (1.1) has no solutions, and thus is said to be *inconsistent*, or *overdetermined*. Here, the possibility of choosing a *generalized solution* x to minimize the norm of the *residual vector*

$$(1.6) \qquad r = Ax - y$$

in Y is considered. Suppose that the set

$$(1.7) \qquad G(A,y) = \{x \mid \|Ax-y\| = \min_{z \in X} \|Az-y\|\}$$

of generalized solutions x is nonempty, as will certainly be the case if equation (1.1) is consistent. If Y is a Hilbert space, then elements $x \in G(A,y)$ are ordinarily called *least-squares solutions* of the linear equation (1.1). Thus, one might require that $A^{\dagger}y \in G(A,y)$ for all $y \in Y$ in addition to property (1). Furthermore, the subset

$$(1.8) \qquad S(A,y) = \{x \mid x \in G(A,y),\ \|x\| = \min_{z \in G(A,y)} \|z\|\}$$

of $G(A,y)$ may be nonempty, and would then consist of the generalized (or least-squares) solutions x of (1.1) of *minimum norm*. The requirement that $A^{\dagger}y \in S(A,y)$ for all $y \in Y$ would then also be a possible additional restriction on the set of generalized inverses of A. If $S(A,y)$ consists of a single point for each $y \in Y$, then the corresponding generalized inverse $A^{\dagger}$ is uniquely determined. In case X and Y are finite-dimensional Euclidean spaces, this generalized inverse $A^{\dagger}$ exists and is the *Moore-Penrose inverse* of A [3, pp. 7, 103-121], which, in addition to (1), satisfies the condition

$$(2) \qquad A^{\dagger}AA^{\dagger} = A^{\dagger} ;$$

that is, $A^{\dagger}$ is also an *outer inverse* of A [25, pp. 12-14], and the symmetry conditions

$$(3) \qquad (AA^{\dagger})^{*} = AA^{\dagger},$$

and

$$(4) \qquad (A^{\dagger}A)^{*} = A^{\dagger}A,$$

where M^{*} denotes the conjugate transpose of the matrix M.

The problem of finding generalized solutions can become delicate in more general spaces, as the set $S(A,y)$ may consist of more than one element or be empty [20,25]; in fact, $G(A,y)$ will be empty if the infimum of the norm of the residual vector is not attained. Of course, there are also many applications of generalized inverses in addition to the solution of linear equations in the singular case [3,21], and this fairly recent subject already has a vast literature [24].

c. The eigenvalue-eigenvector problem.

This problem is posed most naturally in the case $Y = X$ is a Hilbert space with inner product $< , >$. One looks for scalars (real or complex numbers) λ and vectors $x \neq 0$ such that

(1.9) $$Ax = \lambda x.$$

Solutions λ of this problem are called *eigenvalues* of the linear operator A; for each eigenvalue λ, nonzero solutions x of (1.9) are said to be the corresponding *eigenvectors* of A. As equation (1.9) is homogeneous in x, the condition $x \neq 0$ may be replaced, for example, by

(1.10) $$< x,x > = 1 ,$$

or some other *normalization* condition.

From a standpoint of functional analysis, the determination of the eigenvalues of A is a special case of the more general problem of finding the *spectrum* $\sigma(A)$ of A. In a complex Hilbert space X, the set

(1.11) $$\rho(A) = \{\lambda \mid (A - \lambda I)^{-1} \in L(X,X)\}$$

of complex numbers λ is called the *resolvent* of A. Thus, $\lambda \in \rho(A)$ if and only if the operator $A - \lambda I$ has a continuous inverse. The spectrum of A is simply the complement of the resolvent,

(1.12) $$\sigma(A) = C - \rho(A) ,$$

and hence contains any eigenvalues of A.

d. Linear programming.

In order to formulate this problem, suppose that X,Y are real spaces with *partial ordering* relationships denoted by $\leq$. For most applications, X and Y are taken to be finite-dimensional, in which case the partial ordering is the usual componentwise comparison of vectors [26, pp. 155-158]. Also needed is the *dual space* $X^* = L(X,R)$ of X; that is, the space of continuous *linear functionals* defined on X. It is convenient to use the *bracket notation* of Dirac [7, pp. 18-28] for linear functionals. If $c \in X^*$, then define

(1.13) $$< c,x > := c(x) ,$$

which will be consistent with the notation for the inner product if X is a Hilbert space [7, pp. 6-8].

One formulation of the *(primal) linear programming problem* [26, pp. 156-157] is, given $A \in L(X,Y)$, $y \in Y$, $c \in X^*$, find $x \in X$ to *maximize*

(1.14) $$f(x) = < c,x > + \xi$$

subject to

(1.15) $$Ax \leq y , \quad x \geq 0 .$$

The function $f(x)$ defined by (1.14) is called the *objective function* of the problem, and conditions (1.15) are known as *constraints*.

Instead of the primal problem (1.14)-(1.15), one may wish to consider the *dual problem* [26, pp. 188-190] which is to find $z \in Y^*$ to *minimize*

(1.16) $$g(z) = \langle z,y \rangle - \zeta$$

subject to

(1.17) $$A^* z \geq c , \quad z \geq 0 .$$

In (1.17), the operator $A^* \in L(Y^*,X^*)$ is the *adjoint* of A, defined by

(1.18) $$\langle A^* y^*, x\rangle = \langle y^*, Ax \rangle$$

for all $y^* \in Y^*$, $x \in X$. It will also be convenient to write

(1.19) $$y^* A := A^* y^* , \quad y^* \in Y^* ;$$

that is, $y^* A$ is the linear functional on X defined by

(1.20) $$(y^* A) x := y^* (Ax) = \langle y^*, Ax\rangle , \qquad x \in X .$$

This is analogous to the notation frequently used in elementary matrix algebra, with x being considered to be a column vector, and y^* a row vector. The scalar quantity (1.20) will also be denoted by

(1.21) $$\langle y^* Ax \rangle := \langle y^*, Ax \rangle .$$

The subject of perturbation methods and theory has a long history, and there is a vast literature devoted to this topic and its applications. The bibliography at the end of this paper, rather than attempting to be comprehensive, lists only references cited in the text, doubtless at the cost of omitting a number of significant contributions.

2. Perturbed linear problems. Perturbation theory, as applied to the linear problems listed in §1, starts from the assumption that their solutions are known for the given *reference data* $A \in L(X,Y)$, $y \in Y$, $c \in X^*$. The object is to study the behavior of these solutions for various classes of *perturbed data*.

(2.1) $$B = A + \Delta A , \quad z = y + \Delta y , \quad d = c + \Delta c ,$$

where the *perturbations* $\Delta A \in L(X,Y)$, $\Delta y \in Y$, $\Delta c \in X^*$ or appropriate information about them are given. One then desires to calculate or estimate the corresponding changes $\Delta x \in X$, $\Delta A^{-1} \in L(Y,X)$, $\Delta A^{\dagger} \in L(Y,X)$, $\Delta\lambda \in \Lambda$ in the solutions $x \in X$, $A^{-1} \in L(Y,X)$, $A^{\dagger} \in L(Y,X)$, $\lambda \in \Lambda$ of the original problems. Here

(2.2) $$\Delta A^{-1} = B^{-1} - A^{-1}$$

denotes the difference between the inverse, if it exists, of the perturbed operator B and the inverse of the unperturbed operator A, and *not* $(\Delta A)^{-1}$, which may also exist. A similar observation applies to the notation $\Delta A^{\dagger}$.

As an example, the perturbed linear system

(2.3) $$Bw = z$$

can be solved for

(2.4) $$w = x + \Delta x$$

if Δx can be obtained in terms of ΔA and Δy, the solution x of the unperturbed system (1.1) with reference data A, y being assumed to be known.

The motivation behind perturbation methods is that if the perturbations in the data are "small" in some sense, then one might expect the changes in the solutions to be correspondingly small, at least under suitable conditions. What is referred to here as "small" may vary widely, depending on the specific problem, the type of perturbation considered, the computing power available, and perhaps other factors. In the next section, a framework will be developed to characterize the concept of small perturbations more precisely.

The goals of perturbation theory may be either practical or theoretical. Two uses of perturbation methods in actual computation are to find solutions of perturbed problems with economy of effort, and to obtain error estimates. In the first case, computing the solution of a given linear problem might be extremely laborious, but a large amount of information could be generated in the process. One would then hope to be able to use this information to solve perturbations of the reference problem with less work than required when starting from scratch, as indicated in connection with the illustration (2.3)-(2.4) cited above. In the case of error estimation, the perturbations are considered to arise from inaccuracies in the data and from truncation and roundoff errors in the computation. Usually, these perturbations can only be estimated, and one seeks some kind of information about the possible error in the solution. One approach, called *forward error estimation*, starts from assumptions about the perturbations in the data, and obtains a comparison of the solution actually obtained with that of the reference problem if exact data and computation were employed. For *backward error estimation*, as developed by Wilkinson [34], the solution actually obtained is taken to be the exact solution of some perturbation of the reference problem, and estimates are made of the corresponding changes in the data. With the forward method, the computed solution is considered to be acceptable if it can be shown to be "close" to the (unknown) solution of the reference problem, while in the backward procedure, the criterion of acceptability is that the problem actually solved is "close" to the reference problem in some sense. More precise concepts of "closeness" will be introduced in the next section.

Perturbation methods can also be used for theoretical purposes. If a conceptual framework can be developed in which the problems considered can be viewed as perturbations of problems with known theory, then it may be possible to extend this theory from one class to the other. This is the basis, for example, of the classical technique of Erhard Schmidt for obtaining the theory of linear Fredholm integral equations of second kind from the theory of finite linear algebraic systems [4, p. 155]. A more general situation will be described in a later section. Another theoretical use of perturbation methods, closely related to error estimation, is to determine the *sensitivity* of the solution of a linear problem to changes in the data. For example, one may wish to know which components of the solution are affected most strongly by a small change in one of the coefficients of the input data, and which are relatively undisturbed. This kind of analysis can also be used to pursue cause-and-effect relationships in mathematical models of various natural systems and processes.

3. Analytic and algebraic perturbations. For the present purposes, it will be convenient to classify perturbations into two nonexclusive categories; analytic and algebraic. This classification arises from the information available in each case and the methodology used to solve the perturbation problem, as well as an attempt to clarify what is meant by a "small" perturbation. In general, analytic perturbation theory uses metric information, and obtains solutions to perturbation problems in terms of series expansions, or by iterative methods. An objective criterion for a perturbation to be small in this case is that the required series or iterations converge. A more subjective condition is that the convergence be rapid enough to be useful in practice. The satisfaction of this restriction will depend, among other things, on the computing power available and whether the transformations involved can be carried out explicitly, or have to be approximated.

The idea of smallness for algebraic perturbations also depends more or less on outside factors. Here, the perturbations of operators are operators with finite-dimensional range, and vectors and functionals are perturbed by elements belonging to finite-dimensional subspaces of the corresponding spaces. The solution of algebraic perturbation problems will require solving finite algebraic problems of similar type, with the judgment as to what constitutes a "small" finite algebraic problem being again tied up with the resources available for computing. For example, early workers in the theory of linear integral equations knew that replacing them by a corresponding finite linear algebraic system would yield good approximate solutions, but despaired of being able to solve systems of order 10 or 20, as might be required to attain the desired accuracy [18, p. 242]. By contrast, today most computing centers are able to furnish the solutions of well-conditioned linear algebraic systems of order 100 or 200 at nominal cost.

More precise formulations will now be made of the type of information given and expected with each type of perturbation.

a. Analytic perturbations.

The fundamental metric information about vectors in Banach spaces $X, Y, \ldots$ is given, of course, by the respective norms $\| \|_X, \| \|_Y, \ldots$. As confusion is unlikely, the subscripts will usually be dropped. If A is a continuous linear operator from X into Y, that is, if $A \in L(X,Y)$, then the numbers

(3.1)
$$M(A) = \sup_{\|x\|=1} \|Ax\|, \quad m(A) = \inf_{\|x\|=1} \|Ax\|,$$

exist and are finite [2, p. 54; 16, p. 194] . $M(A)$ and $m(A)$ are called the *upper* and *lower bound* of A, respectively. With the natural definitions of addition and scalar multiplication of linear operators, it is well known [38, p. 163] that $L(X,Y)$ is a Banach space for the *operator norm* $\|A\| = M(A)$. In some spaces, this norm is easy to compute, but in others, finding $M(A)$ might require more effort than solving the problem of interest. For numerical purposes, it is often convenient to assign a norm to the linear operator space $L(X,Y)$ which is easier to compute than the operator norm, and is *consistent* with it in the sense that

(3.2)
$$\|A\| \geqq M(A) .$$

For example, if $X = Y = E^n$, (complex) n-dimensional Euclidean space, then $A = (a_{ij})$ is represented by an $n \times n$ matrix with eigenvalues $\lambda_1, \lambda_2, \ldots, \lambda_n$. One has

(3.3)
$$M(A) = \max\{|\lambda_1|, |\lambda_2|, \ldots, |\lambda_n|\},$$

which requires finding the eigenvalue of largest modulus of A. On the other hand, the *Euclidean norm* of A,

(3.4)
$$\|A\| = \left(\sum_{i=1}^{n} \sum_{j=1}^{n} |a_{ij}|^2 \right)^{\frac{1}{2}}$$

is consistent and may be found by a straightforward calculation.

In the case of the adjoint spaces $X^*, Y^*, \ldots$ of continuous linear functionals on $X, Y, \ldots$, the norm will always be defined analogously to the operator norm, that is,

(3.5)
$$\|c\| = \sup_{\|x\|=1} |\langle c,x \rangle|$$

for $c \in X^*$.

Thus, in the perturbed linear system (2.3), one would want a convergent process to calculate Δx, or an estimate for $\|\Delta x\|$ in terms of bounds for $\|\Delta A\|$ and $\|\Delta y\|$, and perhaps also the known quantities $\|A\|, \|x\|, \|y\|$.

b. Algebraic perturbations.

An algebraic perturbation Δy of a vector $y \in Y$ is defined to be an element of a finite-dimensional subspace

(3.6) $$Y_n = span\{y_1, y_2, \ldots, y_n\}$$

of Y consisting of all linear combinations of given independent basis vectors $y_1, y_2, \ldots, y_n$ in Y. A similar definition applies to algebraic perturbations of linear functionals. Ordinarily, algebraic perturbations will be restricted to subspaces with small dimension (in the sense described above). However, if the original spaces are finite-dimensional, it is of course possible to represent an arbitrary perturbation as an algebraic perturbation.

In the case of linear operators, algebraic perturbations are represented by linear operators with finite-dimensional ranges. Such operators are said to be of *finite rank*, or degenerate (in infinite-dimensional spaces). Here, the *dyadic notation* of Dirac [7, pp. 26-28] will be adopted; for $u \in Y$, $v \in X^*$, the symbol $u >< v$ will represent an operator of *rank one* from X into Y, with

(3.7) $$(u >< v)x = u < v,x > = < v,x > u \in Y$$

for $x \in X$. Also, for $y^* \in Y^*$, the transposed operation will be denoted by

(3.8) $$y^*(u >< v) = < y^*,u > v \in X^* ,$$

again consistent with the notation introduced in §1. In these terms, a general algebraic perturbation $\Delta A \in L(X,Y)$ of *rank n* will be written as

(3.9) $$\Delta A = \sum_{i=1}^{n} u_i >< v_i ,$$

where the vectors $u_i \in Y$ and functionals $v_i \in X^*$, $i = 1,2,\ldots,n$, form linearly independent sets. The range of the operator (3.9) is $Y_n = span\{u_1, u_2, \ldots, u_n\}$. In the finite-dimensional case, Y_n could coincide with Y, and arbitrary perturbations of linear operators could be written in the form (3.9).

Algebraic perturbations of vectors and linear operators are sometimes referred to as *finite rank modifications*. This terminology is useful if a clear distinction between analytic and algebraic methods is intended. By the use of algebraic perturbation theory, one would expect to obtain the perturbations in solutions of linear problems in the same form as the perturbations in the data. For example, one would want to express Δx as a linear combination of vectors $x_1, x_2, \ldots, x_n$ to be determined, that is, $\Delta x \in span\{x_1, x_2, \ldots, x_n\} = X_n$, a finite-dimensional subspace of X. Similarly, expressions of the form (3.9) for ΔA^{-1} and $\Delta A^{\dagger}$ would be sought. In other words, algebraic perturbations in the data of linear problems are expected to give rise to finite rank modifications of their solutions.

In contrast to analytic perturbation theory, the use of algebraic methods does not involve restrictions on the norms of the perturbations in the data. However, it is possible that algebraic perturbations can be small in the analytic sense, so that either technique could be employed. Also, as illustrated in the next section, certain problems lend themselves to a combination of algebraic and analytic methods.

4. Compact operators and the Fredholm theory. A theoretical application of perturbation methods, which also has implications for numerical computation, is the extension of the theory of finite linear algebraic systems of n equations in n unknowns to certain types of linear equations (1.1) in infinite-dimensional spaces. An extension of this kind will be obtained here by the use of both analytic and algebraic techniques. First, the alternative structure (1.1a) of the theory of equation (1.1) will be given an explicit formulation for the class of operators to be

Definition 4.1. Linear operators belonging to a class $\mathcal{A} \subset L(X,Y)$ are said to have a *Fredholm theory* if for each $A \in \mathcal{A}$, either (i) the *homogeneous equation*

$$Ax = 0 \tag{4.1}$$

has the unique solution $x = 0$, in which case the inhomogeneous equation (1.1) has a unique solution x for each $y \in Y$, or (ii) equation (4.1) has nonzero solutions, each of which can be expressed as a linear combination of a finite number d linearly independent solutions $x_1, x_2, \ldots, x_d \in X$, in which case the *transposed homogeneous equation*

$$zA = 0 \tag{4.2}$$

likewise has d linearly independent solutions $z_1, z_2, \ldots, z_d \in Y^*$, in terms of which all its nonzero solutions are expressible as linear combinations, and the inhomogeneous equation (1.1) has no solutions unless

$$\langle z_i, y\rangle = 0\ ,\ i = 1,2,\ldots,d. \tag{4.3}$$

If (4.3) is satisfied and x_0 is any solution of (1.1) (sometimes called a *particular solution*), then the *general solution* of the inhomogeneous equation can be written as

$$x = x_0 + \sum_{i=1}^{d} \alpha_i x_i\ , \tag{4.4}$$

with arbitrary scalars $\alpha_1, \alpha_2, \ldots, \alpha_d$.

For the algebraic case $X = Y = R^n$, real n-dimensional space, the class $\mathcal{A}$ of linear operators with Fredholm theory consists of all $n \times n$ real matrices $A = (a_{ij})$, that is, $\mathcal{A} = L(R^n, R^n)$, and the alternatives in Definition 4.1 were known to hold long before 1903, when the Norwegian mathematician Ivar Fredholm [6] established the correspondence between the theories of finite linear algebraic systems and linear integral equations of the form

$$x(s) - \lambda \int_0^1 K(s,t)\ x(t)\,dt = y(s),\quad 0 \leq s \leq 1, \tag{4.5}$$

giving rise to the present name for the theory.

Definition 4.2. A linear operator $K \in L(X,Y)$ is said to be *compact* if, given any $\varepsilon > 0$, there exists a positive integer $n = n(\varepsilon)$ such that

$$K = S + F\ , \tag{4.6}$$

where $\|S\| < \varepsilon$ and F is of finite rank n.

A compact operator may thus be regarded as a small analytic perturbation of an operator of finite rank, or as a finite rank modification of an operator which is small in the analytic sense. It will be shown that the Fredholm theory can be extended to operators which can be expressed as the sum of a linear operator having a continuous inverse and a compact operator. That is, if $\mathcal{J} \subset L(X,Y)$ denotes the class of linear operators J such that $J^{-1} \in L(Y,X)$ exists, $\mathcal{K} \subset L(X,Y)$ the class of compact operators, and $\mathcal{A} = \mathcal{J} \oplus \mathcal{K}$ the class of linear operators of the form

$$A = J + K, \quad J \in \mathcal{J}, \quad K \in \mathcal{K} \tag{4.7}$$

then each $A \in \mathcal{A}$ has a Fredholm theory. This assertion will be proved in the next section by combining results from both analytic and algebraic perturbation theory. First, it will be shown that if $J \in \mathcal{J}$, then one has the well known result that $J + \Delta J \in \mathcal{J}$ for $||\Delta J||$ sufficiently small. Later, the *Fredholm alternative* given in Definition 4.1 will be established for operators which are the sum of invertible linear operators and linear operators of finite rank. The statement that operators of the form (4.7) have a Fredholm theory will then follow from Definition 4.2.

5. Nonsingular linear equations and operators. In this section, the problems of solving linear systems and the inversion of linear operators will be considered for the nonsingular case. Here, alternative (1.1a(i)) holds, and the inverse A^{-1} of the operator A exists.

a. Analytic perturbation of well-posed problems.

Definition 5.1. A problem is said to be *well-posed* if it has a unique solution which depends continuously on the data.

As a general rule, analytic perturbation methods are only successful when applied to well-posed problems. This can require the imposition of additional conditions on the data to insure uniqueness and continuous dependence of the solution, at least in some neighborhood of the solution of the reference problem. For the linear problems considered in this section to be well-posed, the continuity (and hence boundedness) of A^{-1} is required in addition to its existence. Consequently, it will be assumed that $A^{-1} \in L(Y,X)$ in the following discussion of the application of analytic perturbation theory. If A maps X onto Y, then it is well known that $A^{-1} \in L(Y,X)$ if and only if $m(A) > 0$ [2, pp. 145-150]. Lonseth [16, p. 194] has derived the relationship

$$m(A)M(A^{-1}) = M(A)m(A^{-1}) = 1 \tag{5.1}$$

between the upper and lower bounds of a linear operator A with the continuous inverse A^{-1}. Furthermore, $(A + \Delta A)^{-1}$ exists if $M(\Delta A) < m(A)$. Using (5.1), this result may be stated in terms of consistent norms.

Theorem 5.1. If $\|\Delta A\| < \frac{1}{\|A^{-1}\|}$, then $(A+\Delta A)^{-1}$ exists and is given by

(5.2) $$(A+\Delta A)^{-1} = \sum_{n=0}^{\infty} (-A^{-1}\Delta A)^n A^{-1} .$$

Proof: The hypothesis guarantees the convergence of the *Neumann series* on the right side of (5.2). Denoting this series by S, one finds by direct manipulation that $(A+\Delta A)S = I_Y$, the identity operator in Y, and $S(A+\Delta A) = I_X$; hence, $S = (A+\Delta A)^{-1}$. QED

Although the Neumann series expansion (5.2) is useful for theoretical purposes, it is likely to be too slowly convergent for practical computation. The partial sums

(5.3) $$S_k = \sum_{n=0}^{k} (-A^{-1}\Delta A)^n A^{-1}$$

of the Neumann series (5.2) may be obtained by the simple iteration

(5.4) $$S_0 = A^{-1}, \quad S_k = S_0 - (A^{-1}\Delta A)A_{k-1}, \quad k = 1,2,\ldots .$$

From (5.2), for $\theta = \|A^{-1}\Delta A\|$,

(5.5) $$\|(A+\Delta A)^{-1} - S_k\| \leqq \frac{\theta^{k+1}}{1-\theta}\|A^{-1}\|.$$

In order to find a more efficient method, the *Hotelling-Lonseth algorithm* [17] may be adapted to this purpose. In this special case, the iteration process is

(5.6) $$B_0 = A^{-1}, \quad B_k = [1 + (-A^{-1}\Delta A)^{2^{k+1}}]B_{k-1}, \quad k = 1,2,\ldots .$$

It is easy to show by mathematical induction that $B_k = S_{2^k-1}$; hence, from (5.5),

(5.7) $$\|(A+\Delta A)^{-1} - B_k\| \leqq \frac{\theta^{2^k}}{1-\theta} A^{-1},$$

so that the sequence $\{B_k\}$ defined by (5.6) converges quadratically to $(A+\Delta A)^{-1}$. The only additional labor required over the more slowly convergent algorithm (5.4) is the repeated squaring of the small operator $-A^{-1}\Delta A$.

Attention will now be devoted to the estimation of the perturbations ΔA^{-1} and Δx in the inverse of the perturbed operator and the solution of the perturbed linear equation (2.3), respectively [14, 15, 16]. It will be helpful to introduce the notion of the *condition number* of a bounded linear operator. For $A \in L(X,Y)$, the *exact* condition number $\kappa(A)$ of A is defined to be

(5.8) $$\kappa(A) = \frac{M(A)}{m(A)},$$

and is a measure of the distortion of the image in Y of the unit ball in X as transformed by the operator A. If A has a continuous inverse, then $\kappa(A) = M(A)M(A^{-1})$ by (5.1). For computational purposes, it may be expedient to use consistent norms for $L(X,Y)$ and $L(Y,X)$, and the *approximate* condition number

(5.9) $$k(A) = \|A\| \cdot \|A^{-1}\| ,$$

which is an upper bound for $\kappa(A)$. The inequalities given below will be stated in terms of consistent norms and approximate condition numbers, but remain valid if these upper bounds are replaced by their exact values.

First, from (5.2),

(5.10) $$\Delta A^{-1} = (A+\Delta A)^{-1} - A^{-1} = \sum_{n=1}^{\infty} (-A^{-1}\Delta A)^n A^{-1} ,$$

and thus,

(5.11) $$\|\Delta A^{-1}\| \leq \frac{\|A^{-1}\|^2 \, \|\Delta A\|}{1 - \|A^{-1}\| \cdot \|\Delta A\|} .$$

Dividing (5.11) by $\|A^{-1}\|$ and multiplying and dividing $\|\Delta A\|$ on the right hand side by $\|A\|$ gives

(5.12) $$\frac{\|\Delta A^{-1}\|}{\|A^{-1}\|} = \frac{k(A)\frac{\|\Delta A\|}{\|A\|}}{1-k(A)\frac{\|\Delta A\|}{\|A\|}} ,$$

which expresses the relative change in the inverse in terms of the relative perturbation of the reference operator and its (approximate) condition number. A similar expression will now be obtained for the perturbation Δx in the solution of (2.3).

<u>Theorem 5.2.</u> If $\|\Delta A\| < \frac{1}{\|A^{-1}\|}$, then the perturbed linear equation (2.3) has a unique solution $w = x + \Delta x$ for each $z = y + \Delta y$, and

(5.13) $$\frac{\|\Delta x\|}{\|x\|} \leq \frac{k(A)}{1 - k(A)\frac{\|\Delta A\|}{\|A\|}} \left[\frac{\|\Delta A\|}{\|A\|} + \frac{\|\Delta y\|}{\|y\|} \right]$$

provided, of course, that $y \neq 0$.

Proof: By Theorem 5.1, the hypothesis guarantees that $B^{-1} = (A+\Delta A)^{-1} = A^{-1}+\Delta A^{-1}$ exists, which implies the unique solvability of (2.3) for each z. Writing (2.3) as

(5.14) $$(A + \Delta A)(x + \Delta x) = y + \Delta y ,$$

one obtains

(5.15) $$\Delta x = \Delta A^{-1} y + (A+\Delta A)^{-1} \Delta y .$$

As $y = Ax$, from (5.10),

(5.16) $$\Delta A^{-1} y = \sum_{n=1}^{\infty} (-A^{-1}\Delta A)^n x ,$$

so that

$$(5.17)\qquad \| \Delta A^{-1}y\| \leq \frac{\|A^{-1}\| \cdot \|\Delta A\| \cdot \|x\|}{1-\|A^{-1}\| \cdot \|\Delta A\|} = \frac{k(A)}{1-k(A)\frac{\|\Delta A\|}{\|A\|}} \cdot \frac{\|\Delta A\|}{\|A\|} \cdot \|x\|$$

Similarly, from (5.2) and the fact that $\|y\| = \|Ax\| \leq \|A\| \cdot \|x\|$,

$$(5.18)\qquad \|(A+\Delta A)^{-1}\Delta y\| \leq \frac{\|A^{-1}\| \cdot \|\Delta y\|}{1-\|A^{-1}\| \cdot \|\Delta A\|} \cdot \frac{\|A\| \cdot \|x\|}{\|y\|} =$$

$$\frac{k(A)}{1-k(A)\frac{\|\Delta A\|}{\|A\|}} \cdot \frac{\|\Delta y\|}{\|y\|} \cdot \|x\| .$$

Inequality (5.13) now follows directly from (5.15), (5.17), and (5.18). QED

b. <u>Algebraic perturbation of nonsingular linear equations and operators.</u>

The simplest type of algebraic perturbation (2.3) of the linear system (1.1) is with $\Delta A = 0$ and Δy restricted to belong to a finite-dimensional subspace Y_n of Y. Given a basis $\{y_1, y_2, \ldots, y_n\}$ for Y_n, one need only find the corresponding basis vectors

$$(5.19)\qquad x_i = A^{-1}y_i , \qquad i = 1,2,\ldots,n,$$

of the subspace $X_n \subset X$ which will then contain all possible perturbations Δx. Thus, given

$$(5.20)\qquad \Delta y = \alpha_1 y_1 + \alpha_2 y_2 + \ldots + \alpha_n y_n ,$$

it follows that

$$(5.21)\qquad \Delta x = \alpha_1 x_1 + \alpha_2 x_2 + \ldots + \alpha_n x_n .$$

In actual computation, it may be more efficient to solve the n systems $Ax_i = y_i$, $i = 1,2,\ldots,n$, for the basis vectors for X_n, even if X is finite-dimensional [5, p. 77], than to calculate A^{-1}.

To introduce the study of the effect of a finite-rank modification of an operator upon its inverse, the case of rank one perturbation will be considered first, as all the indicated operations can be displayed explicitly. For $\Delta A = u><v$ with $u \in Y$, $v \in X^*$ nonzero, the solvability of the perturbed system (2.3), that is

$$(5.22)\qquad (A + u><v)w = z ,$$

will be investigated for arbitrary z. As A^{-1} is assumed to exist, the equations $A\hat{u} = u$, $A\hat{z} = z$ can be solved uniquely for $\hat{u} = A^{-1}u$, $\hat{z} = A^{-1}z$, respectively. In terms of these solutions, (5.22) may be written as

$$(5.23)\qquad w = \hat{z} - \hat{u}\langle v,w\rangle .$$

The key to the solvability of (5.23), and hence of (5.22), is the determination of the number $\xi = \langle v,w \rangle$. From (5.23),

$$\langle v,w \rangle + \langle v,\hat{u} \rangle \langle v,w \rangle = \langle v,\hat{z} \rangle \tag{5.24}$$

If the *determinant*

$$\delta = 1 + \langle v,\hat{u} \rangle = 1 + \langle vA^{-1}u \rangle \tag{5.25}$$

does not vanish, then (5.24) has the unique solution

$$\langle v,w \rangle = \frac{\langle v,\hat{z} \rangle}{\delta} = \frac{\langle v A^{-1} z \rangle}{1+\langle v A^{-1} u \rangle}, \tag{5.26}$$

where the notation (1.21) has been used in (5.25) and (5.26). Substitution of (5.26) into (5.23) yields

$$w = A^{-1}z - A^{-1}u \frac{\langle vA^{-1}z \rangle}{\delta} = \left(A^{-1} - \frac{A^{-1}u\rangle\langle vA^{-1}}{1 + \langle vA^{-1}u \rangle}\right) z, \tag{5.27}$$

so that

$$(A + u\rangle\langle v)^{-1} = A^{-1} - \frac{A^{-1}u\rangle\langle vA^{-1}}{1 + \langle vA^{-1}u \rangle}, \tag{5.28}$$

provided $\delta \neq 0$. Hence, the inverse of a rank one modification of an invertible oper-operator, if it exists, is a rank one modification of the inverse of the reference operator. The symmetry of (5.28), sometimes called the *Sherman-Morrison-Woodbury formula* [11, pp. 123-124; 35, 46], is appealing.

Using (5.28), the solution $w = x + \Delta x$ of (5.22) is

$$x + \Delta x = x + A^{-1}\Delta y - \frac{\langle v, x+A^{-1}\Delta y \rangle}{1+\langle vA^{-1}u \rangle} A^{-1}u, \tag{5.29}$$

or

$$\Delta x = A^{-1}\Delta y - \frac{\langle v, x+A^{-1}y \rangle}{1+\langle vA^{-1}u \rangle} A^{-1}u. \tag{5.30}$$

Thus, the perturbation Δx is a linear combination of $A^{-1}\Delta y$ and the vector $\hat{u} = A^{-1}u$. If Δy is an algebraic perturbation of the form (5.20), and $\hat{u}$ is independent of the vectors $x_i = A^{-1}y_i$, $i = 1,2,\dots,n$, then Δx will lie in the $(n+1)$-dimensional subspace $X_{n+1} = \mathit{span}\,\{\hat{u},x_1,x_2,\dots,x_n\}$ of X; otherwise $\Delta x \in X_n = \mathit{span}\,\{x_1,x_2,\dots,x_n\}$.

Before going to the general case, two applications of algebraic perturbation theory will be given which involve rank one modifications. The first is to the Fredholm integral equation (4.5) in which the *kernel* $K(s,t)$ has the special form

$$K(s,t) = \begin{cases} u(t)v(s), & 0 \leqq t \leqq s \leqq 1, \\ u(s)v(t), & 0 \leqq s \leqq t \leqq 1. \end{cases} \tag{5.31}$$

This type of kernel arises in applications: for example, as a Green's function determined by a two-point boundary value problem [32]. Given the representation (5.31) for $K(s,t)$, the integral equation (4.5) may be written as

(5.32) $$x(s) - \lambda \int_0^s L(s,t)\, x(t)dt - \lambda \int_0^1 u(s)v(t)x(t)dt = y(s) ,$$

where

(5.33) $$L(s,t) = u(t)v(s) - u(s)v(t), \qquad 0 \leq t \leq s \leq 1 .$$

Equation (5.32) is of the form $(A - \lambda u >< v)x = y$, where $A = I - \lambda L$ is a linear Volterra integral operator of second kind with kernel (5.33), and $u >< v$ is a Fredholm integral operator of first kind and rank one with kernel $u(s)v(t)$. The inverse $A^{-1} = (I - \lambda L)^{-1}$ of the Volterra operator of second kind exists for all λ [30, pp. 52-53], and thus the linear Volterra integral equation

(5.34) $$\hat{w}(s) - \lambda \int_0^s L(s,t)\, \hat{w}(t)dt = w(s)$$

can be solved for arbitrary $w(s)$; in particular, one obtains $\hat{w}(s) = \hat{u}(s)$ for $w(s) = u(s)$, and $\hat{w}(s) = \hat{y}(s)$ for $w(s) = y(s)$. Corresponding to (5.25), if the *Fredholm determinant*

(5.35) $$\delta = 1 - \lambda < v,\hat{u} > = 1 - \lambda \int_0^1 v(t)\hat{u}(t)dt$$

does not vanish, then, from (5.27),

(5.36) $$x(s) = \hat{y}(s) + \frac{\lambda}{\delta}\, \hat{u}(s) \int_0^1 v(t)\hat{y}(t)dt$$

is the unique solution of (5.32). Hence, the solution of the Fredholm integral equation (4.5) with the kernel (5.31) can be obtained by solving the Volterra integral equation (5.34) with right-hand sides $w(s) = u(s)$ and $w(s) = y(s)$, followed by the calculation of the inner product integrals in (5.35) and (5.36).

The second application to be considered for rank one modification of a linear operator is to backward error analysis in the solution of linear equations. Suppose that one attempts to solve the linear equation (1.1) and obtains, instead of x , an approximate solution w such that

(5.37) $$A\, w = y + r ,$$

with nonzero *residual* r , The Hahn-Banach theorem [38, p. 186] guarantees the existence of a linear functional $w^* \in X^*$ such that $\| w^* \| = 1$ and $<w^*,w> = \|w\|$. Thus, w is the *exact* solution of the linear equation

(5.38) $$\left(A - \frac{r >< w^*}{\| w \|} \right) w = y$$

with perturbed operator and desired right-hand side. An analytic bound for the perturbation of A is thus

(5.39) $$\| \Delta A \| = \frac{\| r >< w^* \|}{\| w \|} = \frac{\| r \|}{\| w \|} .$$

Returning to the study of general algebraic perturbations, note that the equivalence of (5.22) and the single scalar equation (5.24) establishes that (5.22) has a Fredholm theory, because (5.24) does. In the case $\delta = 0$, the homogeneous equation $(A + u >< v)w = 0$ is satisfied by (and only by) vectors $w = \alpha\hat{u}$ with α arbitrary. The inhomogeneous equations (5.22) and (5.24) then have solutions only if

$$(5.40) \qquad < v,\hat{z}> = < vA^{-1}z > = < \hat{v},z > = 0 ,$$

where $\hat{v} = vA^{-1}$ satisfies the transposed homogeneous equation

$$(5.41) \qquad \hat{v}(A + u >< v) = 0 .$$

Theorem 5.3. If $\mathfrak{J} \subset L(X,Y)$ denotes the class of all invertible linear operators, and $\mathfrak{F} \subset L(X,Y)$ the class of all linear operators of finite rank, then all linear operators belonging to the class $\mathcal{Q} = \mathfrak{J} \oplus \mathfrak{F}$ have a Fredholm theory.

Proof: If $B \in \mathcal{Q} = \mathfrak{J} \oplus \mathfrak{F}$, then there is an ivertible linear operator $A \in \mathfrak{J}$ for which B can be written as

$$(5.42) \qquad B = A + \sum_{j=1}^{n} u_j >< v_j ,$$

where $u_j \in Y$, $v_j \in X^*$, $j = 1,2,\ldots,n$, are linearly independent sets of vectors and functionals, respectively. Equation (2.3) in this case is equivalent to

$$(5.43) \qquad w = \hat{z} - \sum_{j=1}^{n} \hat{u}_j < v_j,w > ,$$

where $\hat{z} = A^{-1}z$, $\hat{u}_j = A^{-1}u_j$, $j = 1,2,\ldots,n$. Applying the functionals $v_1,v_2,\ldots,v_n$ to (5.43) in turn gives the equivalent finite linear algebraic system of equations

$$(5.44) \qquad \xi_i + \sum_{j=1}^{n} \alpha_{ij}\xi_j = \zeta_i , \qquad i = 1,2,\ldots,n$$

for $\xi_i = < v_i,w >$, where $\zeta_i = < v_i,\hat{z}>$ and $\alpha_{ij} = < v_i,\hat{u}_j >$, $i,j = 1,2,\ldots,n$. As the Fredholm alternative applies to (5.44), it follows that operators of the form (5.42) have a Fredholm theory. QED

In the nonsingular case, an expression can be obtained for B^{-1} as a finite rank modification of A^{-1}. Let

$$(5.45) \qquad M = (\delta_{ij} + \alpha_{ij})$$

debite the matrix of coefficients of the linear system (5.44), where δ_{ij} is the Kronecker delta: $\delta_{ij} = 0$ if $i \neq j$, $\delta_{ii} = 1$. As the determinant δ of M is assumed to be nonzero, the inverse of M may be written

$$(5.46) \qquad M^{-1} = \frac{1}{\delta}(\beta_{ij}) ,$$

and thus

$$(5.47) \qquad \xi_i = < v_i,w> = \frac{1}{\delta}\sum_{j=1}^{n} \beta_{ij}\zeta_j = \frac{1}{\delta}\sum_{j=1}^{n} \beta_{ij} < v_j,\hat{z}> ,$$

$i = 1,2,\ldots,n$. Using (5.43) and the fact that $< v_j,\hat{z}> = < v_jA^{-1}z> = < \hat{v}_j,z >$ for

$\hat{v}_j = v_j A^{-1}$, $i = 1,2,\ldots,n$, one obtains the solution w of (2.3) in this case as

$$(5.48) \qquad w = \left(A^{-1} - \frac{1}{\delta} \sum_{i=1}^{n} \sum_{j=1}^{n} A^{-1} u_i > \beta_{ij} < v_j A^{-1} \right) z .$$

By taking appropriate linear combinations $\tilde{u}_1, \tilde{u}_2, \ldots, \tilde{u}_n$ of $u_1, u_2, \ldots, u_n$ and $\tilde{v}_1, \tilde{v}_2, \ldots, \tilde{v}_n$ of $v_1, v_2, \ldots, v_n$ (for example, by an *LU-decomposition* of M^{-1} [5, pp. 27-32]), (5.48) may be put in the form

$$(5.49) \qquad w = \left(A^{-1} - \frac{1}{\delta} \sum_{j=1}^{n} A^{-1} \tilde{u}_j >< \tilde{v}_j A^{-1} \right) z ,$$

from which

$$(5.50) \qquad \left(A + \sum_{j=1}^{n} u_j >< v_j \right)^{-1} = A^{-1} - \frac{1}{\delta} \sum_{j=1}^{n} A^{-1} \tilde{u}_j >< \tilde{v}_j A^{-1} ,$$

which is analogous to (5.28).

Another way to find the inverse of the perturbed operator (5.42) is the method of *successive rank one modifications*, which does not require obtaining M^{-1} explicitly. Set

$$(5.51) \qquad B_0 = A , \quad B_0^{-1} = A^{-1} ,$$

and then the algorithm

$$(5.52) \qquad \begin{cases} B_k = B_{k-1} + u_k >< v_k , \\ B_k^{-1} = B_{k-1}^{-1} - \dfrac{B_{k-1}^{-1} u_k >< v_k B_{k-1}^{-1}}{1 + < v_k B_{k-1}^{-1} u_k >} , \end{cases}$$

$k = 1,2,\ldots,n$ will give $B^{-1} = B_n^{-1}$ if none of the *intermediate determinants*

$$(5.53) \qquad \delta_k = 1 + < v_k B_{k-1}^{-1} u_k > , \quad k = 1,2,\ldots,n ,$$

vanish.

It should be noted again that it is not necessary to obtain A^{-1} to solve the equation (2.3) for

$$(5.54) \qquad w = \hat{z} - \sum_{j=1}^{n} \xi_j \hat{u}_j ,$$

as given by (5.43). What is required is to solve equation (1.1) for the $n+1$ right-hand sides $y = z, u_1, u_2, \ldots, u_n$ for $x = \hat{z}, \hat{u}_1, \hat{u}_2, \ldots, \hat{u}_n$, calculate the coefficients of the system (5.44) of n equations for the n unknowns $\xi_1, \xi_2, \ldots, \xi_n$, solve this system, and then form the linear combination (5.54).

An important application of the above technique of algebraic perturbation is to the numerical solution of partial differential equations by what is called the *capacitance matrix method* [36, 42]. The basic problem is to solve, for example, the Poisson or Helmholtz equation on a region Ω, with information given on its boundary $\partial\Omega$ (see Figure 5.1). The use of finite-difference methods will lead to a linear

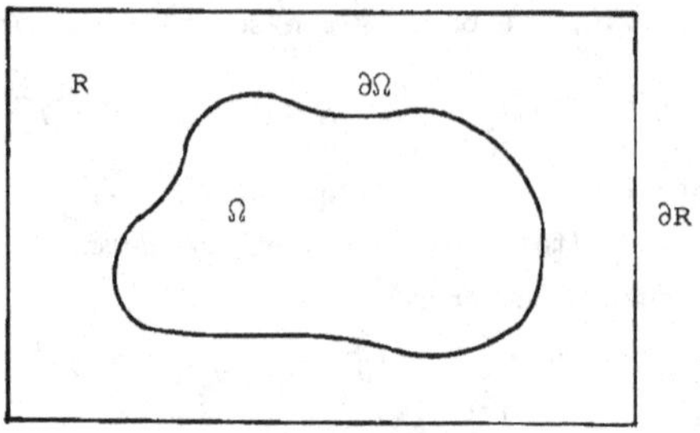

Figure 5.1

algebraic system $B\,w = z$ which may be very laborious to solve. On the other hand, rapid and effective methods may be available for the algebraic system $Ax = y$ corresponding to the same finite-difference approximation to the problem posed on an enclosing rectangle R with boundary ∂R. By regarding the algebraic system obtained for Ω as a finite rank perturbation of the easily solved system arising from the approximate problem on R, a considerable reduction in effort may be possible. Typically, if the order of the systems (1.1) and (2.3) is about n^2, then the rank of the perturbations ΔA and Δy will be approximately n.

The Fredholm theory will now be shown to apply to operators which are the sum of a continuously invertible operator and a compact operator.

Theorem 5.4. Operators A belonging to the class $\mathcal{A}$ defined by (4.7) have a Fredholm theory.

Proof: Choose $\varepsilon < 1/\|J^{-1}\|$. According to Definition 4.2, the compact operatio K may be written as

$$K = S + \sum_{j=1}^{n} u_j >< v_j\,, \tag{5.55}$$

where $n = n(\varepsilon)$ is finite. Thus,

$$A = J + S + \sum_{j=1}^{n} u_j >< v_j\,, \tag{5.56}$$

and Theorem 5.1 guarantees the existence of the inverse operator $(J+S)^{-1} \in L(Y,X)$. It follows from Theorem 5.3 that A has a Fredholm theory. QED

Theorem 5.4 provides a basis for the "kernel splitting" method due to Erhard Schmidt [4, p. 155] for proving the Fredholm Alternative Theorem [6] for the linear integral equation (4.5). Suppose that $K(s,t)$ is continuous, or at least can be approximated sufficiently well by a kernel of finite rank so that one can write

(5.57) $$K(s,t) = S(s,t) + \sum_{j=1}^{n} u_j(s)v_j(t) ,$$

where $S(s,t)$ is the kernel of a linear integral operator S with $\| S \| < \frac{1}{|\lambda|}$ in the appropriate norm. Then, the linear integral operator $I - \lambda K$ in (4.5) has the form

(5.58) $$I - \lambda K = I - \lambda S - \lambda \sum_{j=1}^{n} u_j >< v_j ,$$

where $(I - \lambda S)^{-1} = T(\lambda)$ exists by Theorem 5.1. Applying this operator to the equation $(I - \lambda K)x = y$, the integral equation (4.5) is seen to be equivalent to the linear equation

(5.59) $$(I - \lambda \sum_{j=1}^{n} T(\lambda)u_j >< v_j)x = T(\lambda)y,$$

and thus $I - \lambda K$ has a Fredholm theory by Theorem 5.3. This approach regards $I-\lambda K$ as an algebraic perturbation of the invertible operator $I-\lambda S$.

On the other hand, suppose that

(5.60) $$(I - \lambda \sum_{j-1}^{n} u_j >< v_j)^{-1} = I + \frac{\lambda}{\delta} \sum_{j=1}^{n} \tilde{u}_j >< \tilde{v}_j = Z(\lambda)$$

exists, where the notation (5.50) has been used. Then, $I - \lambda K$ is an analytic perturbation of an invertible operator, and (4.5) is equivalent to the equation

(5.61) $$(I - \lambda Z(\lambda) S)x = Z(\lambda)y .$$

From Theorem 5.1, if $(I - \lambda K)^{-1}$ exists and $\| \lambda S\| < 1/\|(I- \lambda K)^{-1}\|$, then $Z(\lambda)$ exists, so all sufficiently good finite rank approximations (5.57) to $K(s,t)$ will lead to a solvable perturbed equation

(5.62) $$(I - \lambda \sum_{j=1}^{n} u_j >< v_j)w = y$$

which is equivalent to a finite linear algebraic system of the form (5.44). Conversly, if the inverse operator $Z(\lambda)$ exists and $\| \lambda Z(\lambda) S \| < 1$, then it follows from the same theorem that $(I - \lambda K)^{-1}$ exists. If $u_1, u_2, \ldots, u_n$ and $v_1, v_2, \ldots, v_n$ are chosen so that all the inner products required can be calculated explicitly, then this gives a method for concluding the existence and uniqueness of the solution of the integral equation (4.5) on the basis of a finite set of algebraic computations, as well as a technique to obtain approximate solutions. An error analysis for (5.62) may be carried out by the analytic methods of §5a with $\Delta A = -\lambda S$, $\Delta y = 0$. A similar approach can be used on (5.59) with $T(\lambda)$ replaced by

(5.63) $$T_k(\lambda) = I + \lambda S + \lambda^2 S^2 + \ldots + \lambda^k S^k .$$

Setting $z = T_k(\lambda)y$, the perturbed equation

(5.64) $$(I - \lambda \sum_{j=1}^{n} T_k(\lambda)u_j >< v_j)w = z$$

may be analyzed by the same technique, with

(5.65) $$\Delta A = \lambda \sum_{j=1}^{n} (\lambda S)^{k+1} T(\lambda)u_j >< v_j ,$$

and

(5.66) $$\Delta T(\lambda)y = -(\lambda S)^{k+1} T(\lambda)y .$$

As $\|T(\lambda)\| \leqq 1/(1-\|\lambda S\|)$, and for

(5.67) $$\lambda F = \lambda \sum_{j=1}^{n} u_j >< v_j ,$$

one has

(5.68) $$\|\Delta A\| \leqq \frac{\|\lambda S\|^{k+1}}{1-\|\lambda S\|} \|\lambda F\| , \quad \|\Delta T(\lambda)y\| \leqq \frac{\|\lambda S\|^{k+1}}{1-\|\lambda S\|} \|y\|$$

with

(5.69) $$\|\lambda F\| \leqq |\lambda| \sum_{j=1}^{n} \|u_j\| \cdot \|v_j\|$$

from (5.67).

Another analytic approach to the approximate solution of (4.5) which also yields error bounds is to solve (5.61) by iteration, as $\|\lambda Z(\lambda)S\| < 1$ [29].

6. The singular case and generalized inverses. Attention will now be devoted to linear problems which are ill-posed because the linear operator involved does not have a bounded inverse. As the solutions, if any, of ill-posed problems do not depend on the data in a continuous fashion, it might be expected in this situation that analytic perturbation methods will be of little utility, or can be applied only under very restrictive conditions. For example, there is an inherent limitation as to how well an operator $B \in L(X,Y)$ without a continuous inverse can be approximated by an operator A belonging to the class $\mathcal{J} \subset L(X,Y)$ of operators with continuous inverses $A^{-1} \subset L(Y,X)$. From Theorem 5.1,

(6.1) $$\|B - A\| = \|\Delta A\| \geqq \frac{1}{\|A^{-1}\|} ,$$

otherwise, the assumption that $B \notin \mathcal{J}$ would be contradicted. Also, from (6.1),

(6.2) $$\|A^{-1}\| \geqq \frac{1}{\|B-A\|} = \frac{1}{\|\Delta A\|} ,$$

so that $\|A^{-1}\|$ and the approximate condition number

(6.3) $$k(A) \geqq \frac{\|A\|}{\|\Delta A\|} \geqq \left| \frac{\|B\|}{\|\Delta A\|} - 1 \right|$$

grow without limit as $\|\Delta A\| \to 0$. Clearly, computational difficulties can be expecpected in the calculation of A^{-1} or in the solution of the linear equation (1.1) if A is very close in the analytic sense to an operator B which does not have a continuous inverse.

Theorem 6.1. If $\{A_n\} \subset \mathcal{J}$ is any sequence of linear operators such that $\lim_{n\to\infty} \|A_n - B\| = 0$, then $B \notin \mathcal{J}$ if and only if (6.2) holds for each $A = A_n$, $n = 1,2,\ldots$.

Proof: If $B \notin \mathcal{J}$, then it has already been shown that (6.2) holds for each A_n. To show the converse, suppose that $B \in \mathcal{J}$, and choose n sufficiently large so that $\|\Delta A_n\| = \|A_n - B\| < 1/2\|B^{-1}\|$. It then follows from (5.2) that

$$\|A_n^{-1}\| \leq \frac{\|B^{-1}\|}{1-\|\Delta A_n\|\cdot\|B^{-1}\|} < \frac{1}{\|\Delta A_n\|}, \tag{6.4}$$

a contradiction of (6.2) which proves the theorem. QED

An evident drawback of analytic perturbation theory is that, in general, no conclusions can be drawn from the existence of $A^{-1} \in L(Y,X)$ about the invertibility or noninvertibility of any operator B for which inequality (6.1) holds. The algebraic theory, on the other hand, states that if B is the finite rank modification (5.42) of an invertible linear operator $A \in \mathcal{J}$, then B^{-1} exists if and only if

$$\delta = \det(\delta_{ij} + \langle v_i A^{-1} u_j \rangle) \neq 0 . \tag{6.5}$$

Of course, one would still expect computational difficulty if B is nearly singular, especially if the inner products $\alpha_{ij} = \langle v_i A^{-1} u_j \rangle$, $i,j = 1,2,\ldots,n$, can only be calculated approximately.

The algebraic approach also provides information in the singular case. Supposing that $\delta = 0$, consider the transposed homogeneous equation

$$t(A + \sum_{i=1}^{n} u_i >< v_i) = 0 \tag{6.6}$$

for $t \in X^*$. Using the technique of §5b, this is equivalent to the finite linear algebraic system

$$\tau_j + \sum_{i=1}^{n} \tau_i \alpha_{ij} = 0, \quad j = 1,2,\ldots,n, \tag{6.7}$$

for $\tau_j = \langle t, u_j \rangle$. The system of equations (6.7) is the transposed homogeneous system corresponding to (5.44). If $\delta = 0$, then (6.7) has d linearly independent solutions

$$\tau^{(k)} = (\tau_1^{(k)}, \tau_2^{(k)}, \ldots, \tau_n^{(k)}), \quad k = 1,2,\ldots,d, \tag{6.8}$$

and, corresponding to these, equation (6.6) also has d linearly independent solutions

$$t^{(k)} = \sum_{i=1}^{n} \tau_i^{(k)} v_i A^{-1} = \sum_{i=1}^{n} \tau_i^{(k)} \hat{v}_i , \tag{6.9}$$

$k = 1,2,\ldots,d$. Likewise, the homogeneous system

$$\xi_i + \sum_{j=1}^{n} \alpha_{ij}\xi_j = 0, \quad i = 1,2,\ldots,n, \tag{6.10}$$

has d linearly independent solutions

$$\xi^{(k)} = (\xi_1^{(k)}, \xi_2^{(k)}, \ldots, \xi_n^{(k)})^T, \qquad k = 1,2,\ldots,d \ , \tag{6.11}$$

from which are obtained the corresponding linearly independent solutions

$$w^{(k)} = \sum_{j=1}^{n} \xi_j^{(k)} A^{-1} u_j = \sum_{j=1}^{n} \xi_j^{(k)} \hat{u}_j \ , \tag{6.12}$$

$k = 1,2,\ldots,d$, of the homogeneous equation

$$(A + \sum_{j=1}^{n} u_j >< v_j) w = 0 \ . \tag{6.13}$$

Representing the right-hand sides of the system (5.44) as the vector

$$\zeta = (\zeta_1, \zeta_2, \ldots, \zeta_n)^T = (\langle v_1 A^{-1} z\rangle, \langle v_2 A^{-1} z\rangle, \ldots, \langle v_n A^{-1} z\rangle)^T \ , \tag{6.14}$$

it is seen immediately that the conditions for the solvability of the finite inhomogeneous system (5.44) and the equivalent inhomogeneous equation (2.3) for the case $\delta = 0$ are

$$\langle \tau^{(k)}, \zeta \rangle = \langle \sum_{i=1}^{n} \langle \tau_i^{(k)} \hat{v}_i, z \rangle = \langle t^{(k)}, z \rangle = 0 \ , \tag{6.15}$$

$k = 1,2,\ldots,d$; that is, z must be orthogonal to all solutions of the homogeneous equation (6.6). If (6.15) is satisfied, then the *general solution* of (2.3) may be written as

$$w = \hat{w} + \sum_{k=1}^{d} \alpha_k w^{(k)} \ , \tag{6.16}$$

where $\hat{w}$ is some *particular solution* of (2.3), and the *complementary vectors*

$$\tilde{w} = \tilde{w}(\alpha_1, \alpha_2, \ldots, \alpha_k) = \sum_{k=1}^{d} \alpha_k w^{(k)} \tag{6.17}$$

satisfy the homogeneous equation (6.13) for arbitrary $\alpha_1, \alpha_2, \ldots, \alpha_k$.

Usually, in actual computational solution of linear equations, the distinction between the singular and nonsingular cases is not as clear-cut as in the alternatives (1.1a) or the Fredholm theory. In practice, an objective or subjective standard is set for what constitutes an "acceptable" (approximate) solution, and one of the following situations is observed:

(6.18)
- (i) An acceptable solution is obtained,

  or

- (ii) either no solution at all is found, or the computed solution is unacceptable.

In the *computationally singular* case (6.18ii), the method used to solve (1.1) or invert A may break down because A does not have a bounded inverse, or is analytically close to an operator $B \notin \mathcal{J}$. On the other hand, the algorithm employed may actually be trying to solve the system (5.44) with $\delta = 0$ and without (6.15) holding to the desired degree of accuracy. This will be called an *algebraic catastrophe of type I* . In the second situation described in (6.18ii), the acceptable particular solution $\hat{w}$ may be contaminated by a complementary vector (6.17) to the extend that the resulting solution is unacceptable. This *algebraic catastrophe of type II* can occur in the numerical solution of differential equations by the use of approximating difference equations. For example, the difference equation

(6.19) $$3u_{n+1} + 8u_n - 3u_{n-1} = 0$$

with the initial conditions

(6.20) $$u_0 = 1, \quad u_1 = \frac{1}{3},$$

has the bounded solutions

(6.21) $$u_n = (\tfrac{1}{3})^n, \qquad n = 0,1,2,\ldots,$$

which may be the ones considered to be acceptable. However, a slight perturbation of (6.20), such as rounding $\frac{1}{3}$ to eight decimal places,

(6.22) $$w_0 = 1, \quad w_1 = 0.33333333$$

gives the corresponding solutions w_n of $3w_{n+1} + 8w_n - 3w_{n-1} = 0$ as

(6.23) $$w_n = (0.999999999)(\tfrac{1}{3})^n + (0.000000001)(-3)^n,$$

$n = 1,2,\ldots,$ and the second term on the right-hand side of (6.23) will eventually wreak havoc with the accuracy of the approximation of u_n by w_n.

As indicated in §1b, if the operator A is singular, then a generalized inverse $A^\dagger$ of A having certain useful properties may be sought, for example, to give a solution of (1.1) in the form (1.4) if (1.1) is consistent. As (1.5) indicates, the vector $x = A^\dagger y$ will be a particular solution of (1.1) for any inner inverse $A^\dagger$ of A. An algebraic perturbation method may be used to obtain inner inverses of singular operators which have a Fredholm theory, under the technical assumption that the space Y is reflexive, that is, $Y^{**} = Y$ [38, p.192]. In this case, if

(6.24) $$U^* = \{u_1^*, u_2^*, \ldots, u_d^*\} \subset Y^*$$

is a set of linearly independent functionals on Y, then the Hahn-Banach theorem guarantees the existence of a set of d linearly independent vectors in Y to which the Gram-Schmidt orthonormalization process [38, p. 116] may be applied, if necessary, to obtain the set

(6.25) $$U = \{u_1, u_2, \ldots, u_d\} \subset Y$$

for which

(6.26) $$\langle u_i^*, u_j \rangle = \langle u_i, u_j^* \rangle = \delta_{ij}, \qquad i,j = 1,2,\ldots,d,$$

where δ_{ij} again denotes the Kronecker delta. Similarly, given a set of linearly independent vectors

(6.27) $$V = \{v_1, v_2, \ldots, v_d\} \subset X,$$

a set of functionals

(6.28) $$V^* = \{v_1^*, v_2^*, \ldots, v_d^*\} \subset X^*$$

exists such that

(6.29) $$\langle v_i^*, v_j \rangle = \delta_{ij}, \qquad i,j = 1,2,\ldots,d,$$

whether X is reflexive or not.

Theorem 6.2. Suppose that $A \in L(X,Y)$ has a Fredholm theory, and

(6.30) $$u^* A = Av = 0$$

if and only if $u^* \in \mathit{span}\,\{u_1^*, u_2^*, \ldots, u_d^*\} \subset Y^*$ and $v \in \mathit{span}\{v_1, v_2, \ldots, v_d\} \subset X$, where the *defect* d of A is positive. Then, for $u_k \in U$ and $v_k^* \in V^*$, $k = 1,2,\ldots,d$, where U and V^* are defined by (6.24)-(6.29), the operator

(6.31) $$B = A - \sum_{k=1}^{d} u_k >< v_k^*$$

is invertible, and

(6.32) $$AB^{-1}A = A ,$$

so that $A^\dagger = B^{-1}$ is an inner inverse of A.

Proof: To show that B is invertible, consider the homogeneous equation $Bz = 0$, which is equivalent to

(6.33) $$Az = \sum_{k=1}^{d} u_k \langle v_k^*, z \rangle .$$

As this equation is solvable if and only if the right-hand side is orthogonal to $u_1^*, u_2^*, \ldots, u_d^*$ because A has a Fredholm theory, it follows from (6.26) that

(6.34) $$\langle v_k^*, z \rangle = 0 , \quad k = 1,2,\ldots,d ,$$

and thus $Az = 0$. This means that z is of the form $z = \alpha_1 v_1 + \alpha_2 v_2 + \ldots + \alpha_d v_d$, where the coefficients α_k are given by (6.34), and hence $z = 0$ is the unique solution of the homogeneous equation $Bz = 0$, which implies the existence of B^{-1}. To prove (6.32), note that from (6.29), (6.30), and (6.31),

(6.35) $$Bv_i = -\sum_{k=1}^{d} u_k \langle v_k^*, v_i \rangle = -u_i, \qquad i = 1,2,\ldots,d ,$$

hence

(6.36) $$B^{-1}u_k = -v_k, \qquad k = 1,2,\ldots,d,$$

and

(6.37) $$B^{-1}A = B^{-1}(B + \sum_{k=1}^{d} u_k >< v_k^*) = I - \sum_{k=1}^{n} v_k >< v_k^* ,$$

and (6.32) follows directly from (6.30). QED

Instead of (6.36), one could also use the relationships

(6.38) $$v_k^* B^{-1} = -u_k^* , \quad k = 1,2,\ldots,d,$$

to establish (6.32). The operator $B^{-1} = A^\dagger$ obtained from (6.31) is called *Hurwitz pseudoinverse* of A [31, pp. 165-168; 12], which goes back to 1912.

By the same reasoning as above, any operator of the form

(6.39) $$A^\dagger = (A - \sum_{k=1}^{d} u_k > \beta_k < v_k^*)^{-1}$$

for $\beta_1, \beta_2, \ldots, \beta_d$ such that $\beta_1\beta_2\ldots\beta_d \neq 0$ will be an inner inverse of A. However, as these operators are invertible, they cannot satisfy condition (2) of §1b which characterizes outer inverses; consequently, the construction (6.39), while useful for some purposes, only gives a partial solution to the problem of finding generalized inverses.

Another matter of computational importance relates to the calculation of generalized inverses of perturbations of operators with known generalized inverses. Suppose, for example, that one has an efficient technique to obtain the Moore-Penrose generalized inverse $A^\dagger$ of A [27], and then would like to use the result to obtain the generalized inverses of perturbed operators $B = A + \Delta A$ with less effort than calculating $B^\dagger$ *ab initio*, or error bounds for the approximation of $B^\dagger$ by $A^\dagger$. As $A^\dagger$ is not a continuous function of A in general, it would be expected that analytic perturbation methods apply only under restrictive conditions, as even for $\|\Delta A\|$ arbitrarily small, one of the algebraic catastrophes that the rank of B is greater or less than the rank of A could occur. Most applications of analytic perturbation theory to the above problems are carried out under assumptions that ensure $\mathit{rank}(B) = \mathit{rank}(A)$, or that the change in rank is known [23, pp. 333-351]. Algebraic perturbation methods, on the other hand, are not necessarily subject to this kind of limitation. For rank one modifications of A, C. D. Meyer, Jr. [19; 23, pp. 351-352] has obtained formulas of the type

$$(A + u\rangle\langle v)^\dagger = A^\dagger + G \tag{6.40}$$

for all six possible cases, where G depends on $A^\dagger$ and the data. More general finite-rank modifications (5.42) of A can then be handled by the method of successive rank one modifications corresponding to (5.51)-(5.52). This latter algorithm was originated by Greville [9] for the recursive calculation of the Moore-Penrose generalized inverse of a matrix. Formula (6.40) reduces to (5.28) in the special case that A is invertible, as for any generalized inverse of A, $A^\dagger = A^{-1}$ for all $A \in \mathcal{I}$. This suggests the computational strategy of using a method for generalized inversion on an operator which is suspected of being singular or nearly singular. If the operator or the perturbed operator actually involved in the calculation is nonsingular, then this technique will yield its inverse, whereas a straightforward inversion method might fail.

Another approach to ill-posed problems is to approximate them by a perturbed problem which is well conditioned. An example is the technique of *regularization*, due to A. N. Tihonov [39, 40], which has close connections with the subject of generalized inverses [22]. If the operator A in (1.1) does not have a bounded inverse, then the smallest perturbation Δy in the data can cause an enormous change Δx in the solution of the perturbed problem (2.3) as compared to the solution of the reference problem. A typical situation in which problems of this type arise in applications is that X and Y are Hilbert spaces, and $A = K$ is a compact operator. The prototype of the resulting equation

(6.41) $$Kx = y\ ,\quad K \in \mathcal{K}\ ,$$

is the linear Fredholm integral equation of the *first kind* ,

(6.42) $$\int_0^1 K(s,t)x(t)\,dt = y(s)\ ,\qquad 0 \leq s \leq 1\ .$$

As perturbations in (6.42) in actual practice are inevitable, due to errors of measurement, discretization, and computation, direct numerical solution of (6.42) by standard techniques that work well for the integral equation (4.5) of second kind are rarely successful. The same observation may be made for (6.41) as compared to

(6.43) $$(\alpha I - K)x = y$$

for $\alpha \neq 0$. In order to find an acceptable approximate solution of the perturbed version of (6.41), the method of regularization consists of finding an element $w(\alpha) \in X$ which minimizes the functional

(6.44) $$f(w;\alpha) = \|Kw - z\|^2 + \alpha^2\|w\|^2\ .$$

Thus, (6.44) represents a trade-off between the fidelity with which the perturbed equation $Kw = z$ is satisfied, and the size of the norm of the corresponding solution. The parameter α (or sometimes α^2) in (6.44) is called the *regularization parameter* . The crucial problem in this field is the determination of the *optimal* regularization parameter, for which the value of $f(w;\alpha)$ is minimum, or at least a method for obtaining good approximations to the optimal value. A significant recent advance in this area is the application by Grace Wahba [41] of the method of weighted cross-validation to the case that the perturbation is due to discretization of the data with random errors of the type known as "white noise".

7. The eigenvalue-eigenvector problem. As stated in §1c, this problem is to find eigenvalues λ and *right eigenvectors* $x \neq 0$ satisfying (1.9), where $A \in L(X,X)$, X a Hilbert space. It follows that one is interested in the values of λ for which the linear operator

(7.1) $$T(\lambda) = A - \lambda I$$

is singular, and one may also want to find the *left eigenvectors* $y \neq 0$ of A corresponding to the eigenvalue λ which satisfy the homogeneous equation

(7.2) $$y(A - \lambda I) = 0\ .$$

The additional assumption will be made that the values of λ considered are restricted to those for which $T(\lambda)$ has a Fredholm theory. This condition does not exclude any λ in the finite-dimensional algebraic case; however, for Fredholm integral operators of the first kind or compact operators in general, it is customary to formulate the eigenvalue-eigenvector problem in terms of the reciprocal eigenvalues $\mu = 1/\lambda$, as the operator

$$(7.3) \qquad S(\mu) = I - \mu K, \quad K \in \mathcal{K},$$

will have a Fredholm theory for all scalars $\mu \in \Lambda$ by Theorem 5.3. This is equivalent to excluding $\lambda = 0$ from consideration in (7.1) if A is compact.

In order to contemplate the application of analytic perturbation methods to the eigenvalue-eigenvector problem, it is essential to determine conditions under which this problem is well-posed, as the operator $T(\lambda)$ will be singular if λ is an eigenvalue. One way to do this is to convert equation (1.9) and the normalization condition (1.10) into the nonlinear system

$$(7.4) \qquad P(q) := \begin{pmatrix} Ax - \lambda x \\ \\ \frac{1}{2} - \frac{1}{2} < x,x > \end{pmatrix} = 0$$

in the product space $Q = X \times \Lambda$ of vectors $q = (x,\lambda)^T$, $x \in X$, $\lambda \in \Lambda$. Suppose that $q_1 = (x_1,\lambda_1)^T$ is a solution of (7.4); that is, λ_1 is an eigenvalue of A, and x_1 is a corresponding normalized eigenvector. Then, the implicit function theorem [10] guarantees continuous dependence of the solution of (7.4) on the data if the linear operator $P'(q_1) \in L(Q,Q)$ has a bounded inverse, where $P'(q)$ is the *Frechet derivative*

$$(7.5) \qquad P'(q) = \begin{pmatrix} A - \lambda I & -x > \\ \\ - < x & 0 \end{pmatrix}$$

of the operator P at q [30, pp. 97-100]. The formulation (7.4), while not the most general [1], has the advantage that if A is *Hermitian* ($A^* = A$ [38, pp. 324-327]), then so is $P'(q)$. The following theorem gives an explicit formulation of the inverse operator $[P'(q_1)]^{-1}$ in this case if the defect of $T(\lambda)$ is equal to one, that is, if all solutions x of the homogeneous equation $T(\lambda_1)x = 0$ are scalar multiples of the normalized eigenvector x_1, making use of the fact that the right and left eigenvectors of an Hermitian operator can be identified.

<u>Theorem 7.1.</u> If A is Hermitian, $q_1 = (x_1,\lambda_1)^T$ satisfies (7.4), and the defect of $T(\lambda_1)$ is equal to one, then

$$(7.6) \qquad [P'(q_1)]^{-1} = \begin{pmatrix} B_1^{-1} - x_1 >< x_1 & -x_1 > \\ \\ - < x_1 & 0 \end{pmatrix}$$

where

$$(7.7) \qquad B_1^{-1} = (A - \lambda_1 I - x_1 >< x_1)^{-1}$$

is the Hurwitz pseudoinverse of $A - \lambda_1 I$.

Proof: It follows by direct calculation and the use of (6.36) and (6.38) that

$$(7.8)\quad P'(q_1)\begin{pmatrix} B_1^{-1} - x_1 >< x_1 & -x_1 > \\ -< x_1 & 0 \end{pmatrix} = \begin{pmatrix} B_1^{-1} - x_1 >< x_1 & -x_1 > \\ -< x_1 & 0 \end{pmatrix} P'(q_1) =$$

$$= \begin{pmatrix} I & 0 \\ 0 & 1 \end{pmatrix},$$

the identity operator in $Q = X \times \Lambda$. QED

By the use of Theorem 6.2, formula (7.6) can be extended immediately to the non-Hermitian case $y_1 T(\lambda_1) = T(\lambda_1)x_1 = 0$, provided the defect of $T(\lambda_1)$ remains equal to one [1, §3]. Under these circumstances, results are available by the methods of analytic perturbation theory similar to those for nonsingular linear equations (1.1) [1, §5].

For the finite-dimensional case, perturbation methods and error analysis for the algebraic eigenvalue problem have been presented in great detail in the comprehensive work by J. H. Wilkinson [44, pp. 62-188]. Just one of these results will be cited here, which fits into the framework of algebraic perturbation theory. Suppose that w is a unit vector, and $p = (w,\mu)^T$ is an approximate solution of (7.4), so that

$$(7.9)\qquad (A - \mu I)w = r ,$$

with residual vector r. From equation (5.38), it follows that

$$(7.10)\qquad (A - r >< w^* - \mu I)w = 0 ,$$

so that w is an exact eigenvector of the perturbed operator

$$(7.11)\qquad B = A - r >< w^*$$

corresponding to the eigenvalue μ [44, pp. 170-171]. The perturbed operator B is simply a rank one modification of the reference operator A.

Another application of algebraic perturbation theory to the eigenvalue-eigenvector problem has been given by W. Stenger [37] to find inequalities between eigenvalues of perturbed and reference integral operators.

8. Linear programming. The solution of linear programming problems as formulated in §1d is one of the primary tools for decision making in government and commerce at the present time [8]. The number of variables involved is typically large, and a lot of computer time is expended for this purpose. Thus, an application of perturbation theory which would increase efficiency could result in substantial savings. Once again, the fact that the solutions do not depend continuously on the data in general limits the applicability of analytic perturbation techniques. A necessary and

sufficient condition for continuous dependence of the solution of the primal and dual linear programming problems in a neighborhood of solvable reference problems has been given recently by S. M. Robinson [34]. Studies of what is called *parametric programming* give conditions under which the solution of the reference problem remains unchanged under perturbation of the data [8, pp. 144-154]. On the subject of error estimation, P. Wolfe [45] has contributed a method for error analysis and control in the solution of linear programming problems.

Although changes in the objective function (1.14) are not usually difficult to deal with, perturbations in the constraints (1.15), as would result, for example, by the introduction of a new technology in an industry, may require the complete restarting of the solution method used. Consequently, the following problem may be of practical interest.

Problem 8.1. Given the solution x of (1.14)-(1.15) and the associated information, such as the choice of pivots in the simplex algorithm [45], find an efficient method for solving

(8.1) $$\text{minimize } f(w) := \langle d,w \rangle + \eta$$

subject to

(8.2) $$B\,w \leq z\ , \quad w \geq 0\ ,$$

where all perturbations in the reference data are of finite rank which is small compared to the size of the reference problem.

9. Nonlinear problems. Although this survey has been concerned mainly with linear problems, it should be mentioned that perturbation methods are widely applied to the solution of nonlinear operator equations

(9.1) $$P(x) = 0\ ,$$

where P maps X into Y, and also *fixed point problems* in X of the form

(9.2) $$x = H(x)\ .$$

(It is evident that (9.2) is a special case of (9.1); conversely, there are many ways to convert (9.1) into an equivalent fixed point problem.)

These problems are well-posed in the neighborhood of a solution x_0 if, for example, H is continuous and contractive [30, Chapter 2], or, more restrictively, if P is differentiable and

(9.3) $$\Gamma_0 = [P'(x_0)]^{-1} \in L(Y,X).$$

Depending on the smoothness of P, in this case one can base analytic perturbation techniques on the implicit function theorem [10], Newton's method and its variants, Taylor series expansions, inversion of power series, and so on [30, Chapter 4]. These methods are all essentially derived from the corresponding ideas of elementary scalar calculus.

Recently, W. Rheinboldt has given generalizations of the condition numbers (5.8) and (5.9) for nonlinear operators for which (9.3) holds, and a corresponding generalization of the perturbation formula (5.13) for error estimation [33].

Algebraic perturbation methods for nonlinear operator equations are less well investigated. A nonlinear operator F with range belonging to the finite-dimensional space

(9.4) $$Y_n = \mathit{span}\,\{y_1, y_2, \dots, y_n\}$$

will be of the form

(9.5) $$F(\cdot) = \sum_{j=1}^{n} y_j > f_j(\cdot) ,$$

where $f_1(\cdot), f_2(\cdot), \dots, f_n(\cdot)$ are (generally nonlinear) functionals on X. The perturbed operator equation

(9.6) $$Q(x) = 0 ,$$

where $Q = P - F$, is equivalent to the equation

(9.7) $$P(x) = \sum_{j=1}^{n} \xi_j y_j ,$$

where

(9.8) $$\xi_j = f_j(x) , \qquad j = 1,2,\dots,n .$$

Suppose, and this is the *big assumption*, that the equation $P(x) - y$ is solvable for $y \in Y_n$, that is, an operator G is known which gives

(9.9) $$x = G(\xi_1, \xi_2, \dots, \xi_n)$$

if $P(x) = y$ is of the form (9.7). Then, applying $f_1, f_2, \dots, f_n$ in turn to (9.9) yields the nonlinear system

(9.10) $$\xi_i = h_i(\xi_1, \xi_2, \dots, \xi_n) , \qquad i = 1,2,\dots,n ,$$

where $h_1 = f_1 G$, $h_2 = f_2 G, \dots, h_n = f_n G$, which is a finite-dimensional fixed-point problem in Λ^n of the form (9.2). On the basis of the *additional assumption* that (9.10) is solvable, the substitution of its solutions $\xi_1, \xi_2, \dots, \xi_n$ into (9.9) provides a solution x of the nonlinear operator equation (9.6). As an example of this approach, the *Hammerstein integral equation* with kernel (5.31)

(9.11) $$x(s) - \int_0^1 K(s,t)\phi(t,x(t))\,dt = 0$$

is a rank one modification of the *nonlinear Volterra integral equation*

(9.12) $$x(s) - \int_0^s L(s,t)\phi(t,x(t))\,dt = 0$$

with kernel (5.33). Thus, if one can solve

(9.13) $$x(s) - \int_0^s L(s,t)\phi(t,x(t))\,dt = \xi u(s) ,$$

where

(9.14) $$\xi = \int_0^1 v(t)\phi(t,x(t))\,dt .$$

for $x(s) = g(s;\xi)$, then from (9.14), the system (9.10) is equivalent to the scalar fixed point problem

$$\xi = h(\xi) := \int_0^1 v(t)\phi(t,g(t;\xi))dt , \tag{9.15}$$

which is one nonlinear equation in one unknown [32, §5].

Although quite a bit is known about nonlinear systems (9.10) in finite-dimensional spaces [28], the theory and practice of their solution is far from the highly developed technology available for finite linear systems (5.44). There is also the ever-present big assumption. Even though (9.9) is not obtainable explicitly, the form of the problem (9.7) suggests iteration: Solve (9.7) for given $\xi_1^{(0)},\xi_2^{(0)},\ldots,\xi_n^{(0)}$, substitute into (9.10) to obtain

$$\xi_i^{(1)} = h_i(\xi_1^{(0)},\xi_2^{(0)},\ldots,\xi_n^{(0)}), \quad i = 1,2,\ldots,n , \tag{9.16}$$

and so on. In the case that (9.6) is a boundary-value problem for a nonlinear differential equation, this is called "shooting" [13, Chapter 2, also §6.1]. Of course, this iteration may not converge, and some other method for solving (9.6) may be more appropriate .

This section will also conclude with an important problem, as much more work needs to be done.

Problem 9.1. For differentiable P , develop existence theory and find effective techniques for computing solutions x_0 of the nonlinear operator equation (9.1) in the case that $P'(x_0)$ does not have a bounded inverse.

REFERENCES

1. Anselone, P.M. and Rall, L. B., The solution of characteristic value-vector problems by Newton's method, Numer. Math. 11 (1968), 38-45. MR 36, #4795.

2. Banach, S., Théorie des Opérations Linéaires, Warsaw, 1932. Reprinted by Chelsea, New York, 1963.

3. Ben-Israel, A. and Greville, T. N. E., Generalized Inverses: Theory and Applications, John Wiley & Sons, New York, 1974. MR 53, #469.

4. Courant, R. and Hilbert, D., Methods of Mathematical Physics, Vol. 1., Interscience, New York, 1953.

5. Forsythe, G. E. and Moler, C. B., Computer Solution of Linear Algebraic Systems, Prentice-Hall, Englewood Cliffs, N. J., 1967. MR 36, #2306.

6. Fredholm, I., Sur une classe d'équations fonctionnelles, Acta Math. 27 (1903), 365-390.

7. Friedman, Bernard, Principles and Techniques of Applied Mathematics, John Wiley & Sons, New York, 1956. MR 18, 43.

8. Gass, S. J., Linear Programming. Methods and Applications, 3rd Ed., McGraw-Hill, New York, 1969. MR 42, #1509.

9. Greville, T. N. E., Some applications of the pseudoinverse of a matrix, SIAM Rev. 2 (1960), 15-22. MR 22, #1067.

11. Householder, A. S., The Theory of Matrices in Numerical Analysis, Blaisdell, New York, 1964. MR 30, #5475.

12. Hurwitz, W. A., On the pseudo-resolvent to the kernel of an integral equation, Trans. Amer. Math. Soc. 13 (1912), 405-418.

13. Keller, H. B., Numerical Methods for Two-Point Boundary-Value Problems, Blaisdell, Waltham, Massachusetts, 1968. MR 37, #6038.

14. Lonseth, A. T., Systems of linear equations with coefficients subject to error, Ann. Math. Statistics 13 (1942), 332-337. MR 4, 90.

15. Lonseth, A. T., On relative errors in systems of linear equations, Ann. Math. Statistics 15 (1944), 323-325. MR 6, 51.

16. Lonseth, A. T., The propagation of error in linear problems, Trans. Amer. Math. Soc. 62 (1947), 193-313. MR 9, 192.

17. Lonseth, A. T., An extension of an algorithm of Hotelling, Proceedings of the Berkeley Symposium on Mathematical Statistics and Probability, 1945, 1946, pp. 353-357. University of California Press, Berkeley and Los Angeles, 1949. MR 10, 627.

18. Lonseth, A. T., Sources and applications of integral equations, SIAM Rev. 19 (1977), 241-278.

19. Meyer, C. D., Jr., Generalized inversion of modified matrices, SIAM J. Appl. Math. 24 (1973), 315-323. MR 47, #5010.

20. Nashed, M. Z., Generalized inverses, normal solvability, and interation for singular operator equations, Nonlinear Functional Analysis and Applications, ed. by. L. B. Rall, pp. 311-359, Academic Press, New York, 1971. MR 43, #1003.

21. Nashed, M. Z. (Editor), Generalized Inverses and Applications, Academic Press, New York, 1976.

22. Nashed, M. Z., Aspects of generalized inverses in analysis and regularization, Generalized Inverses and Applications, ed. by M. Z. Nashed, pp. 193-244, Academic Press, New York, 1976.

23. Nashed, M. Z., Perturbations and approximations for generalized inverses and linear operator equations, Generalized Inverses and Applications, ed. by M. Z. Nashed, pp. 325-396, Academic Press, New York, 1976.

24. Nashed, M. Z. and Rall, L. B., Annotated bibliography on generalized inverses and applications, Generalized Inverses and Applications, ed. by M. Z. Nashed, pp. 771-1041, Academic Press, New York. 1976.

25. Nashed, M. Z. and Votruba, G. F., A unified operator theory of generalized inverses, Generalized Inverses and Applications, ed. by M. Z. Nashed, pp 1-109, Academic Press, New York, 1976.

26. Noble, B., Applied Linear Algebra, Prentice-Hall, Englewood Cliffs, N.J., 1969. MR 40, #153.

27. Noble, B., Methods for computing the Moore-Penrose generalized inverse, and related matters, Generalized Inverses and Applications, ed. by M. Z. Nashed, pp. 245-301, Academic Press, New York, 1976.

28. Ortega, J. M. and Rheinboldt, W. C., Iterative Solution of Non-linear Equations in Several Variables, Academic Press, New York, 1970. MR 42, #8686.

29. Rall, L. B., Error bounds for iterative solution of Fredholm integral equations, Pacific J. Math. 5 (1955), 977-986. MR 18, 72.

30. Rall, L. B., Computational Solution of Nonlinear Operator Equations, John Wiley & Sons, New York, 1969. MR 39, #2289.

31. Rall, L. B., The Fredholm pseudoinverse--an analytic episode in the history of generalized inverses, Generalized Inverses and Applications, ed. by M. Z. Nashed, pp. 149-173, Academic Press, New York, 1976.

32. Rall, L. B., Resolvent kernels of Green's function kernels and other finite rank modifications of Fredholm and Volterra kernels, J. Optimization Theory Appl. (to appear). Preprint: MRC TSR#1670, University of Wisconsin-Madison, 1976.

33. Rheinboldt, W. C., On measures of ill-conditioning for nonlinear equations, Math. Comp. 30 (1976), 104-111.

34. Robinson, S. M., A characterization of stability in linear programming, Operations Res. (to appear). Preprint: MRC TSR#1542, University of Wisconsin-Madison, 1975.

35. Sherman, J. and Morrison, W. J., Adjustment of an inverse matrix corresponding to a change in one element of a given matrix, Ann. Math. Statistics 21 (1950), 124-127. MR 11, 693.

36. Shieh, A. S., On the numerical solution of Poisson's equation by a capacitance matrix method, Proceedings of the 1977 Army Numerical Analysis and Computers Conference, U. S. Army Research Office, Research Triangle Park, N.C. (to appear).

37. Stenger, W., On perturbations of finite rank, J. Math. Anal. Appl. 28 (1969), 625-635.

38. Taylor, A. E., Introduction to Functional Analysis, John Wiley & Sons, New York, 1958. MR 20, #5411.

39. Tihonov, A. N., On the solution of ill-posed problems and the method of regularization (Russian), Dokl. Akad. Nauk SSSR 151 (1963), 501-504. MR 28, #5576.

40. Tihonov, A. N., On the regularization of ill-posed problems (Russian), Dokl. Akad. Nauk SSSR 153 (1963), 49-52. MR 28, #5577.

41. Wahba, Grace, Practical approximate solutions to linear operator equations when the data are noisy, SIAM J. Numer. Anal. 14 (1977) (to appear). Preprint: Department of Statistics Technical Report #430, University of Wisconsin-Madison, 1975.

42. Widlund, O. and Proskurowski, W., On the numerical solution of Helmholtz's equation by the capacitance matrix method, ERDA Rep. C00-3077-99, Courant Institute of Mathematical Sciences, New York University, New York, 1975.

43. Wilkinson, J. H., Rounding Errors in Algebraic Processes, Prentice-Hall, Englewood Cliffs, N. J., 1963. MR 28, #4661.

44. Wilkinson, J. H., The Algebraic Eigenvalue Problem, Claredon Press, Oxford, 1965. MR 32, #1894.

45. Wolfe, P., Error in the solution of linear programming problems, Error in Digital Computation, Vol. II, ed. by L. B. Rall, pp. 271-284, John Wiley & Sons, New York, 1965. MR 32, #4830.

46. Woodbury, M. A., Inverting modified matrices, Statistical Research Group, Memo. Rep. no. 42, Princeton University, Princeton, N. J., 1950. MR 12, 361.

Difference Approximations to Boundary Value Problems with Deviating Arguments

by

G. W. Reddien

Mathematics Department
Vanderbilt University
Nashville, Tennessee 37235

1. Introduction.

In this paper we present a numerical method for computing approximations to the solutions of a class of boundary value problems for ordinary differential equations that have both delays and advances in the argument of the solution. Problems with just delays have been considered in [1-5], and mixed problems have been treated apparently just in [6]. Mixed problems arise quite naturally as the necessary conditions for certain optimal control problems with delays in the state equation.

An example of such a problem is the following. Let A and B be constant $s \times s$ matrices and C a constant $s \times m$ matrix. Let ϕ be a vector in R^s, let G and Q be positive semidefinite symmetric $s \times s$ matrices and let R be a positive definite symmetric $m \times m$ matrix. The following optimal control problem (P) was studied in detail in [7]: Find an R^m-valued control u^* in $L_2[0,t_f]$ which minimizes the performance criterion

$$J(u) = \tfrac{1}{2}[x(t_f)^T G x(t_f)] + \tfrac{1}{2}\int_0^{t_f} \{x(t)^T Q x(t) + u(t)^T R u(t)\} dt \tag{1.1}$$

subject to the constraints

$$\dot{x}(t) = Ax(t) + Bx(t-r) + Cu(t), \quad t > 0, \quad x(t) = \phi(t), \quad t \le 0. \tag{1.2}$$

The following maximum principle gives necessary conditions for the problem (P) and was given in [7].

Theorem (Maximum Principle). Let u^* be the optimal control for the problem (P). Then there exists a scalar $\alpha \leq 0$ and an R^s-values function ψ in $L_2[0,t_f + r]$ such that $|\alpha| + \|\psi\|_{L_2} \neq 0$, $\psi(t) = 0$ for t in $(t_f,t_f + r]$, and such that

(i) ψ is absolutely continuous on $[0,t_f]$ and on this interval satisfies the equation

$$\dot{\psi}(t)^T = (-\alpha Qx(t))^T - \psi(t)^T A - \psi(t + r)^T B;$$

(ii) ψ satisfies the transversality condition

$$\psi(t_f) = \alpha Gx(t_f),$$

(iii) u^* satisfies the pointwise maximum principle

$$\tfrac{1}{2}\alpha u^*(t)^T Ru^*(t) + \psi(t)^T Cu^*(t) = \max_{v\in R^m} \{\tfrac{1}{2}\alpha v^T Rv + \psi(t)^T Cv\},$$

for almost all t in $[0,t_f]$.

There are some cases when it can be shown that necessarily $\alpha \neq 0$. Then without loss of generality one can set α equal to -1 and the necessary conditions become

$$\begin{aligned} \dot{x}(t) &= Ax(t) + Bx(t - r) - CR^{-1}C^T\psi(t) \\ \dot{\psi}(t)^T &= (Qx(t))^T + \psi(t)^T A - \psi(t + r)^T B, \end{aligned} \tag{1.3}$$

with boundary conditions

$$\begin{aligned} x(t) &= \phi(t) \quad \text{for} \quad -r \leq t \leq 0 \\ \psi(t) &= 0 \quad \text{for} \quad t_f < t \leq t + r \\ \psi(t_f) &= -Gx(t_f). \end{aligned} \tag{1.4}$$

It is for boundary value problems of the form (1.3)-(1.4) that we will develop a numerical method. Numerical methods that have been studied for the problem (P) generally involve discretizing (1.1)-(1.2) directly, see [8,9]. Discretizing (1.3)-(1.4) does not appear to be straightforward. First, (1.3)-(1.4) is a coupled vector system of equations with both an advance and a delay which will make the matrices arising out of discretizations have an unusual structure. And second, the solution to (1.3)-(1.4) is not going to be smooth. For example, it is usually the case that $\ddot{X}$ will have a jump discontinuity at $t = r$ and $\dot{\psi}$ will have a jump discontinuity at $t_f - r$. We present a numerical method for (1.3)-(1.4) that takes this into account. This method was analyzed in [6] under a restrictive assumption on the boundary conditions. The results given here generalize and improve those earlier results.

Section 2 contains a statement of the problem to be treated and a definition of the numerical method. Also, needed hypotheses are made. Section 3 contains a convergence analysis and Section 4 contains a numerical example.

2. Problem and Method.

Let r and t_f be positive numbers with $t_f > r$. We consider the boundary value problem to solve for an absolutely continuous function u on $[0,t_f]$ satisfying a.e. the equation

$$(2.1)\qquad \dot{u}(t) = A(t)u(t) + B(t)u(t - r) + C(t)u(t + r) + f(t), \quad 0 < t < t_f,$$

subject to the boundary conditions

$$(2.2)\qquad Mu(0) + Nu(t_f) = \theta,$$

where A, B and C are $k \times k$ matrix valued functions with B zero for $t < r$ and C zero for $t > t_f - r$, M and N are $k \times k$ matrices, f and u are R^k-valued functions, and θ is a vector in R^k. We will assume (H1) that A, B and C are piecewise continuous matrix valued functions and that f is a piecewise continuous vector valued function over $[0,t_f]$, each with only a finite number of jump discontinuities. Since f is only piecewise continuous, formulation of problems such as (1.3)-(1.4) into the form (2.1) poses no difficulty since the boundary behavior outside $[0,t_f]$ can be built into eqn (2.1) through f. The remaining boundary condition can be handled through (2.2). Define $u = (x\ \psi)^T$, $M = \begin{pmatrix} I & 0 \\ 0 & 0 \end{pmatrix}$, $N = \begin{pmatrix} 0 & 0 \\ G & I \end{pmatrix}$ and $\theta = \begin{pmatrix} \phi(0) \\ 0 \end{pmatrix}$ where all submatrices and vectors have order s. Then the boundary condition (2.2) simply says that $x(0) = \phi(0)$ and $Gx(t_f) + \psi(t_f) = 0$. Without loss of generality, we will throughout simply take $\theta = 0$.

We remark here that one can treat more general problems than (2.1)-(2.2) with the analysis to be given here. The addition of multiple delay terms or even more general functional behavior such as terms of the form $\int_{t-r}^{t} u(s)ds$ can be made quite easily. The more general equation causes only notational and not mathematical difficulties. These more general problems are treated by a direct minimization method in [9].

The difference method to be used is defined over a mesh $\{t_j^n\}_{j=0}^{n}$ where $t_0^n = 0$, $t_n^n = t_f$, $t_j^n - t_{j-1}^n = h_j^n > 0$, and $\Delta_n = \max_j h_j^n$. We will assume that any sequence of meshes considered here is quasi-uniform. In order to simplify notation, we will omit the superscript n on the mesh designations. Our numerical method uses centered differences to solve for u_i in R^k as an approximation to u at t_i. Namely, we solve the system

$$(2.3)\quad \frac{u_j - u_{j-1}}{h_j} = A(t_{j-\frac{1}{2}})\frac{u_j + u_{j-1}}{2} + B(t_{j-\frac{1}{2}})(\theta_j u_{\beta(j)} + (1-\theta_j)u_{\beta(j)+1})$$

$$+ C(t_{j-\frac{1}{2}})(\varepsilon_j u_{\gamma(j)} + (1-\varepsilon_j)u_{\gamma(j)+1}) + f(t_{j-\frac{1}{2}}), \qquad 1 \le j \le n,$$

subject to the boundary conditions

(2.4) $$Mu_0 + Nu_n = 0$$

where we have used the following notations: $t_{j-\frac{1}{2}} = t_j - h_j/2$, $\beta(j)$ is defined so that $t_{\beta(j)}$ is the first mesh point less than or equal to $t_{j-\frac{1}{2}} - r$ if $t_{j-\frac{1}{2}} - r \geq 0$, $\gamma(j)$ is defined so that $t_{\gamma(j)+1}$ is the first mesh point greater than or equal to $t_{j-\frac{1}{2}} + r$ if $t_{j-\frac{1}{2}} + r \leq t_f$, $\theta_j = (t_{\beta(j)+1} - (t_{j-\frac{1}{2}} - r))/h_{\beta(j)+1}$, and $\varepsilon_j = (t_{\gamma(j)+1} - (t_{j-\frac{1}{2}} + r))/h_{\gamma(j)+1}$. Recall that $B(t_{j-\frac{1}{2}}) = 0$ for $t_{j-\frac{1}{2}} < r$ and $C(t_{j-\frac{1}{2}}) = 0$ for $t_{j-\frac{1}{2}} > t_f - r$ so that the definitions of ε_j, θ_j, $\beta(j)$ and $\gamma(j)$ are complete. Note also that the mesh spacings are not assumed to be uniform. Our discretization (2.3)-(2.4) of (2.1)-(2.2) is an adaptation of the centered difference method for ordinary differential equations as studied in [10,11].

We next make several additional assumptions. In general, it is too restrictive to assume $(M + N)^{-1}$ exists, although this assumption is satisfied for the specific control problem given earlier. This hypothesis was made in [6]. Rather, we assume (H2) that for some matrix D, the equation $\dot{u} + Du = 0$ on $[0,t_f]$ and the boundary conditions $Mu(0) + Nu(t_f) = 0$ imply that u is identically zero. We assume (H3) that the problem (2.1)-(2.2) has a unique solution for any piecewise continuous function f. Finally, let Σ denote the set of discontinuities of A, B, C and the given function f in (2.1). We require (H4) that the partitions $\{t_j\}$ of $[0,t_f]$ that are used contain Σ. From the form of eqn (2.1) and hypothesis (H1), it follows that all the discontinuities of $\dot{u}$ will be included in the mesh points. Later, to achieve a high rate of convergence, we will want Σ to contain the points of discontinuity of $\ddot{u}$ and $\dddot{u}$. These points can be identified a priori based on the form of (2.1).

3. Convergence Results.

Because (2.1) involves both an advance and a delay, the matrix arguments of [1] or [10] seem unlikely to apply here. The matrices are not symmetric and although sparse, the advance and the delay introduce nonzero entries well off the diagonal. We give an operator theoretic argument using the theory of collectively compact operators. An alternative approach can be developed using the ideas in [5]. We first establish some lemmas.

Lemma 3.1. The matrix $(M + Ne^{-D})^{-1}$ exists. Moreover, define

$$G(t,s) = \begin{cases} -e^{-tD}(M + Ne^{-D})^{-1}Ne^{(s-1)D} + e^{(s-t)D} & s \leq t \\ -e^{-tD}(M + Ne^{-D})^{-1}Ne^{(s-1)D} & t < s, \end{cases}$$

and let w be a piecewise continuous mapping from $[0,t_f]$ to R^k. Then $\dot{u} = w$ on $[0,t_f]$ and $Mu(0) + Nu(t_f) = 0$ if and only if $u(t) = \int_0^{t_f} G(t,s)w(s)ds$.

Proof. Let $(M + Ne^{-D})d = 0$, d in R^k and $d \neq 0$. Define $u(t) = e^{-tD}d$. Then $u = -De^{-tD}d$ and $-De^{-tD}d + De^{-tD}d = 0$. Thus u solves $\dot{u} + Du = 0$. Since $Md + Ne^{-D}d = 0$, the boundary conditions are satisfied by u. This contradicts (H2), and so $d = 0$, i.e. $(M + Ne^{-D})^{-1}$ exists. It is a straightforward computation to see that if $\dot{u} + Du = w$ plus $Mu(0) + Nu(t_f) = 0$, then $u(t) = -e^{-tD}(M + Ne^{-D})^{-1}N \int_0^t e^{(s-1)D}w(s)ds + \int_0^t e^{(s-t)D}w(s)ds$ which is equivalent to $u(t) = \int_0^{t_f} G(t,s)w(s)ds$.

We next establish a discrete version of the preceding lemma.

Lemma 3.2. Let w be continuous on $[0,t_f]$. Then the equation

$$\frac{u_i - u_{i-1}}{h_i} + D\left(\frac{u_i + u_{i-1}}{2}\right) = w_{i-\frac{1}{2}}, \quad i = 1, \ldots, n, \tag{3.4}$$

with boundary conditions

$$(3.5) \qquad Mu_0 + Nu_n = 0$$

is equivalent to

$$u_i = -e^{t_i D}((M + N)e^{-D})^{-1} N \sum_{j=1}^{n} e^{t_{j-\frac{1}{2}} D} w_{j-\frac{1}{2}} h_j$$

$$(3.6) \qquad + e^{-t_i D} \sum_{j=1}^{i} e^{t_{j-\frac{1}{2}} D} w_{j-\frac{1}{2}} h_j + (B_n \{w_{j-\frac{1}{2}}\}_{j=1}^{n})_i$$

$$= \sum_{j=1}^{n} G(t_i, t_{j-\frac{1}{2}}) w_{j-\frac{1}{2}} h_j + (B_n \{w_{j-\frac{1}{2}}\})_i, \quad i = 0, \ldots, n,$$

where B_n is given in (3.8) as a matrix mapping R^{kn} into $R^{k(n+1)}$, $\|B_n\{w_{j-\frac{1}{2}}\}\| \le c(\Delta_n)\|\{w_{j-\frac{1}{2}}\}\|$, $c(\Delta_n) \to 0$ as $\Delta_n \to 0$, and $\|\cdot\|$ is the maximum norm.

Proof. We assume Δ_n is small enough so that both $(I + \frac{h_i}{2} D)^{-1}$ and $(I - \frac{h_i}{2} D)^{-1}$ exist. Then (3.4) is equivalent to

$$u_i = (I + \frac{h_i}{2} D)^{-1}(I - \frac{h_i}{2} D)u_{i-1} + (I + \frac{h_i}{2} D)^{-1} w_{i-\frac{1}{2}} h_i .$$

Define $P_i = (I + \frac{h_i}{2} D)^{-1}$ and $Q_i = (I - \frac{h_i}{2} D)$. Then it follows that

$$u_i = \prod_{j=1}^{i} P_j Q_j \; u_0 + \sum_{j=1}^{i} \left(\prod_{s=j+1}^{i} P_s Q_s \right) P_j w_{j-\frac{1}{2}} h_j .$$

Consider the matrix differential equation $\dot{Z} = -DZ$, $Z(0) = I$, on $[0,t_f]$. The midpoint difference scheme for this problem is $\frac{Z_i - Z_{i-1}}{h_i} = -D(\frac{Z_i + Z_{i-1}}{2})$, $Z_0 = I$, and so $Z_i = \prod_{j=1}^{i} P_j Q_j$. The solution to the matrix differential equation is $Z(t) = e^{-tD}$. Since the midpoint rule is known to be stable for initial value problems, it follows that $\|Z(t_i) - Z_i\| \le c\Delta_n^2$ for some constant $c > 0$. By the Banach lemma, we thus have (for Δ_n small) that $(M + N(\prod_{j=1}^{n} P_jQ_j))^{-1}$ exists. Thus (3.4) and (3.5) hold if and only if

$$u_i = -((\prod_{j=1}^{i} P_jQ_j)(M + N(\prod_{j=1}^{n} P_jQ_j))^{-1}N(\sum_{j=1}^{n} (\prod_{s=j+1}^{n} P_sQ_s)P_jw_{j-\frac{1}{2}}h_j)$$

(3.8)

$$+ \sum_{j=1}^{i} (\prod_{s=j+1}^{i} P_sQ_s)P_jw_{j-\frac{1}{2}}h_j \overset{\text{def}}{=} (F_n\{w_{j-\frac{1}{2}}\})_i \,, \quad i = 0, \ldots, n,$$

where the empty product is taken to equal one. Define

$$(B_n\{w_{j-\frac{1}{2}}\})_i = e^{t_iD}((M + N)e^{-D})^{-1}Ne^{-D} \sum_{j=1}^{i} e^{t_{j-\frac{1}{2}}D} w_{j-\frac{1}{2}}h_j$$

$$-e^{-t_iD} \sum_{j=1}^{i} e^{t_{j-\frac{1}{2}}D} w_{j-\frac{1}{2}}h_j + (F_n\{w_{j-\frac{1}{2}}\})_i, \; i = 0,\ldots,n,$$

The estimate $\|e^{-t_iD} - \prod_{j=1}^{i} P_jQ_j\| \le c\Delta_n^2$ gives $\|B_n\| = o(\Delta_n)$. This result and the definition of G give the claimed equivalence.

Eqn (2.1) may be written first as

$$\dot{u}(t) + Du(t) = Au(t) + Du(t) + Bu(t - r) + Cu(t + r) + f(t)$$

and then using Lemma 3.1 as

$$u(t) = \int_0^{t_f} G(t,s)((A(s) + D)u(s) + B(s)u(s - r) + C(s)u(s + r))ds$$

$$+ \int_0^{t_f} G(t,s)f(s)ds.$$

Define $Gf = \int_0^{t_f} G(t,s)f(s)ds$ and $Hu = \int_0^{t_f} G(t,s)((A + D)u(s) + Bu(s - r) + Cu(s + r))ds$. Then (2.1)-(2.2) becomes

$$u = Hu + Gf. \tag{3.9}$$

Note that Gf is defined with f only piecewise continuous. If we think of u as having values equal to zero outside of $[0,t_f]$, then eqn (3.9) may be studied in the space $C[0,t_f]$.

Next, we first write (2.3) in the form

$$\frac{u_j - u_{j-\frac{1}{2}}}{h_j} + D(\frac{u_j + u_{j-1}}{2}) = (A(t_{j-\frac{1}{2}}) + D)(\frac{u_j + u_{j-1}}{2})$$

$$+ B(t_{j-\frac{1}{2}})(\theta_j u_{\beta(j)} + (1 - \theta_j)u_{\beta(j)+1}) + C(t_{j-\frac{1}{2}})(\varepsilon_j u_{\gamma(j)} + (1 - \varepsilon_j)u_{\gamma(j)+1}) \tag{3.10}$$

$$+ f_{j-\frac{1}{2}}, \quad 1 \leq j \leq n.$$

Define $\{R_n\{u_j\}\}$ to be an n vector with i^{th} component equal to $(A(t_{i-\frac{1}{2}}) + D)(\frac{u_i + u_{i-1}}{2}) + B(t_{i-\frac{1}{2}})(\theta_i u_{\beta(i)} + (1 - \theta_i)u_{\beta(i)+1})$ $+ C(t_{i-\frac{1}{2}})(\varepsilon_i u_{\gamma(i)} + (1 - \varepsilon_i)u_{\gamma(i)+1})$. Then we use Lemma 3.2 to write (3.10) in the form

$$u_j = \sum_{i=1}^{n} G(t_j, t_{i-\frac{1}{2}})(R_n\{u_j\})_i + \sum_{i=1}^{n} G(t_j, t_{i-\frac{1}{2}})f_{i-\frac{1}{2}}h_i$$

$$+ (B_n\{f_{i-\frac{1}{2}}\})_j + (B_n\{R_n\ u_j\}\})_j, \quad j = 0, 1, \ldots, n. \tag{3.11}$$

Define $(H_n\{u_i\})_j$ to be the first term on the right in (3.11) and $(G_n\{f_{i-\frac{1}{2}}\})_j$ to be the next two terms. Then in vector form, (3.11) becomes

$$\{u_j\} = \{H_n\{u_i\}\} + \{G_n\{f_{i-\frac{1}{2}}\}\} + B_n\{R_n\{u_j\}\}.$$

Next define Q_n to be the mapping from $C[0,t_f]$ into $R^{k(n+1)}$ given by $Q_n u = \{u_i\}_{i=0}^{n}$ where $u_i = u(t_i)$, P_n to be the piecewise linear interpolation mapping at $\{t_i\}$ to $C[0,t_f]$, S_n to be the discretization mapping given by

$S_n u = \{u(t_{i-\frac{1}{2}})\}_{i=1}^n$, and finally $\bar{u}_n$ to be the piecewise linear and continuous interpolant at the mesh points of u_i. Then since a piecewise linear function over the mesh $\{t_i\}_{i=0}^n$ is uniquely determined by its values at the mesh points, (3.11) is equivalent to the equation

$$\bar{u}_n = P_n H_n Q_n \bar{u}_n + P_n G_n S_n f + P_n B_n R_n Q_n \bar{u}_n. \tag{3.12}$$

We will apply the theory of collectively compact operators to eqn (3.9) and eqn (3.12) as developed by Anselone [12] and his coworkers. Bounded linear operators $\{K_n\}_{n\geq 1}$ and K operating on a Banach space X are said to be collectively compact if (A1) K is compact, (A2) $K_n y \to Ky$ for each y in X as $n \to \infty$, and (A3) the set $\{K_n y: n \geq 1, \|y\| \leq 1\}$ has compact closure in X. We next establish that these hypotheses are satisfied for the operators introduced here. We let X be the Banach space $C[0,t_f]$ with the usual maximum norm.

<u>Lemma 3.3</u>. $P_n H_n Q_n + P_n B_n R_n Q_n$ converges pointwise on X as $\Delta_n \to 0$.

<u>Proof</u>. The matrices R_n are easily seen to satisfy $\|R_n\| \leq c$ for some constant $c > 0$. Using Lemma 3.2 and the fact that $\|P_n\| = 1$, it follows that $P_n B_n R_n Q_n$ converges in norm to zero. The lemma will thus be established if we show $P_n H_n Q_n$ converges pointwise to H on X. First note that $\|P_n H_n Q_n u - Hu\| \leq \|P_n\| \cdot \|H_n Q_n u - Hu\| + \|P_n Hu - Hu\|$. By the usual Arzela-Ascoli theorem argument, H is a compact operator. Since $P_n \to I$ on X as $\Delta_n \to 0$ and $\|P_n\| = 1$ for each n, then $\|P_n Hu - Hu\| \to 0$ uniformly on bounded sets in u as $\Delta_n \to 0$. Now $H_n Q_n u$ may be viewed as a special Riemann sum approximating Hu for each t. Thus $H_n Q_n u$ converges to Hu as $\Delta_n \to 0$ for each t in $[0,t_f]$. To complete the proof of this lemma, we need to show that this convergence is uniform in t. Using the definition of $H_n Q_n u$ given in (3.10), it follows that for $\xi < \eta$ in $[0,t_f]$,

$$|(H_nQ_nu)(\xi) - (H_nQ_nu)(\eta)| = |\sum_{\substack{i \\ \xi \le t_{i-\frac{1}{2}} < \eta}} w_i h_i|$$

where

$$w_i = (A(t_{i-\frac{1}{2}}) + D)(\frac{u_i + u_{i-1}}{2}) + B(t_{i-\frac{1}{2}})(\theta_i u_{\beta(i)} + (1 - \theta_i)u_{\beta(i)+1})$$

$$+ C(t_{i-\frac{1}{2}})(\varepsilon_i u_{\gamma(i)} + (1 - \varepsilon_i)u_{\gamma(i)+1}).$$

Then for some constant c independent of n,

$$|(H_nQ_nu)(\xi) - (H_nQ_nu)(\eta)| \le c \sum_{\substack{i \\ \xi \le t_{i-\frac{1}{2}} < \eta}} h_i \le c(\xi - \eta + 2\Delta_n).$$

Now let $\varepsilon > 0$ be given. Let N be chosen so that for $n \ge N$, $\Delta_n < \varepsilon/8c$. Since $(Hu)(t)$ is uniformly continuous in t, we may choose an increasing sequence of points $\{z_i\}$ in $[0,t_f]$ so that $|(Hu)(z_i) - (Hu)(z_{i+1})| \le \varepsilon/4$ for each i. Increase this set of points if necessary so that also $|z_i - z_{i+1}| \le \varepsilon/4c$. Since H_nQ_nu converges to Hu pointwise in t, we may increase N if necessary so that $|(H_nQ_nu)(z_i) - (Hu)(z_i)| \le \varepsilon/4$ for each z_i. Now for any t in $[0,t_f]$ and $n \ge N$, let z_i be the closest z_j to t. Then

$$|(H_nQ_nu)(t) - (Hu)(t)| \le |(H_nQ_nu)(z_i) - (Hu)(z_i)| + |(H_nQ_nu)(t) - (H_nQ_nu)(z_i)|$$

$$+ |(Hu)(z_i) - (Hu)(t)| \le \varepsilon/4 + c(|t - z_i| + 2\Delta_n) + \varepsilon/4 \le \varepsilon.$$

Hence the convergence is uniform, completing the proof of the lemma.

Lemma 3.4. $\{P_nH_nQ_n + P_nB_nR_nQ_n\}_{n\ge1}$ and H form a collectively compact family of operators on X as $\Delta_n \to 0$.

Proof. We first note that each of the operators $P_nH_nQ_n$ and $P_nB_nR_nQ_n$ is bounded linear and of finite rank, and so compact. Since $P_nB_nR_nQ_n$ converges in norm to zero, it follows from Proposition 3.1 of [12] that it will be sufficient to show $\{P_nH_nQ_n\}_{n\geq 1}$ and H form a collectively compact family. Now choose u in X and define

$$r_n(t_i,u) = ((A(t_{i-\frac{1}{2}}) + D)(\frac{u_i + u_{i-1}}{2}) + B(t_{i-\frac{1}{2}})(\theta_i u_{\beta(i)} + (1-\theta_i)u_{\beta(i)+1})$$

$$+ C(t_{i-\frac{1}{2}})(\varepsilon_i u_{\gamma(i)} + (1-\varepsilon_i)u_{\gamma(i)+1}))h_i$$

where $\{u_i\} = Q_n u$. Since A, B, C and D are bounded, it follows that $|r_n(t_i,u)| \leq ch_i$ for some constant c and all partitions $\{t_i\}$ as u ranges over the unit ball in X. Now let ξ and η be arbitrary in $[0,t_f]$ and satisfy $\xi < \eta$. Then arguing as in the proof of Lemma 3.3, it follows that

$$(3.13)\quad |\sum_{i=1}^{n} G(\xi,t_{i-\frac{1}{2}})r_n(t_i,u) - \sum_{i=1}^{n} G(\eta,t_{i-\frac{1}{2}})r_n(t_i,u)| = |\sum_{\substack{i \\ \xi\leq t_{i-\frac{1}{2}}<\eta}} r_n(t_i,u)|.$$

Now let $\xi = t_{i-\frac{1}{2}}$ for some i and let $t_{i-\frac{1}{2}} < \eta < t_{i+\frac{1}{2}}$. Then the right hand side of eqn (3.13) becomes less than or equal to ch_i where c is independent of t_i and u in the unit ball in X. Thus the line segment connecting

$$(t_{i-1}, \sum_{j=1}^{n} G(t_{i-1},t_{j-\frac{1}{2}})r_n(t_j,u)) \text{ to } (t_i, \sum_{j=1}^{n} G(t_i,t_{j-\frac{1}{2}})r_n(t_j,u)),$$

i.e. $P_nH_nQ_nu$, has slope bounded by c. Thus it follows that $|(P_nH_nQ_nu)(\xi) - (P_nH_nQ_nu(\eta)| \leq c|\xi - \eta|$ for all $\{t_i\}$ and ξ, η in $[0,t_f]$. By Lemma 3.3 and the uniform boundedness principle, $P_nH_nQ_n$ are uniformly bounded in norm. Then using Ascoli's theorem, the proof is complete.

It follows from hypothesis (H3) that $(I-H)^{-1}$ exists. Since H is completely continuous, then $(I-H)^{-1}$ is bounded. Then from Theorem 1.11 of [12], it follows that there exist constants $\delta > 0$ and $M > 0$ so that $(I - P_nH_nQ_n - P_nB_nR_nQ_n)^{-1}$ exists and $\|(I - P_nH_nQ_n - P_nB_nR_nQ_n)^{-1}\| \leq M$ for all partitions with $\Delta_n \leq \delta$. This result uses the collectively compact property established in Lemma 3.4. We can now give our basic convergence result. We say that a function f is in $PC^p\ [0,t_f]$ if $f, f', \ldots, f^{(p)}$ are piecewise continuous, each with only a finite number of jump discontinuities.

Theorem 3.5. Let hypotheses (H1), (H2), (H3) and (H4) hold. Then solutions to eqns (2.3)-(2.4) exist and are unique for all partitions $\{t_i\}$ with Δ_n sufficiently small. Let u, the solution of (2.1)-(2.2) be in $PC^p\ [0,t_f]$, $p \geq 1$. Finally, let all of the partitions used contain the discontinuities of $u, u', \ldots, u^{(p)}$. Then there exists a constant $c > 0$ so that

$$(3.14) \qquad \|u - \bar{u}_n\| \leq \begin{cases} c\Delta_n^{p-1}\ \omega(u^{(p)},\Delta_n) & \text{if } p = 1,2 \\ c\Delta_n^2 & \text{if } p = 3 \end{cases}$$

for all Δ_n sufficiently small where $\bar{u}_n$ is the piecewise linear interpolant of $\{u_j\}_{j=0}^n$ and $\omega(u^{(p)},\Delta_n)$ is the usual modulus of continuity function taken over every interval of continuity of $u^{(p)}$ in $[0,t_f]$.

Proof. For Δ_n small, $(I - P_nH_nQ_n - P_nB_nR_nQ_n)^{-1}$ exists and so the existence and uniqueness of $\{u_i\}$ follows from the representation of eqns (2.3)-(2.4) as $\bar{u}_n = P_nH_nQ_n\bar{u}_n + P_nB_nR_nQ_n\bar{u}_n + P_nG_nS_nf$. Now define $z_j = u(t_j)$, $j = 0, 1, \ldots, n$. Then using Taylor's theorem and the boundedness of A, B and C, $\{z_j\}$ will satisfy eqns (2.3)-(2.4) up to a truncation error term τ_j, $j = 1, \ldots, n$, for the jth equation where τ_j has order given by the right hand side of (3.14) as a function of t for some constant c independent of j and Δ_n. Define $\bar{z}_n$ to be the piecewise linear interpolant of $\{z_j\}$ and $\bar{\tau}_n$ to be the piecewise constant function with value on the open i^{th} subinterval of τ_i. It then follows as before that $\bar{z}_n = P_nH_nQ_n\bar{z}_n + P_nB_nR_nQ_n\bar{z}_n + P_nG_nS_nf + P_nG_nS_n\bar{\tau}_n$. Thus $\bar{z}_n - \bar{u}_n =$ $P_nH_nQ_n(\bar{z}_n - \bar{u}_n) + P_nB_nR_nQ_n(\bar{z}_n - \bar{u}_n) + P_nG_nS_n\bar{\tau}_n$, and so $\bar{z}_n - \bar{u}_n =$

$(I - P_nH_nQ_n - P_nB_nR_nQ_n)^{-1}P_nG_nS_n\overline{\tau}_n$. The operators $P_nG_nS_n$ and $(I - P_nH_nQ_n - P_nB_nR_nQ_n)^{-1}$ are uniformly bounded in norm, and the proof is complete.

4. Examples.

We first note that for a two-point boundary value problem of the form $\ddot{u} + qu(t - r) = f$ on $[0,t_f]$ with boundary conditions $u(0) = u(t_f) = 0$, the corresponding matrix $M + N$ for this problem when written in first order vector form is singular. Thus the device of adding Du to both sides of the equation enables us to apply Theorem 3.5.

We next present the results of some numerical experiments. Consider the simple problem

$$\min J(u) = \tfrac{1}{2}\gamma(x(3))^2 + \tfrac{1}{2}\int_0^3(u(t))^2dt$$

subject to

$$\dot{x}(t) = x(t - 1) + u(t), \quad 0 < t < 3,$$

$$x(t) \equiv 1, \quad -1 \leq t \leq 0.$$

The necessary conditions can be explicitly solved in this case with

$$u^*(t) = \psi(t) = \begin{cases} \delta\{-(t - 2)^2/2-3/2 & 0 \leq t \leq 1 \\ \delta(t - 3) & 1 < t \leq 2 \\ -\delta & 2 < t \leq 3 \end{cases}$$

where $\delta = 37\gamma/[6(1 + 319\gamma/30)]$. Note that $\dot{\psi}$ has jump discontinuities. We report the results of applying the method given in (2.3)-(2.4) to this problem with $\gamma = 3$ in Table 1. Assuming the method converges like $c\Delta^\beta$, the observed β is also presented in Table 1 for both the approximations to u^* and the associated response x^*. The parameter β should be approximately equal to 2 in this example from Theorem 3.5. We used uniform mesh spacings with $\Delta = 1/n$.

In Table 2, we present a comparison of a computed value for $J(u_n)$ using the midpoint rule with the exact value $J(u^*)$. Our computed value for $J(u_n)$ is designated as $J_n(u_n)$.

This same problem was treated in [7] using their method of averaging projections. We present their approximation, J_n, to $J(u^*)$, in Table 3. The parameter n refers to a discretization of the interval into subintervals of length $1/n$ so that our two methods involve the same number of variables.

They also obtained magnitudes for $|u^*(t) - u_9(t)|$ at selected points on the order of 1.4×10^{-3}. Their method appears to be linearly convergent whereas ours is second order. In addition, our absolute error estimates are smaller.

Table 1

n	$\lvert u_n(1)-u^*(1)\rvert$	β	$\lvert u_n(2)-u^*(2)\rvert$	β	$\lvert x_n(1)-x^*(1)\rvert$	β
5	7.8×10^{-5}	-	3.9×10^{-5}	-	2.10×10^{-4}	-
9	2.4×10^{-5}	2.0	1.2×10^{-5}	2.0	0.66×10^{-4}	2.0
13	1.1×10^{-5}	2.1	0.6×10^{-5}	1.9	0.32×10^{-4}	1.97

Table 2

n	$J_n(u_n)$	$\lvert J_n(u_n)-J(u^*)\rvert$	β
5	1.734606	$8.17\ 10^{-4}$	-
9	1.734042	$2.53\ 10^{-4}$	2.0
13	1.733910	$1.21\ 10^{-4}$	2.0

Table 3

n	$\lvert J_n-J(u^*)\rvert$	β
9	$1.89\ 10^{-2}$	-
13	$1.32\ 10^{-2}$	0.98
17	$1.01\ 10^{-2}$	1.00
20	$8.60\ 10^{-3}$	0.99

References

1. C. W. Cryer, The numerical solution of boundary value problems for second order functional differential equations by finite differences, Numer. Math. 20, 288-299 (1973).

2. K. deNevers and K. Schmitt, An application of the shooting method to boundary value problems for second order delay equations, J. Math. Anal. Appl. 36, 588-597 (1971).

3. G. W. Reddien and C. C. Travis, Approximation methods for boundary value problems of differential equations with functional arguments, J. Math. Anal. Appl. 46, 62-74 (1974).

4. G. W. Reddien and G. F. Webb, Boundary value problems for functional differential equations with L^2 initial functions, Trans. Am. Math. Soc. 223, 305-321 (1976).

5. G. W. Reddien, Difference approximations of boundary value problems for functional differential equations, J. Math. Anal. Appl., to appear.

6. F. H. Mathis and G. W. Reddien, Numerical solution of hereditary control problems using necessary conditions, Comp. and Maths. with Appls. 2, 193-200 (1976).

7. H. T. Banks, J. A. Burns, E. M. Cliff and P. R. Thrift, Numerical solutions of hereditary control problems via an approximation technique, CDS Technical Report 75-6, Division of Applied Mathematics, Brown University (1975).

8. H. T. Banks and J. A. Burns, An abstract framework for approximate solutions to optimal control systems governed by hereditary systems, International Conference on Differential Equations (H. A. Antosiewicz, ed.) 10-25. Academic Press, New York (1975).

9. F. H. Mathis and G. W. Reddien, Difference approximations to control problems with functional arguments, SIAM J. Control, to appear.

10. H. B. Keller, Accurate difference methods for linear ordinary differential systems subject to linear constraints, SIAM J. Numer. Anal. 6, 8-30 (1969).

11. H. B. Keller, Accurate methods for nonlinear two-point boundary value problems, SIAM J. Numer. Anal. 11, 305-320 (1974).

12. P. M. Anselone, Collectively Compact Operator Approximation Theory. Prentice-Hall, New Jersey (1971).

APPLICATIONS OF BANACH SPACE INTERPOLATION TO FINITE ELEMENT THEORY *

Ridgway Scott
Applied Mathematics Department
Brookhaven National Laboratory
Upton, New York 11973

1. Introduction

Our purpose is two fold. First, we give an introduction to the most basic technique, operator interpolation, for obtaining estimates for the finite element error for nonsmooth solutions (section 4). Then we move on to more complicated applications (sections 5 and 6) to problems first considered by the author in [18]. These problems find their natural setting in the Banach space interpolation framework, and we are able to settle affirmatively an open question from [18] using these more powerful techniques. In the following section (#2), we recall the abstract results from Banach space interpolation theory to be used, and in section 3, we give examples of interpolation spaces and the sort of functions that one expects to find in them. In the last section (#7), we mention an open problem in interpolation theory related to proving L^p estimates for the error arising from finite element approximation.

2. Definitions and basic results from interpolation theory

Let B_0 and B_1 be Banach spaces such that there is a continuous inclusion $B_1 \subset B_0$. For $u \in B_0$ and $t>0$, define

$$K(u,t) = \inf\{\|u-v\|_{B_0} + t\|v\|_{B_1} \; : \; v \in B_1\}.$$

---

*Work performed under the auspices of the ERDA.

For $\theta\in]0,1[$ and $p\in[1,\infty]$, $B_{\theta,p} = [B_0,B_1]_{\theta,p}$ is defined to be the Banach space with norm

$$\|u\|_{B_{\theta,p}} = \begin{cases} \left(\int_0^\infty K(u,t)^p t^{-\theta p-1}dt\right)^{1/p} & \text{if } p<\infty \\ \sup\{K(u,t)t^{-\theta}: t>0\} & \text{if } p=\infty. \end{cases}$$

If also $\lambda\in]0,1[$ and $q\in[1,\infty]$, then we say

$$(\theta,p) < (\lambda,q) \quad \text{if} \begin{cases} \theta<\lambda \\ \text{or} \\ \theta=\lambda \quad \text{and} \quad p>q. \end{cases}$$

If $(\theta,p) \leq (\lambda,q)$, then there exist the following continuous inclusion mappings:

(2.1) $$B_0 \subset B_{\theta,p} \subset B_{\lambda,q} \subset B_1.$$

In particular, we have for any Banach space B

(2.2) $$[B,B]_{\theta,p} \cong B.$$

The <u>iteration theorem</u> states that if $\lambda,\mu,\nu\in]0,1[$ and $p,q,r\in[1,\infty]$, then

(2.3) $$[[B_0,B_1]_{\lambda,p}, [B_0,B_1]_{\mu,q}]_{\nu,r} \cong [B_0,B_1]_{\theta,r}$$

where $\theta = (1-\nu)\lambda + \nu\mu$ (note the lack of dependence on p and q).

Now suppose we have two interpolation pairs (A_0,A_1) and (B_0,B_1) as above, and a linear map $T:A_i \to B_i$ having a finite norm M_i, $i=0,1$. In a picture, we assume that

$$T: \begin{cases} A_0 \to B_0 & \text{with norm } M_0 \\ A_1 \to B_1 & \text{with norm } M_1. \end{cases}$$

Then the <u>operator interpolation property</u> says that T may be viewed as a bounded linear map of $A_{\theta,p} \to B_{\theta,p}$ having norm $\le 2M_0^{1-\theta}M_1^{\theta}$ for any $\theta \in]0,1[$ and $p \in [1,\infty]$:

$$T: [A_0,A_1]_{\theta,p} \to [B_0,B_1]_{\theta,p} \quad \text{with norm} \le 2M_0^{1-\theta}M_1^{\theta}.$$

The above results may be found, e.g., in the book by Butzer and Berens [7]. We now give a more special result, cf. [21]. Let $\{S_\epsilon : 0<\epsilon\le 1\}$ be a family of subspaces of B_0 such that

$$\inf_{v\in S_\epsilon} \|u-v\|_{B_0} \le \epsilon \|u\|_{B_1}$$

for all $u \in B_1$ and ϵ in the range $0<\epsilon\le 1$. If $u \in B_{\theta,p}$ for some $\theta \in]0,1[$ and some <u>finite</u> $p(1\le p<\infty)$, then

$$\lim_{\epsilon \downarrow 0} \epsilon^{-\theta} \inf_{v\in S_\epsilon} \|u-v\|_{B_0} = 0. \tag{2.4}$$

3. Examples of interpolation spaces

Our main interest lies with the Sobolev spaces $H^k(\Omega)$ defined, for a bounded open set $\Omega \subset \mathbb{R}^n$ and, initially, for a nonnegative integer k, as the Hilbert space with norm

$$\|u\|_{H^k(\Omega)} = \Big(\sum_{|\alpha|\le k} \int_\Omega |D^\alpha u(x)|^2 dx \Big)^{\frac{1}{2}} \tag{3.1}$$

(for details, see e.g. the book by Stein [23]). The importance of interpolation spaces comes from the fact (cf. Magenes [14]) that

(3.2) $$[H^k(\Omega),\ H^m(\Omega)]_{\theta,2} \simeq H^{(1-\theta)k+\theta m}(\Omega)$$

provided $(1-\theta)k + \theta m$ is an integer. For this reason, the space $H^{k+\theta}(\Omega)$, $0<\theta<1$, is <u>defined</u> as follows:

(3.3) $$H^{k+\theta}(\Omega) \equiv [H^k(\Omega), H^{k+1}(\Omega)]_{\theta,2}.$$

In view of (3.2), the iteration theorem (2.3) implies that we could equally well have chosen any pair of integers as interpolation end-points; that is, with (3.3) as the definition of the fractional space, (3.2) is now valid for k and m any nonnegative <u>real</u> numbers. In fact, with the negative Sobolev spaces defined by duality,

(3.4) $$H^{-S}(\Omega) = (H^S(\Omega))^* \quad \text{for } S\geq 0,$$

(3.2) becomes valid for all $k,m\in\mathbb{R}$. This result may be found in Lions and Peetre [13]; the case when k and m have opposite signs has some important applications, as we shall see later.

We now give an example to show that interpolation spaces are really necessary for studying practical problems. We say a function f is <u>piecewise smooth on Ω</u> if there is a closed set $\omega \subset \Omega$ such that

i) $f\in H^1(\Omega-\omega)$ and $f\in L^\infty(\Omega)$

and ii) meas $\{x\in\Omega: \text{dist}(x,\omega\cup\partial\Omega)<h\} = O(h)$.

Here "meas" means Lebesgue measure and "dist" means Euclidean distance. Condition ii) simply requires that ω and $\partial\Omega$ be reasonably smooth and that $\dim \omega \leq n-1$.

Theorem 1. If f is piecewise smooth on Ω, then $f \in [H^0(\Omega), H^1(\Omega)]_{\frac{1}{2},\infty}$.

Proof. Let ρ be a smooth, nonnegative function supported in the unit ball such that $\int \rho(x)dx = 1$. Extend f outside Ω by zero and define $f_h \equiv f*\rho_h$, where $\rho_h(x) = h^{-n}\rho(x/h)$, $h>0$. Standard estimates yield

$$\|f-f_h\|_{H^s(\Omega^h)} \le ch^s\|f\|_{H^1(\Omega-\omega)}, \quad s=0,1,$$

where $\Omega^h = \{x\in\Omega : \text{dist}(x,\omega\cup\partial\Omega) > h\}$. Further,

$$\|f*\rho_h\|_{L^\infty(\Omega)} + h\|\nabla(f*\rho_h)\|_{L^\infty(\Omega)} \le c\|f\|_{L^\infty(\Omega)},$$

so our assumptions on ω and $\partial\Omega$ imply that

$$\|f-f_h\|_{H^0(\Omega)} + h\|f_h\|_{H^1(\Omega)} \le c(h^{\frac{1}{2}}\|f\|_{L^\infty(\Omega)} + h\|f\|_{H^1(\Omega-\omega)}).$$

Thus (see Section 2) $K(f,h) \le Ch^{\frac{1}{2}}$ for $h \le h_0$, and since $K(f,h) \le \|f\|_{H^0(\Omega)}$ for any $h>0$, we have $K(f,h) \le C'h^{\frac{1}{2}}$ for all $h>0$ as required.

Remarks. Assumption i) on f above is perhaps too strong. However, the theorem is sharp in the sense that $f \notin [H^0,H^1]_{\theta,p}$ for $(\theta,p) > (\frac{1}{2},\infty)$ if it is really discontinuous, as we now show by example. Suppose f is +1 for $x_n > 0$ and -1 for $x_n < 0$, at least for $|x| \le 1$. Choose a mesh of cubes of side length h with vertices having coordinates $(j+\frac{1}{2})h$, $j\in Z$. Let S^h be the linear space of piecewise constants on this mesh. Then, for any $h>0$ and any $u\in H^1$,

$$\inf_{v\in S^h}\|u-v\|_{H^0} \le ch\|u\|_{H^1},$$

for example, as a consequence of the Bramble-Hilbert lemma [6]. Thus we may apply (2.4) (take $\varepsilon = ch$): if $f\in[H^0,H^1]_{\theta,p}$ for $(\theta,p) > (\frac{1}{2},\infty)$, then, by (2.1), $f\in[H^0,H^1]_{\frac{1}{2},q}$ for some $q<\infty$, and hence

$$\inf_{v\in S^h}\|f-v\|_{H^0} = o(h^{\frac{1}{2}}).$$

However, the best approximation to f in the cubes containing $x_n = 0$ will be zero, since f is +1 in one half of each such cube and -1 in the other half. Thus, there is a pointwise error of ½ in a strip of width h, and so

$$\inf_{v\in S^h}\|f-v\|_{H^0} \geq ch^{\frac{1}{2}},$$

a contradiction. The reason that such simple functions as these have a fractional number of derivatives is that we are measuring them in the mean square, rather than pointwise, sense; their simple behavior is obfuscated by our elaborate measuring apparatus.

4. Basic applications to finite element methods

The typical framework for the application of "finite elements" to a time independent problem is as follows. One has a bilinear form $a(\ ,\ )$ defined (and continuous) on $H^m(\Omega) \times H^m(\Omega)$, and a variational problem to find $u\in V$ such that

$$a(u,v) = (f,v) \quad \text{for all } v\in V, \tag{4.1}$$

where V is some closed subspace of $H^m(\Omega)$ and $f\in V^*$ is data. For example, V might be the space $H_0^m(\Omega)$ of functions in $H^m(\Omega)$ vanishing to m-th order on $\partial\Omega$; then (4.1) corresponds to a Dirichlet problem. The existence-uniqueness of u depends on the coercivity of $a(\ ,\ )$ on V, cf. Agmon [1].

A finite element method for approximating u may be described loosely as follows. One has a finite dimensional space of functions S^h and we choose $u^h \epsilon S^h$ such that

$$(4.2) \qquad a_h(u^h,v) = (f,v)_h \quad \text{for all } v \epsilon S^h,$$

where $a_h(\ ,\ ) \sim a(\ ,\ )$ and $(\ ,\ )_h \sim (\ ,\ )$ in some suitable fashion. Typically, S^h is (or at least contains) a space of piecewise polynomials on a mesh of size h. The choice of all these ingredients is made with one end in mind, namely, to obtain an error estimate, say of the form

$$(4.3) \qquad \|u-u^h\|_{H^m} \leq c \inf_{v \epsilon S^h} \|u-v\|_{H^m} + \text{"other terms"}.$$

For specific examples, one may refer to the surveys by Bramble [5], Nitsche [15], and the author [22], and to the books by Babuška and Aziz [2], Ciarlet [9], Oden and Reddy [16] and Strang and Fix [25], and references therein.

To complete the error estimates for $u - u^h$, there are two steps, one to bound the "other terms" if any, and then to construct a $v \epsilon S^h$ that approximates u well. The first step is sometimes crucial, but we do not discuss it here. The typical solution to the second problem is to interpolate u (at selected points in Ω, not in the Banach space sense), obtaining (cf. the books above or Ciarlet and Raviart [10])

$$(4.4) \qquad \inf_{v \epsilon S^h} \|u-v\|_{H^m} \leq ch^{k-m} \|u\|_{H^k}$$

for some integer k (≡ the approximation degree of S^h). The problem we wish to address here is what to do if u is not smooth enough to define the interpolant or other constructed approximation. We are assuming $u \in H^k$ is sufficient for our approximation construction, although this is not always so: consider Lagrange interpolation of piecewise linear functions in $\mathbb{R}^4$ (here k=2 and $H^2(\mathbb{R}^4) \not\subset C^0(\mathbb{R}^4)$). But given an approximation for $u \in H^k$, we can obtain one for less smooth u by interpolating, in the Banach space sense!

Denote by P the projection of $H^m(\Omega)$ onto S^h:

$\inf_{v \in S^h} \|u-v\|_{H^m} = \|u-Pu\|_{H^m}$. Then P is a continuous linear operator on H^m, and hence $T \equiv I - P$ is as well. In fact,

$$\|Tu\|_{H^m} \le \|u\|_{H^m} + \|Pu\|_{H^m} \le 2\|u\|_{H^m}.$$

Combining this with (4.4) thus shows that

$$T: \begin{cases} H^m \to H^m & \text{with norm} \le 2 \\ H^k \to H^m & \text{with norm} \le ch^{k-m}, \end{cases}$$

and we can apply the operator interpolation property to conclude that, for $s \in [m,k]$,

$$T: \quad H^s \to H^m \quad \text{with norm} \le c'h^{s-m}$$

(take $\theta = (s-m)/(k-m)$ and p=2; here $c' \le 2^{2-\theta}c^{\theta}$). Translating back, we have

$$\inf_{v \in S^h} \|u-v\|_{H^m} = \|Tu\|_{H^m} \le c'h^{s-m}\|u\|_{H^s}.$$

Thus we obtain the final error estimate

(4.5) $$\|u-u^h\|_{H^m} \le ch^{s-m}\|u\|_{H^s} ,$$

provided the "other terms" can be so estimated.

However, there are times when they can not. For example, when the method of interpolated boundary conditions [19] is used for a second order (m=1) Dirichlet problem, the "other terms" can be bounded by $ch^s\|u\|_{H^s}$ for $s\in]^3/2,k-^1/2]$, but no such bound is possible for $s \le {}^3/2$, due to the failure of a certain trace theorem. Thus the technique described above is not sufficient for $s \le {}^3/2$. Instead, we make use of the fact that

(4.6) $$\|u^h\|_{H^m} \le c\|f\|_{H^{-m}} ,$$

which holds also for other finite element methods as well. Using elliptic regularity theory (cf. Agmon [1]), we have

(4.7) $$\|u-u^h\|_{H^m} \le ch^{k-m}\|u\|_{H^k} \le ch^{k-m}\|f\|_{H^{k-2m}} .$$

Thus the mapping T given by $Tf = u - u^h$ is ripe for interpolating:

$$T: \begin{cases} H^{-m} \to H^m & \text{with norm} \le c \\ H^{k-2m} \to H^m & \text{with norm} \le ch^{k-m}, \end{cases}$$

and hence $T: H^{s-m} \to H^m$ with norm $\le ch^s$, $s\in[0,k-m]$. More concretely,

(4.8) $$\|u-u^h\|_{H^m} = \|Tf\|_{H^m} \le ch^s\|f\|_{H^{s-m}}.$$

Of course, the same estimate follows directly from (4.5) as well via elliptic regularity when (4.5) holds.

Now consider our example of a piecewise smooth f. From theorem 1 and the inclusion relations (2.1), we know that $f \in H^s$ for all $s < \frac{1}{2}$, and hence $\|u-u^h\|_{H^m} = O(h^r)$ for all $r < m + \frac{1}{2}$. However, it is possible to improve this estimate to $\|u-u^h\|_{H^m} = O(h^{m+\frac{1}{2}})$ as follows. In the derivation of (4.8) (or equivalently (4.5)), instead of using the operator interpolation property with the second index p equal to 2, choose $p=\infty$. In view of (2.2), we obtain

$$(4.9) \qquad \|u-u^h\|_{H^m} \leq ch^{\theta(k-m)}\|f\|_{[H^{-m},H^{k-2m}]_{\theta,\infty}} .$$

From the iteration theorem (2.3), we have

$$[H^{-m},H^{k-2m}]_{(m+\frac{1}{2})/(k-m),\infty} \simeq [H^0,H^1]_{\frac{1}{2},\infty} .$$

Thus $\|u-u^h\|_{H^m} \leq ch^{m+\frac{1}{2}}\|f\|_{[H^0,H^1]_{\frac{1}{2},\infty}}$ in general; hence our claim concerning piecewise smooth f follows from theorem 1.

The estimate (4.5) can be improved in another way, as observed by Babuška and Kellogg [3]. Suppose that (4.4) holds and that we are given a $u \in H^s$ for some $s<k$. Then (2.4) implies that (choose $\varepsilon = ch^{k-m}$)

$$(4.10) \qquad \inf_{v \in S^h} \|u-v\|_{H^m} = o(h^{s-m}),$$

instead of just "big O" as proved above using operator interpolation (recall H^s interpolates H^k and H^m with the second index $p=2<\infty$). Thus $\|u-u^h\|_{H^m} = o(h^{s-m})$ for methods where the "other terms" have a similar bound, e.g., the method of interpolated boundary conditions, where an extra power of h is available. (For $s \in]1,3/2]$, the "other terms" can be bounded by $c_r h^r \|f\|_{H^{s-2}}$ for $r < 3(s-1)$, hence r can be chosen to be larger than s-1. The proof of this is again by interpolation, but it requires a nonlinear interpolation technique, hence is beyond our scope here. Such an estimate obviates the need for the argument using estimate (4.6), but the latter is more general in applicability as well as being more elementary in its derivation.)

Another application of interpolation is to obtain estimates in lower norms. By solving a "dual problem" similar to $u \to u^h$, one often obtains estimates of the form

$$(4.11) \qquad \|u-u^h\|_{H^{2m-k}} \le ch^{2(k-m)} \|u\|_{H^k}.$$

If we also have, e.g.,

$$(4.12) \qquad \|u-u^h\|_{H^m} \le ch^{k-m} \|u\|_{H^k},$$

we can interpolate the operator $Tu = u - u^h$ to get

$$(4.13) \qquad \|u-u^h\|_{H^s} \le ch^{k-s} \|u\|_{H^k}, \quad s \in [2m-k, m].$$

Here, we have interpolated with the range of T varying and the domain fixed, where before the range was fixed and the domain varied. One can vary both simultaneously to obtain estimates of the form $\|u-u^h\|_{H^s} \le ch^{r-s} \|u\|_{H^r}$, as well as "little o" estimates using (2.4), but we leave the details as an exercise for the true fanatics. The point here is simply that to bound $\|u-u^h\|_{H^s}$ for s near m by the standard duality approach requires solving a dual problem with non-smooth data, and this can be avoided by interpolation. Note that the derivation of (4.13), and (4.8) as well, involves interpolation between Sobolev spaces of negative and positive order.

5. Error estimates for a polygonal approximation

Suppose that $\Omega \subset \mathbb{R}^2$ is convex and that $\{\Pi_h : 0<h\leq 1\}$ is a family of polygonal domains inscribed in Ω such that each segment of $\partial\Pi_h$ has length $\leq h$. Let $\gamma>0$ be a fixed constant, and for each h let T_h be a triangulation of Π_h such that each triangle in T_h is contained in a ball of radius h and contains a ball of radius γh. Let $\bar{S}^h$ be the set of all piecewise quadratic functions on T_h and let S^h be the subspace of $\bar{S}^h$ consisting of those functions vanishing on $\partial\Pi_h$. The significance of S^h is that it is an admissible space for a Dirichlet problem on Ω:

$$S^h \subset H_0^1(\Pi_h) \subset H_0^1(\Omega),$$

where the subscript "0" denotes the subspace of functions vanishing on the boundary and the second inclusion map is "extension by zero." For simplicity, let $a(v,w) \equiv \int_\Omega \nabla v \cdot \nabla w + vw \, dx$ and let $u \in H_0^1(\Omega)$ satisfy

$$a(u,v) = (f,v) \quad \text{for all } v \in H_0^1(\Omega).$$

Define $u^h \in S^h$ by

$$a(u^h,v) = (f,v) \quad \text{for all } v \in S^h;$$

again, we think of $v \in S^h$ as extended by zero outside Π_h, hence all integrations are restricted to Π_h. If u is smooth enough, then $\|u-u^h\|_{H^1(\Pi_h)} = O(h^{3/2})$, and this is the best possible rate regardless of the smoothness of u (Strang and Berger [24] and Thomée [26]. Let $|\cdot|$ denote the norm for the space $[H^2(\Omega),H^3(\Omega)]_{\frac{1}{2},\infty}$. Then the optimal result one could expect would be

$$\|u-u^h\|_{H^1(\Pi_h)} \le ch^{3/2}|u|.$$

Note, however, that this can not follow directly from the techniques of the previous section, since we do not have a higher approximation rate for smoother u to interpolate from. We show below how to get around this, and thus we obtain the optimal rate of convergence for, say, piecewise smooth data f (see theorem 1--$f\in[H^0(\Omega),H^1(\Omega)]_{\frac{1}{2},\infty}$ implies $u\in[H^2(\Omega),H^3(\Omega)]_{\frac{1}{2},\infty}$ simply by interpolating the standard elliptic regularity theory [1]), solving a problem posed by the author in [18].

Theorem 2. Suppose $u\in[H^2(\Omega),H^3(\Omega)]_{\frac{1}{2},\infty}$ (in particular, this holds if f is piecewise smooth). Then

$$\|u-u^h\|_{H^1(\Pi_h)} \le ch^{3/2}|u|,$$

where $|u|$ is the norm of u in $[H^2,H^3]_{\frac{1}{2},\infty}$.

Proof. Define $I_h:H^2(\Omega)\to\bar{S}^h$ by the requirement that $u - I_hu$ vanish at the nodes (each vertex and edge midpoint) of T_h. Then, for k=2 and 3,

$$\|u-I_hu\|_{H^1(\Pi_h)} \le ch^{k-1}\|u\|_{H^k(\Omega)}$$

(cf. Ciarlet and Raviart [10]).

Applying the operator interpolation property to $Tu \equiv u - I_h u$ with $(\theta, p) = (\frac{1}{2}, \infty)$, we thus find that

$$\|u - I_h u\|_{H^1(\Pi_h)} \leq ch^{3/2}|u|.$$

Now define $v \in S^h$ by the requirement that $v = I_h u$ at all the nodes of T_h not lying on $\partial\Pi_h$. Then the only nodes at which $I_h u - v$ is nonzero are the edge midpoint nodes on $\partial\Pi_h$, which we denote by $\{z_j\}$. Let $\varphi_j \in \bar{S}^h$ be 1 at z_j and 0 at all other nodes in T_h. Each φ_j is supported in the triangle containing z_j, and $\|\varphi_j\|_{H^1(\Pi_h)} \leq c(\gamma)$ (cf. Strang and Fix [25], Chapter 3). Thus

$$\begin{aligned} \|I_h u - v\|_{H^1(\Pi_h)} &= \|\sum_j u(z_j)\varphi_j\|_{H^1(\Pi_h)} \\ &\leq ch^{-\frac{1}{2}} \sup_{z \in \partial\Pi_h} |u(z)|, \end{aligned} \tag{5.1}$$

since there are at most ch^{-1} nodes z_j. As $\text{dist}(\partial\Pi_h, \partial\Omega) \leq ch^2$ ($\partial\Pi_h$ is a piecewise linear interpolant of $\partial\Omega$) and $u|_{\partial\Omega} = 0$,

$$\begin{aligned} \sup_{z \in \partial\Pi_h} |u(z)| &\leq ch^2 \sup_{x \in \Omega} |\nabla u(x)| \\ &\leq c'h^2 \|u\|_{H^{9/4}(\Omega)} \\ &\leq c''h^2 |u|, \end{aligned} \tag{5.2}$$

via Taylor's theorem, Sobolev's inequality [12], and the containment relations (2.1). Therefore

$$\|I_h u - v\|_{H^1(\Pi_h)} \le ch^{3/2}|u|,$$

and we have

$$\|u-v\|_{H^1(\Pi_h)} \le \|u - I_h u\|_{H^1(\Gamma_h)} + \|I_h u - v\|_{H^1(\Pi_h)}$$

$$\le ch^{3/2}|u|.$$

As shown in Strang and Berger [24], u^h minimizes the error in approximating u from S^h (because $S^h \subset H_0^1(\Omega)$)*:

$$\|u-u^h\|_{H^1(\Gamma_h)} = \inf_{v^h \in S^h} \|u-v^h\|_{H^1(\Pi_h)} \le ch^{3/2}|u|,$$

and this completes the theorem.

Remarks. 1) It is necessary to consider the error $u-u^h$ only on Π_h, since

$$\|u-u^h\|_{H^1(\Omega-\overline{\Pi}_h)} = \|u\|_{H^1(\Omega-\overline{\Pi}_h)} = O(h).$$

2) Theorem 2 can be generalized to n dimensions, $n \ge 3$, but the results are no longer optimal. The major changes occur in (5.2) and the derivation of (5.1). In the latter, we have $\|\varphi_j\|_{H^1(\Pi_h)} \le ch^{n/2-1}$, but $O(h^{1-n})$ points z_j, so (5.1) remains the same.

---

*For completeness: $a(u-u^h,w) = 0$ for any $w \in S^h$, so if $a_h(v,w) \equiv \int_{\Pi_h} \nabla v \cdot \nabla w + vw\,dx$, then $a_h(u-u^h,w) = a(u-u^h,w) = 0$, hence $\|u-u^h\|^2_{H^1(\Pi_h)} = a_h(u-u^h,u-v) \le \|u-u^h\|_{H^1(\Pi_h)} \|u-v\|_{H^1(\Gamma_h)}$, $v \in S^h$.

However, (5.2) is no longer valid since u is not smooth enough. Using the results of [21] one has

$$\sup_{x\in\partial\Pi_h} |u(z)| \le \begin{cases} ch^2\|u\|_{[H^1,H^{n+1}]_{\frac{1}{2},1}} \\ \text{and} \\ ch^0\|u\|_{[H^0,H^n]_{\frac{1}{2},1}} \end{cases},$$

and we can interpolate to get

$$\sup_{x\in\partial\Pi_h} |u(z)| \le \begin{Bmatrix} c_\varepsilon h^{2-\varepsilon} & \text{if } n=3 \\ \\ c\ h & \text{if } n=4 \end{Bmatrix} |u|.$$

Therefore, one obtains

$$\|u-u^h\|_{H^1(\Pi_h)} \le \begin{Bmatrix} c_\varepsilon h^{3/2-\varepsilon} & \text{if } n=3 \\ \\ c\ h^{\frac{1}{2}} & \text{if } n=4 \end{Bmatrix} |u|.$$

When $n\ge5$, there are unbounded functions in $[H^2,H^3]_{\frac{1}{2},\infty}$ (cf. [21]), so the above techniques yield no convergence rate at all.

6. Error estimates when f is the Dirac δ-function

For a 2m-th order elliptic problem in variational form, it is typically possible to define a Galerkin approximation u^h only when $f\in H^{-m}$. However, for finite element spaces $S^h \subset C^{m-1}$, it is possible to extend this definition to the Dirac δ-function (or even its derivatives of order $\le m-1$):

$$\text{(6.1)} \qquad a(g^h,v) = (\delta,v) = v(x) \quad \text{for all } v\in S^h,$$

where δ is supported at the point $x \in \Omega$. The Green's function g satisfies a similar relation:

$$a(g,v) = v(x) \quad \text{for all } v \in \tilde{V}, \tag{6.2}$$

where $\tilde{V}$ is a smoothed version of V. (The natural framework is to consider a(,) as a bilinear form on $W_1^m \times W_\infty^m$ where W_p^m is the Sobolev space of functions having m derivatives in L^p.) Since g may not lie in H^m, standard estimates for the error $g - g^h$ do not apply. However, it is possible to obtain optimal error estimates for $g - g^h$ in suitable lower norms, as we describe below, by using interpolation theory techniques.

First we show that, in some cases, one can derive the right error estimates simply by using the operator interpolation property and a special result of Babuška and Aziz [2]. They showed that it is possible to extend the usual range of validity of the finite element error estimate

$$\|u-u^h\|_{H^0} \le ch^{2m+s}\|f\|_{H^s} \tag{6.3}$$

to allow any $s > -m - \frac{1}{2}$, rather than just $s \ge -m$. The δ-function lies in the interpolation space $[H^{-n},H^0]_{\frac{1}{2},\infty}$ (cf. [21]) and the iteration theorem (2.3) implies that $[H^{-n},H^0]_{\frac{1}{2},\infty} \cong [H^{-n/2-1/4},H^0]_{\theta,\infty}$ with $\theta = 1/(2n+1)$. Thus if $n \le 2m$, we can interpolate (6.3) between s=0 and $s = -n/2-1/4$ to get

$$\|u-u^h\|_{H^0} \le ch^{2m-n/2}\|f\|_{[H^{-n},H^0]_{\frac{1}{2},\infty}} \tag{6.4}$$

in general; hence for the particular case of $f=\delta$,

$$\|g-g^h\|_{H^0} = O(h^{2m-n/2}).$$

The above technique, due to Babuška and Kellogg [4], does not work when n > 2m. However, it is possible in the case f=δ to use more specialized ideas ([18], [20]) from interpolation theory to prove that

$$(6.5) \qquad \|g-g^h\|_{H^{-s}} \le ch^{2m-s+n/2} \quad \text{for } s\in\,]n/2-2m,k-2m],$$

where k is the degree of approximation for smooth u: $\|u-u^h\|_{H^{2m-k}} \le ch^{2(k-m)}\|u\|_{H^k}$. Note that no error estimate is obtained if $n/2 \ge k$, and that a general result such as (6.4) is not proved; the special character of δ is exploited in proving (6.5). For details, the reader is referred to [18] and [20]. (In [20], the results are stated only for n=2, but the techniques easily generalize.)

7. An open problem

Recently there has been much activity concerning L^∞ estimates for finite element methods, by J. Frehse, F. Natterer, J. Nitsche, R. Rannacher, A. Schatz, L. Wahlbin, and the author. The principle result has been to show that, for the case p=∞,

$$(7.0) \qquad \|u-u^h\|_{L^p} \le ch^k\|u\|_{W^k_p},$$

where k is the degree of approximation of S^h and W^k_p is the Sobolev space of functions with k (weak) derivatives in L^p. (Presently, k has to be larger than 2 for second order problems to avoid a log h term in the estimate.) As described in Section 4, (7.0) is also known for p=2 ($W^k_2 = H^k$). Thus one would hope to prove (7.0) for

$p \in]2,\infty[$ by interpolating the map $Tu = u - u^h$. It is true that

$$(7.1) \qquad [L^2, L^\infty]_{1-2/p,p} \simeq L^p ,$$

so the operator interpolation property implies that

$$\|u-u^h\|_{L^p} \le ch^k \|u\|_{[W_2^k, W_\infty^k]_{1-2/p,p}}$$

for $p \in]2,\infty[$. While it is clear that $[W_2^k, W_\infty^k]_{1-2/p,p} \subset W_p^k$, the reverse inclusion needed here is not known, except in one dimension where $(1+i\frac{d}{dx})^k$ provides an isomorphism between W_p^k and L^p for all $p \in [1,\infty]$ (Grisvard [11]). So the open problem we propose is to show that

$$W_p^k \cong [W_2^k, W_\infty^k]_{1-2/p,p} \qquad \text{for } p \in]2,\infty[,$$

or, more generally in view of (2.3), that

$$(?) \qquad W_p^k \cong [W_1^k, W_\infty^k]_{1-1/p,p} \qquad \text{for } p \in]1,\infty[.$$

We remark that the estimate (7.0) has been proved for all $p \in [1,\infty]$ by the author [20] ($n=2$ and $k \ge 3$ or $k=2$ with a power of $|\log h|$ as an added factor), but of course by a different technique. The basic tool is a representation of the error via the continuous and discrete Green's functions, plus the interpolation result (7.1), i.e., the Riesz-Thorin theorem. As far as we know, this is the only technique presently available to derive L^p estimates. Also, the interpolation spaces between W_p^k and W_q^k for $1<p, q<\infty$ have been calculated (Calderón [8] and Peetre [17]), and one has the expected result $[W_p^k, W_q^k]_{\theta,r} = W_r^k$

with $\theta = (1/r-1/p)/(1/q-1/p)$. Thus the problem (?) concerns only the limit cases.

REFERENCES

[1] S. Agmon, Lectures on Elliptic Boundary Value Problems, Van Nostrand, 1965.

[2] I. Babuška and A. K. Aziz, The Mathematical Foundations of the Finite Element Method, Academic Press, 1972.

[3] I. Babuška and B. Kellogg, "Nonuniform estimates for the finite element method," SIAM J. Num. Anal. 12 (1975), 868-875.

[4] I. Babuška and B. Kellogg, private communication.

[5] J. H. Bramble, "A survey of some finite element methods proposed for treating the Dirichlet problem," Advances in Math. 16 (1975), 187-196.

[6] J. H. Bramble and S. R. Hilbert, "Estimation of linear functionals on Sobolev spaces with applications to Fourier transforms and spline interpolation," SIAM J. Num. Anal. 7 (1970), 113-124.

[7] P. L. Butzer and H. Berens, Semi-Groups of Operators and Approximation, Springer-Verlag, 1967.

[8] A. P. Calderón, "Lebesgue spaces of differentiable functions and distributions," Proc. Symposia in Pure Math. IV (1961), C. B. Morrey, Jr., ed., Amer. Math. Soc., 33-49.

[9] P. G. Ciarlet, Numerical Analysis of the Finite Element Method, to appear.

[10] P. G. Ciarlet and P.-A. Raviart, "General Lagrange and Hermite interpolation in $\mathbb{R}^n$ with applications to finite element methods," Arch. Rat. Mech. Anal. 46 (1972), 177-199.

[11] P. Grisvard, private communication.

[12] J. L. Lions and E. Magenes, Non-homogeneous boundary value problems and applications, v. 1, Springer-Verlag, 1972.

[13] J. L. Lions and J. Peetre, "Sur une classe d'espaces d'interpolation," Pub. Math. IHES 19 (1964), 5-68.

[14] E. Magenes, "Spazi di interpolazione ed equazioni a derivate parziali," Atti del VII Congresso dell'Unione Matematica Italiana Genova 1963, Roma (1964), Edizioni Cremonese, 134-197.

[15] J. Nitsche, "On projection methods for the plate problem," to appear.

[16] J. T. Oden and J. N. Reddy, Variational Methods in Theoretical Mechanics, Springer-Verlag, 1976.

[17] J. Peetre, "Espaces d'interpolation et théorème de Soboleff," Ann. Inst. Fourier (Grenoble) 16 (1966), 279-317.

[18] R. Scott, "Finite element convergence for singular data," Numer. Math. 21 (1973/74), 317-327.

[19] R. Scott, "Interpolated boundary conditions in the finite element method," SIAM J. Num. Anal. 12 (1975), 404-427.

[20] R. Scott, "Optimal L^{∞} estimates for the finite element method on irregular meshes," Math. Comp. 30 (1976), 681-697.

[21] R. Scott, "A sharp form of the Sobolev trace theorems," J. Func. Anal. 25 (1977), 70-80.

[22] R. Scott, "A survey of displacement methods for the plate bending problem," Formulations and Computational Algorithms in Finite Element Analysis, M.I.T. Press, to appear.

[23] E. M. Stein, Singular Integrals and Differentiability Properties of Functions, Princeton Univ. Press, 1970.

[24] G. Strang and A. E. Berger, "The change in solution due to change in domain," Proc. AMS Summer Inst. on Partial Differential Equations, 1971.

[25] G. Strang and G. J. Fix, An Analysis of the Finite Element Method, Prentice-Hall, 1973.

[26] V. Thomée, "Polygonal domain approximation in Dirichlet's problem," J. Inst. Maths Applics 11 (1973), 33-44.

A MINIMAX PROBLEM IN PLASTICITY THEORY

Gilbert Strang
Massachusetts Institute of Technology

ABSTRACT

We discuss a specific example of infinite-dimensional optimization. It arises in the limit analysis of a perfectly plastic structure, when the loads are increased until they are no longer balanced by admissible stresses. This moment of collapse can be determined, both analytically and numerically, as the extreme value either of an optimization problem in the stresses or of its dual in terms of displacements. In our example the dual (but not, so far, the primal) can actually be solved. We hope that the example will suggest the right choice of function spaces in the general theory, and also provide a good test case for the numerical analysis.

I am grateful for the support of the National Science Foundation, under contract MCS 76-22289.

Our goal is to find a limit analysis problem which, without being trivial, can nevertheless be solved. Ignoring for a moment its engineering interpretation, the problem will have the following form: to find the minimum norm solution of a given differential equation $L\sigma = f$. Geometrically, we want to find the distance between a point and a subspace--between a particular solution σ_p and the subspace of solutions to the homogeneous equation. This subspace, the kernel of L, ought to be infinite-dimensional.

One possible choice is the equation

$$\operatorname{div} \sigma = \frac{\partial \sigma_1}{\partial x} + \frac{\partial \sigma_2}{\partial y} = f . \tag{1}$$

In this case the homogeneous equation is easy to solve; for any ϕ we can take $\sigma_1 = \partial\phi/\partial y$ and $\sigma_2 = -\partial\phi/\partial x$, so the kernel is very large. We specify $f \equiv 1$, and work in a square domain $\Omega \subset R^2$. This leaves open only the choice of norm. Certainly L_2 would be convenient, but for our purposes it is too convenient; the optimization problem leads just to a linear equation (corresponding to elasticity theory). We need a norm which matches the pointwise conditions of plasticity.

The natural norm is L_∞, but at each point the vector $\sigma(x,y)$ is measured in L_2:

$$||\sigma|| = \sup_{\Omega} |\sigma(x,y)| = \sup_{\Omega} (\sigma_1^2(x,y) + \sigma_2^2(x,y))^{\frac{1}{2}}.$$

This is not the only possibility, and the alternative of a pure L_∞ norm $||\sigma|| = \sup(|\sigma_1(x,y)|, |\sigma_2(x,y)|)$ is also interesting; it would turn our optimization problem into infinite-dimensional linear programming. But this norm introduces preferred directions, parallel to the coordinate axes, whereas the first choice is correct when the material is isotropic; $\sigma_1^2 + \sigma_2^2$ is invariant under rotation in the $x - y$ plane. This choice leads to convex programming.

In fact the linear and convex programs are just restatements of the original problem; instead of solving $\operatorname{div} \sigma = f$ with minimum norm, we solve $\operatorname{div} \sigma = \lambda f$, require $||\sigma|| \leq 1$, and make λ as large as possible. This is the usual statement within plasticity theory, and we adopt it from now on:

$$\textit{Maximize } \lambda \textit{ subject to } \operatorname{div} \sigma = \lambda f \textit{ and } ||\sigma|| \leq 1 .$$

If σ^* and λ^* are optimal in this new formulation, then σ^*/λ^* was the solution of the earlier problem and the minimum norm was $1/\lambda^*$. The requirement $||\sigma|| \leq 1$ is always a convex constraint; for the pure L_∞ norm it amounts to a family of linear constraints $-1 \leq \sigma_1(x,y) \leq 1$, $-1 \leq \sigma_2(x,y) \leq 1$. Computationally, it is conceivable that also the convex constraint $\sigma_1^2 + \sigma_2^2 \leq 1$ should be made piecewise linear, replacing the unit circle by an approximating polygon; but the numerical evidence seems opposed to this approach. Apparently a direct nonlinear algorithm is more efficient.

So far our equation is only analogous to the equations of plastic collapse, without itself having a physical interpretation. However, to his delight, the author learned that it does represent a true problem in mechanics--known as <u>anti-plane</u> or <u>longitudinal shear</u>, and quite easy to visualize. Imagine an infinitely long vertical pipe with square cross-section Ω, filled with a plastic material. The external force λf is gravity, acting vertically; it is balanced by the shear stresses $\sigma_1 = \sigma_{xz}$ and $\sigma_2 = \sigma_{yz}$, and the correct equilibrium equation is exactly $\operatorname{div} \sigma = \lambda f$. The constraint $||\sigma|| \leq 1$ is the von Mises (and in this case also the Tresca) yield condition. With the material fixed at the edges of the square, there is no boundary condition on σ; instead the displacement is zero. Eventually, as λ increases, the weight λf will be unsupportable--and collapse will occur. We want to find the maximal λ^*, and the mechanism by which collapse occurs.

Solutions have been found to some related problems, in which the body force f is absent but $\sigma \cdot n = \lambda g$ is specified on the boundary $\partial\Omega$; see Hult-McClintock [1] and Rice [2]. Physically, a part of the pipe's surface is pushed up and another part pulled down; eventually this shearing action splits the pipe, and the two parts slide away from one another. Mathematically, the problem becomes:

$$\text{Maximize } \lambda \text{ subject to } \operatorname{div} \sigma = 0 \text{ in } \Omega, \quad \sigma \cdot n = \lambda g \text{ on } \partial\Omega,$$
$$\text{and } \|\sigma\| \le 1.$$

There is a constraint on the traction g, coming from the identity $\iint \operatorname{div} \sigma = \int \sigma \cdot n \, ds$; this implies that $\int g \, ds = 0$, which prevents a translation of the whole pipe.

A more general case would admit both f and g at the same time, but we concentrate instead on the original problem in a square--and we approach it indirectly, through duality.

The Dual Problem

The dual can be found either geometrically, or formally, or more slowly through a minimax theorem. The geometry is the easiest: when the primal problem minimizes the distance from a point to a subspace, its dual admits all hyperplanes through the subspace--and maximizes their distance from the point. (Given a point in 3-space, the minimum distance to a line is the largest distance to planes through the line.) Formally, the dual is constructed by reversing almost everything in the primal--the L_∞ norm changes to L_1, the divergence operator is replaced by its transpose the gradient, and the unknowns are the displacements instead of the stresses. The result is a translation into analytical terms of the dual found geometrically. But it is worthwhile to go more slowly, and reach the dual in several steps, in order to recognize which minimax theorem is required.

First we write the maximization as a maximin:

$$\max_{\substack{\|\sigma\|\leq 1\\ \mathrm{div}\,\sigma=\lambda f}} \lambda = \max_{\|\sigma\|\leq 1} \min_{\int uf=1} \int u \,\mathrm{div}\,\sigma \quad .$$

The dual variable u is to vanish outside the square Ω, and all integrals extend over Ω. One can check that the inner minimum is $-\infty$ unless div σ is proportional to f. Then the outer maximum is interested only in this case div $\sigma = \lambda f$, and this leads back to the original problem.

Now, if we can, we want to reverse the order of optimization:

$$\max_{\|\sigma\|\leq 1} \min_{\int uf=1} \int u \,\mathrm{div}\,\sigma = \min_{\int uf=1} \max_{\|\sigma\|\leq 1} \int u \,\mathrm{div}\,\sigma \ .$$

It is certainly possible that we need to write sup and inf for max and min; the choice of function spaces will decide whether or not the extreme values are attained. And it is also possible that the reversal is completely false, and that not even a sup-inf theorem is correct. An example of this "duality gap" is given in [3]: if σ and u are kept in C^∞ but f is not, then div $\sigma = \lambda f$ can only hold if $\lambda = 0$, whereas the dual solution coming next may well be non-zero.

To reach the dual, we integrate by parts and then maximize: the minimax finally becomes

$$\min_{\int uf=1} \max_{\|\sigma\|\leq 1} -\int \sigma\cdot \mathrm{grad}\, u = \min_{\int uf=1} \int |\mathrm{grad}\, u| \quad .$$

The maximum occurred when at each point the vector σ is chosen proportional to grad u. Under the constraint $\|\sigma\| \leq 1$ the best we could do at each point was

$$\sigma = \frac{-\text{grad } u}{|\text{grad } u|}, \text{ giving } -\sigma\cdot\text{grad } u = |\text{grad } u| = \left(\left(\frac{\partial u}{\partial x}\right)^2 + \left(\frac{\partial u}{\partial y}\right)^2\right)^{\frac{1}{2}}.$$

This is the integrand in the dual (notice the L_1 norm, and the appearance of the gradient). Because of the homogeneity, the dual can also be written as

$$\min_{\int uf=1} \int|\text{grad } u| = \min \frac{\int|\text{grad } u|}{\int uf}.$$

Now we can comment further on the space of admissible u. Among the Sobolev spaces, the natural choice seems to be $\mathring{W}_1^1$ --the gradient is in L_1, and $u = 0$ on $\partial\Omega$. <u>This is a poor choice if we want the minimum to be attained.</u> On an interval, the space W_1^1 fails to include a step function just as L_1 fails to include a delta-function. And, at least on the interval $[0,1]$, these excluded functions are exactly the extrema. If $f = 1$ and the maximum of u is taken to be 1 (the ratio is homogeneous anyway), then since u has to go up to 1 and down again,

$$\min \frac{\int|u'|}{\int u} = \min \frac{2}{\int u}.$$

The denominator is as large as possible when u stays at its maximum over the whole interval. Therefore the minimum ratio is 2, and <u>the minimizing</u> u^* <u>is the characteristic function of</u> $[0,1]$. It is not in $\mathring{W}_1^1$; its gradient is the delta-function at $x = 0$ minus the delta-function at $x = 1$. However duality holds, since the solution to the primal problem is also

$$\max_{\substack{\|\sigma\|\leq 1 \\ \sigma'=\lambda}} \lambda = 2, \text{ attained at } \sigma^* = 2x - 1.$$

We expect something similar in two dimensions.

For the space of admissible u, the most natural choice is the functions of bounded variation. In one dimension this means that $\int |u'| < \infty$. For functions of two variables, the early literature contains at least seven inequivalent definitions [4]. The right one is the last of the seven, given by Tonelli; later Cesari and DeGiorgi studied its extension to discontinuous u (which must certainly be admitted). The essential part of the definition is exactly this, that the numerator in our ratio should be finite:

$$\|u\|_{BV} = \int_{\Omega} |\text{grad } u| < \infty \quad .$$

In other words, the gradient of u is a signed measure, of finite variation; this is an exact analogue of the one-dimensional case. (Except that there may be consequences which are special to one dimension, such as the splitting $u = u^+ - u^-$ into non-decreasing parts.) Our force f is assumed to belong to L_∞, so the full definition of the admissible space BV requires that u lie in L_1 and that support $(u) \subset \bar{\Omega}$, as well as $\|u\|_{BV} < \infty$.

For the minimax theorem itself--which is mathematically the most crucial and the most delicate point--the reader is referred to [5] and to our paper with Christiansen and Matthies [3]. Here we assume that the minimum of the dual equals the maximum of the primal, and try to compute it.

The key is to recognize that the optimal u^* is again the characteristic function of some set E. There are heuristic arguments which make this seem reasonable; but, even better, it is a theorem of Fleming [6]. The extreme points of the unit ball in BV are multiples of characteristic functions of nice sets--the boundary of E is a simple closed curve of finite length. For such a characteristic function, the gradient is a delta-function distributed along the boundary of E, and the ratio we want to minimize becomes (with $f = 1$)

$$\frac{\int|\text{grad } u|}{\int u} = \frac{\text{perimeter of } E}{\text{area of } E} .$$

Therefore we face the classical isoperimetric problem, under the extra constraint $E \subset \Omega$.

The optimal u^* is an extreme point of BV for this reason: if we keep $\int|\text{grad } u| \leq 1$, then the problem amounts to maximizing the linear functional $\int u$ over this unit ball--in other words, to translating a hyperplane until it just touches the ball. We expect the extreme value to be attained in BV by Federer's theorem [7, 4.5.9], which expresses $\int|\text{grad } u|$ as an integral over the perimeters $P(s)$ of the sets $E = \{x:u(x) > s\}$. This is a kind of substitute, in our example, for the Krein-Milman theorem.

We are left with the constrained isoperimetric problem, to minimize perimeter/area. Certainly the optimal set E must reach to the edges of the square, or we could scale E by some factor $\alpha > 1$; our ratio P/A would be multiplied by α/α^2, and would decrease. On the other hand, the optimal E is not necessarily the whole square (or the inscribed circle--they both give the same value $P/A = 2$ for a square of side 2). We do know that away from the edges, where the constraints are inactive, the boundary ∂E must be a circular arc; the calculus of variations applies. Therefore ∂E is a union of circular arcs and pieces of the square $\partial\Omega$.

Furthermore, the arcs are tangent to the square. Otherwise we could remove the corner at the point where ∂E leaves $\partial\Omega$; a shortcut of length ε would reduce the perimeter by $O(\varepsilon)$ and the area by only $O(\varepsilon^2)$, and thus reduce P/A. (This may be the first clue towards a regularity theory for minimax problems. It depends on the smoothness of the constraints $\sigma_1^2 + \sigma_2^2 \leq 1$ at each point; for the linear program in L_∞ with $|\sigma_1| \leq 1$ and $|\sigma_2| \leq 1$, E will leave Ω more abruptly and ∂E is not C^1.) By symmetry, there must be four equal arcs, one in each corner, which remove the four corners of the square from the optimal set E.

If the arcs have radius ρ, then the area of E is $4 - 4\rho^2 + \pi\rho^2$ --the whole square minus four small squares in the corners plus four quadrants. The perimeter is $8 - 8\rho + 2\pi\rho$. The smallest ratio is found to be

$$\lambda^* = \min \frac{P(E)}{A(E)} = 1 + \frac{\sqrt{\pi}}{2} .$$

This occurs when the arcs have radius $\rho = 1/\lambda^* \sim .53$; the optimal ∂E stays with the square along slightly less than half of its edge.

Remark: We have discovered a short note [8] on the same isoperimetric problem, discussing also the case of triangles--where again all three arcs have radius equal to the maximum ratio A/P. The references lead back ultimately to Steiner [9], with links to Minkowski's theorem about convex sets which contain no lattice points and to problems of Besicovitch. There are some fascinating experiments described by Garvin [10], who dropped mercury into Ω and measured the arcs in the corners.

Duality has solved most of the original problem. We know that the collapse factor λ^* is $1 + \sqrt{\pi}/2$, and we know that ∂E is the "collapse line" for the plastic material in the pipe; the part inside E falls through while the part outside remains in the corners. We do not know the stress distribution σ, except for the fact that (by the minimax argument) it assumes its maximum along the collapse line; $\sigma \cdot n = 1$ along ∂E. To find σ in the whole square we turn to the computer.

Numerical Analysis

There are two different approaches to the numerical computations. In the first, we gradually increase the load factor λ and watch the growth of σ and u; this embeds the problem into incremental plasticity, for which large-scale finite element systems have been developed over the last ten years. We hope to borrow one, and to try it on our example. At every load increment the mechanical flow law yields a constrained minimization--a variational inequality--and collapse occurs when the constraints can no longer be satisfied and the minimization is over an empty set. The problem will be entirely elastic up to the value λ_0 at which the stresses reach $|\sigma| = 1$; this happens first at the mid-sides of the square. Then separate plastic regions should begin to grow around these points, and when they meet there is collapse.

The alternative is a direct approximation of the programming problem. One possibility is to work with the primal--to construct a finite-dimensional family of stresses σ_h, with $\operatorname{div} \sigma_h$ proportional to f, and to maximize the proportionality factor subject to $\|\sigma_h\| \leq 1$. The optimal λ_h is a lower bound for the true λ^*, and is therefore "safe." The upper bound approach works with the dual, minimizing $\int |\operatorname{grad} u_h|$ subject to $\int f u_h = 1$ and $u_h = 0$ on $\partial\Omega$. Between these two lies a mixed method, in which stresses and displacements are represented separately, and the equilibrium constraint $\operatorname{div} \sigma_h = \lambda f$ is not imposed. The approximation to λ^* is determined from a convex program in finite dimensions:

$$\lambda_h^* = \max_{\|\sigma_h\| \leq 1} \; \min_{\int f u_h = 1} \; \int u_h \operatorname{div} \sigma_h \; .$$

For this problem the minimax theorem does hold.

We want to give a brief analysis of the convergence of the mixed method. We write $a(\sigma,u)$ for the bilinear form $\int u \operatorname{div} \sigma$, and we denote the sets of admissible stresses ($\|\sigma\| \leq 1$) and displacements by Σ and U; in the finite-dimensional approximation we have sets $\Sigma_h \subset \Sigma$ and $U_h \subset U$. Then the optimal solutions are saddle-points:

$$a(\sigma,u^*) \leq a(\sigma^*,u^*) = \lambda^* \leq a(\sigma^*,u)$$

$$a(\sigma_h,u_h^*) \leq a(\sigma_h^*,u_h^*) = \lambda_h^* \leq a(\sigma_h^*,u_h)$$

for all σ in Σ, u in U, σ_h in Σ_h, and u_h in U_h. On the first line we make the specific choices $\sigma = \sigma_h^*$ and $u = u_h^*$. Then by subtraction

$$a(\sigma_h-\sigma^*,u_h^*) \leq \lambda_h^* - \lambda^* \leq a(\sigma_h^*,u_h-u^*)$$

for all σ_h and u_h. This is the fundamental error estimate.

To deduce convergence, we need the same two properties as always:

i) <u>Stability</u>, to bound σ_h^* and u_h^*

ii) <u>Consistency</u>, or <u>approximability</u>, to produce a pair σ_h, u_h for which $\sigma_h - \sigma^*$ and $u_h - u^*$ will approach zero.

The rate of convergence depends on the smoothness of the true solution (σ^*,u^*), and on the construction of the approximating sets Σ_h and U_h.

It is natural to construct these sets from finite elements. In other words, after subdividing Ω into triangles or squares, we choose a family of <u>piecewise polynomials</u>. For example, we may choose σ_h linear within each triangle, and require that $\sigma_h \cdot n$ be continuous between triangles and constant along each edge. Then Σ_h has one basic function for every edge of the triangulation, corresponding to that value of $\sigma_h \cdot n$.

Independently, u_h might be chosen piecewise linear and continuous--the basic finite element--or conceivably even piecewise constant, since that leaves u_h in BV. But this choice is not entirely independent of Σ_h, since the calculation will collapse unless it is stable. The stability condition connects Σ_h and U_h

by the requirement that for each u_h,

$$\sup_{\sigma_h \text{ in } \Sigma_h} |\int u_h \operatorname{div} \sigma_h| \geq c\|u_h\|.$$

In fact, since we have to choose bases $\phi_1,\ldots,\phi_n$ for the stresses and $\psi_1,\ldots,\psi_m$ for the displacements, this is a condition on the matrix $A_{ij} = \int \psi_i \operatorname{div} \phi_j$. The matrix is m by n, and its rank must be m. Like the divergence operator, of which it is the discrete form, we expect a nullspace (therefore $m \leq n$ is an absolute requirement) but we also expect a right inverse; furthermore the rows of A must be linearly independent uniformly in h. Then the constant c does not approach zero with h.

Once the spaces are chosen and A is computed, it remains to solve the discrete problem. The primal asks for the minimum norm:

$$\min_{A\sigma=F} \|\sigma\|, \text{ or equivalently } \max_{\substack{A\sigma=\lambda F \\ \|\sigma\|\leq 1}} \lambda$$

Inevitably the dual is closer to ℓ_1:

$$\lambda_h^* = \min_{u^TF=1} \|u\|_{BV} = \min_{u^TF=1} |||A^Tu|||.$$

The vector F has components $\int f\psi_j$, and the unknowns σ in R^n and u in R^m give the weights in the linear combinations $\Sigma\sigma_i\phi_i$ and $\Sigma u_i\psi_i$. A number of algorithms have been proposed for the minimum ℓ_∞ solution of a linear system, and in fact many of them switch (as we did in the continuous case) to the dual; see Barrodale-Roberts [11] and Ascher [12], as well as the references they give. The dual is solved by a variant of the simplex method. But our norm $\|\ \|$ is not quite ℓ_∞, and the dual norm $|||\ |||$ is not quite ℓ_1. Both have to be computed in agreement with the original function space norms of σ

and u.

Surely the ideas in those algorithms can still be used. However there is another approach to mention, adopted by Casciaro-Di Carlo [13] and Hutula [14], which is far from the simplex method. They begin with optimization in ℓ_2 (the elastic problem, which is linear) and progress in three or four steps towards ℓ_∞. The variable is the exponent p in the norm, and it was Polya's conjecture--now proved--that the minimal ℓ_p solutions converge to a minimal solution in ℓ_∞. It is not known whether they converge to the same solution as in incremental plasticity, nor what properties distinguish this solution; the optimal σ is not generally unique. But the numerical experiments reported in [14] look very encouraging.

We hope, working jointly with Matthies and with help from many friends, to report again on both the functional analysis and the numerical analysis of this example.

REFERENCES

1. J. Hult and F.A. McClintock, Elastic-plastic stress and strain distribution around sharp notches under repeated shear, Ninth Congress of Applied Mechanics, 8, 1956.

2. J.R. Rice, Contained plastic deformation near cracks and notches under longitudinal shear, Int. J. of Fracture Mechanics, 2 (1966), 426-447.

3. E. Christiansen, H. Matthies, and G. Strang, Optimization, plastic limit analysis, and finite elements, to appear.

4. C.R. Adams and J.A. Clarkson, On definitions of bounded variation

for functions of two variables, Trans. Amer. Math. Soc. 35 (1933), 824-854.

5. B. Nayroles, Essai de théorie fonctionelle des structures rigides plastiques parfaites, J. de Mécanique 9 (1970), 491-506.

6. W.H. Fleming, Functions with generalized gradient and generalized surfaces, Annali di Matematica 44 (1957), 93-103.

7. H. Federer, Geometric Measure Theory, Springer-Verlag, New York, 1969.

8. T.-P. Lin, Maximum area under constraint, Math. Magazine 50 (1977), 32-34.

9. J. Steiner, Sur le max et le min des figures dans le plan..., J. Reine Angew. Math. 24 (1842), 93-152.

10. A.D. Garvin, A note on DeMar's "A simple approach to isoperimetric problems in the plane" and an epilogue, Math. Magazine 48 (1975), 219-221.

11. I. Barrodale and F.D.K. Roberts, An improved algorithm for discrete ℓ_1-linear approximation, SIAM J. Num. Anal. 10 (1973), 839-848.

12. U. Ascher, Linear programming algorithms for the Chebyshev solution..., MRC Report 617, Univ. of Wisconsin, 1976.

13. R. Casciaro and A. DiCarlo, Mixed finite element models in limit analysis, Computational Methods in Nonlinear Mechanics,

ed. J.T. Oden, Texas Inst. for Comp. Mech., 1974.

14. D. Hutula, Finite element limit analysis of two-dimensional plane structures, Limit Analysis Using Finite Elements, ASME annual meeting (1976), 35-51.

Vol. 609: General Topology and Its Relations to Modern Analysis and Algebra IV. Proceedings 1976. Edited by J. Novák. XVIII, 225 pages. 1977.

Vol. 610: G. Jensen, Higher Order Contact of Submanifolds of Homogeneous Spaces. XII, 154 pages. 1977.

Vol. 611: M. Makkai and G. E. Reyes, First Order Categorical Logic. VIII, 301 pages. 1977.

Vol. 612: E. M. Kleinberg, Infinitary Combinatorics and the Axiom of Determinateness. VIII, 150 pages. 1977.

Vol. 613: E. Behrends et al., L^p-Structure in Real Banach Spaces. X, 108 pages. 1977.

Vol. 614: H. Yanagihara, Theory of Hopf Algebras Attached to Group Schemes. VIII, 308 pages. 1977.

Vol. 615: Turbulence Seminar, Proceedings 1976/77. Edited by P. Bernard and T. Ratiu. VI, 155 pages. 1977.

Vol. 616: Abelian Group Theory, 2nd New Mexico State University Conference, 1976. Proceedings. Edited by D. Arnold, R. Hunter and E. Walker. X, 423 pages. 1977.

Vol. 617: K. J. Devlin, The Axiom of Constructibility: A Guide for the Mathematician. VIII, 96 pages. 1977.

Vol. 618: I. I. Hirschman, Jr. and D. E. Hughes, Extreme Eigen Values of Toeplitz Operators. VI, 145 pages. 1977.

Vol. 619: Set Theory and Hierarchy Theory V, Bierutowice 1976. Edited by A. Lachlan, M. Srebrny, and A. Zarach. VIII, 358 pages. 1977.

Vol. 620: H. Popp, Moduli Theory and Classification Theory of Algebraic Varieties. VIII, 189 pages. 1977.

Vol. 621: Kauffman et al., The Deficiency Index Problem. VI, 112 pages. 1977.

Vol. 622: Combinatorial Mathematics V, Melbourne 1976. Proceedings. Edited by C. Little. VIII, 213 pages. 1977.

Vol. 623: I. Erdelyi and R. Lange, Spectral Decompositions on Banach Spaces. VIII, 122 pages. 1977.

Vol. 624: Y. Guivarc'h et al., Marches Aléatoires sur les Groupes de Lie. VIII, 292 pages. 1977.

Vol. 625: J. P. Alexander et al., Odd Order Group Actions and Witt Classification of Innerproducts. IV, 202 pages. 1977.

Vol. 626: Number Theory Day, New York 1976. Proceedings. Edited by M. B. Nathanson. VI, 241 pages. 1977.

Vol. 627: Modular Functions of One Variable VI, Bonn 1976. Proceedings. Edited by J.-P. Serre and D. B. Zagier. VI, 339 pages. 1977.

Vol. 628: H. J. Baues, Obstruction Theory on the Homotopy Classification of Maps. XII, 387 pages. 1977.

Vol. 629: W. A. Coppel, Dichotomies in Stability Theory. VI, 98 pages. 1978.

Vol. 630: Numerical Analysis, Proceedings, Biennial Conference, Dundee 1977. Edited by G. A. Watson. XII, 199 pages. 1978.

Vol. 631: Numerical Treatment of Differential Equations. Proceedings 1976. Edited by R. Bulirsch, R. D. Grigorieff, and J. Schröder. X, 219 pages. 1978.

Vol. 632: J.-F. Boutot, Schéma de Picard Local. X, 165 pages. 1978.

Vol. 633: N. R. Coleff and M. E. Herrera, Les Courants Résiduels Associés à une Forme Méromorphe. X, 211 pages. 1978.

Vol. 634: H. Kurke et al., Die Approximationseigenschaft lokaler Ringe. IV, 204 Seiten. 1978.

Vol. 635: T. Y. Lam, Serre's Conjecture. XVI, 227 pages. 1978.

Vol. 636: Journées de Statistique des Processus Stochastiques, Grenoble 1977, Proceedings. Edité par Didier Dacunha-Castelle et Bernard Van Cutsem. VII, 202 pages. 1978.

Vol. 637: W. B. Jurkat, Meromorphe Differentialgleichungen. VII, 194 Seiten. 1978.

Vol. 638: P. Shanahan, The Atiyah-Singer Index Theorem, An Introduction. V, 224 pages. 1978.

Vol. 639: N. Adasch et al., Topological Vector Spaces. V, 125 pages. 1978.

Vol. 640: J. L. Dupont, Curvature and Characteristic Classes. X, 175 pages. 1978.

Vol. 641: Séminaire d'Algèbre Paul Dubreil, Proceedings Paris 1976–1977. Edité par M. P. Malliavin. IV, 367 pages. 1978.

Vol. 642: Theory and Applications of Graphs, Proceedings, Michigan 1976. Edited by Y. Alavi and D. R. Lick. XIV, 635 pages. 1978.

Vol. 643: M. Davis, Multiaxial Actions on Manifolds. VI, 141 pages. 1978.

Vol. 644: Vector Space Measures and Applications I, Proceedings 1977. Edited by R. M. Aron and S. Dineen. VIII, 451 pages. 1978.

Vol. 645: Vector Space Measures and Applications II, Proceedings 1977. Edited by R. M. Aron and S. Dineen. VIII, 218 pages. 1978.

Vol. 646: O. Tammi, Extremum Problems for Bounded Univalent Functions. VIII, 313 pages. 1978.

Vol. 647: L. J. Ratliff, Jr., Chain Conjectures in Ring Theory. VIII, 133 pages. 1978.

Vol. 648: Nonlinear Partial Differential Equations and Applications, Proceedings, Indiana 1976–1977. Edited by J. M. Chadam. VI, 206 pages. 1978.

Vol. 649: Séminaire de Probabilités XII, Proceedings, Strasbourg, 1976–1977. Edité par C. Dellacherie, P. A. Meyer et M. Weil. VIII, 805 pages. 1978.

Vol. 650: C*-Algebras and Applications to Physics. Proceedings 1977. Edited by H. Araki and R. V. Kadison. V, 192 pages. 1978.

Vol. 651: P. W. Michor, Functors and Categories of Banach Spaces. VI, 99 pages. 1978.

Vol. 652: Differential Topology, Foliations and Gelfand-Fuks-Cohomology, Proceedings 1976. Edited by P. A. Schweitzer. XIV, 252 pages. 1978.

Vol. 653: Locally Interacting Systems and Their Application in Biology. Proceedings, 1976. Edited by R. L. Dobrushin, V. I. Kryukov and A. L. Toom. XI, 202 pages. 1978.

Vol. 654: J. P. Buhler, Icosahedral Golois Representations. III, 143 pages. 1978.

Vol. 655: R. Baeza, Quadratic Forms Over Semilocal Rings. VI, 199 pages. 1978.

Vol. 656: Probability Theory on Vector Spaces. Proceedings, 1977. Edited by A. Weron. VIII, 274 pages. 1978.

Vol. 657: Geometric Applications of Homotopy Theory I, Proceedings 1977. Edited by M. G. Barratt and M. E. Mahowald. VIII, 459 pages. 1978.

Vol. 658: Geometric Applications of Homotopy Theory II, Proceedings 1977. Edited by M. G. Barratt and M. E. Mahowald. VIII, 487 pages. 1978.

Vol. 659: Bruckner, Differentiation of Real Functions. X, 247 pages. 1978.

Vol. 660: Equations aux Dérivée Partielles. Proceedings, 1977. Edité par Pham The Lai. VI, 216 pages. 1978.

Vol. 661: P. T. Johnstone, R. Paré, R. D. Rosebrugh, D. Schumacher, R. J. Wood, and G. C. Wraith, Indexed Categories and Their Applications. VII, 260 pages. 1978.

Vol. 662: Akin, The Metric Theory of Banach Manifolds. XIX, 306 pages. 1978.

Vol. 663: J. F. Berglund, H. D. Junghenn, P. Milnes, Compact Right Topological Semigroups and Generalizations of Almost Periodicity. X, 243 pages. 1978.

Vol. 664: Algebraic and Geometric Topology, Proceedings, 1977. Edited by K. C. Millett. XI, 240 pages. 1978.

Vol. 665: Journées d'Analyse Non Linéaire. Proceedings, 1977. Edité par P. Bénilan et J. Robert. VIII, 256 pages. 1978.

Vol. 666: B. Beauzamy, Espaces d'Interpolation Réels: Topologie et Géometrie. X, 104 pages. 1978.

Vol. 667: J. Gilewicz, Approximants de Padé. XIV, 511 pages. 1978.

Vol. 668: The Structure of Attractors in Dynamical Systems. Proceedings, 1977. Edited by J. C. Martin, N. G. Markley and W. Perrizo. VI, 264 pages. 1978.

Vol. 669: Higher Set Theory. Proceedings, 1977. Edited by G. H. Müller and D. S. Scott. XII, 476 pages. 1978.